NEURAL NETWORK COMPUTING FOR THE ELECTRIC POWER INDUSTRY

PROCEEDINGS OF THE 1992 INNS SUMMER WORKSHOP

Editor
Dejan J. Sobajic

LEA LAWRENCE ERLBAUM ASSOCIATES, PUBLISHERS
Hillsdale, New Jersey Hove and London

Lawrence Erlbaum Associates, Inc., Publishers
365 Broadway
Hillsdale, New Jersey 07642

Library of Congress Cataloging-in-Publication Data

Neural network computing for the electric power industry : proceedings of the 1992 INNS summer workshop / editor, Dejan J. Sobajic.
p. cm.
Includes bibliographical references.
ISBN 0-8058-1467-1 (pbk. : alk. paper)
1. Neural networks (Computer science)--Congresses. 2. Electric power systems--Data processing--Congresses. I. Sobajic, Dejan J. II. International Neural Network Society.
QA76.87.N476 1993
621.31'0285'63--dc20 93-17104
CIP

Books published by Lawrence Erlbaum Associates are printed on acid-free paper, and their bindings are chosen for strength and durability.

Printed in the United States of America

PREFACE

Welcome to the initial proceedings publication of the International Neural Network Society, co-published with Lawrence Erlbaum Associates, Publishers. This volume is the initial step in a detailed study of the numerous aspects of neural networks. Over the coming years, INNS publications will seek to illuminate artificial and biological neural networks. I hope that these publications will spark the interests and educate the minds of people engaged in activities that may benefit from greater knowledge and use of neural networks.

Established in 1987, the INNS is composed of scientists, engineers, and others who have recognized the importance of neural networks and their enormous potential application in many fields including engineering, medicine and finance. Since 1991, INNS has aggressively expanded its activities, adding workshops, newsletters, and book publications to the internationally renowned journal *Neural Networks.* In addition, the Society conducts the annual World Congress on Neural Networks, an international meeting that brings together the best minds in the field and provides an opportunity to see the latest developments in neural networks. I invite you to look at the back of this volume for more information about the benefits of INNS membership.

I hope you will find the INNS publications of proceedings and other books on neural networks to be of great value to you in your professional lives and research, and that I will see your contribution to our understanding of neural networks in future volumes.

Harold Szu
1993 INNS President

The 1992 INNS Summer Workshop on Neural Network Computing for the Electric Power Industry was conceived as a way to bring together researchers in neural networks to exchange their ideas and findings, and to show that neural networks, either alone or in combination with other techniques, have broad applicability throughout the electric power industry. The scope of the applications described in these proceedings certainly attest to the latter conclusion. The relative maturity and demonstrated success of many of these projects also indicate that the technology of neural networks is now ready and available for inclusion in practical product developments for our industry.

At present, it would appear that the major practical barrier to the greater use of neural networks in the electric power industry is their opacity (i.e., the difficulty of relating the internal structures and operation of the network to the external relations that it models). Either a detailed theoretical basis or a set of "engineering handbook" procedures is needed to guide the design. Ideally, the completed network would then provide a causal (rather than merely statistical) explanation of the phenomena or system it has been trained to emulate.

The Office of Exploratory and Applied Research of the Electric Power Research Institute has been the sponsor, in whole or in part, of many of the research projects reported at these proceedings. EPRI has supported neural network investigations, and will continue to do so, with two goals in mind: (1) Finding innovative and valuable applications of neural network computing specific to the electric power industry, and (2) establishing fundamental theory and/or more reliable procedures for neural net design and development.

A. Martin Wildberger
Executive Scientist, Electric Power Research Institute

Summer Workshop on Neural Network Computing for the Electric Power Industry

Hartley Conference Center, Stanford University
Stanford, California, August 17-19, 1992

PROGRAM COMMITTEE

Chair: Dejan J. Sobajic, Case Western Reserve University
Chairperson, INNS Power Engineering Special Interest Group

James A. Anderson, Brown University
Merwin Brown, Pacific Gas and Electric
Richard F. Daniels, Southern California Edison
T.S. Dillon, La Trobe University, Australia
Alain Germond, Swiss Federal Institute of Technology
G. Lambert-Torres, University of Itajuba, Brazil
C.-C. Liu, University of Washington
John Maulbetsch, EPRI
Paul Mauldin, Pacific Gas & Electric
Jean Mazalerat, Electricite De France
Hiroyuki Mori, Meiji University
G.K.H. Pang, University of Waterloo, Canada
Yoh-Han Pao, Case Western Reserve University
David Rumelhart, Stanford University
Yasuji Sekine, University of Tokyo
Jennie Si, Arizona State University
Harold Szu, Naval Surface Warfare Center
Al Sudduth, Duke Power Company
Hideo Tanaka, Tokyo Electric Power Company
Belle Upadhyaya, University of Tennessee
Robert Uhrig, University of Tennessee
Louis Wehenkel, University of Liege
Bernard Widrow, Stanford University
Martin Wildberger, EPRI
Richard M. Wood, KBTAC, Syracuse University
B.R. (Bob) York, Tennessee Valley Authority

TABLE OF CONTENTS

E. Modeling and Prediction

F. Control

G. Load Forecasting

H. Scheduling and Optimization

I. Fault Diagnosis

FOREWORD

Power system computing with neural networks is one of the fastest growing fields in the history of power system engineering. Since 1988 a considerable amount of work has been done in investigating computing capabilities of neural networks and understanding their relevance to providing efficient solutions for outstanding complex problems of the electric power industry.

A principal objective of a power utility is to provide electric energy to its customers in a secure, reliable and economic manner. Toward this aim utility personnel are engaged in a variety of activities in areas of supervisory control and monitoring, evaluation of operating conditions, operation planning and scheduling, system development, equipment testing etc. Over the past decades significant advances have been made in development of new concepts, design of hardware and software systems and implementation of solid-state devices which all contributed to the steadily improving power system performance that we are experiencing today. Advanced information processing technologies played an important role in these development efforts.

In a short period of time, neural-net computing has moved rapidly from research laboratories to practical use and commercialization. Demonstrated information processing functionalities of neural-net systems including pattern classification, estimation of feature attributes, learning internal representations and cooperative optimization of systems functions provided the opportunity for development of a new generation of methodologies and tools capable of dealing with challenging power system problems.

Members of the Special Interest Group for Power Engineering of the INNS have recognized the need for bringing together leading researchers in the field of neurocomputing with experts from power utilities and manufacturing companies to assess the current state of affairs and to explore the directions of further research and practice. Sponsored by INNS and the Electric Power Research Institute (EPRI) and with the cooperation of IEEE, these efforts resulted in organizing *The Summer Workshop on Neural Network Computing for the Electric Power Industry.*

The Workshop was held in the Hartley Conference Center on the campus of the Stanford University, August 17-19, 1992. It brought together approximately forty specialists with backgrounds in power engineering, system operation and planning, neural network theory and AI systems design. An informal and highly inspiring atmosphere of the Workshop facilitated open discussion and exchange of expertise between the participants.

These Proceedings, reviewed by the Workshop Technical Committee, present the written materials submitted in support to authors' presentations.

I would like to take this opportunity to express my gratitude to the authors, invited speakers, session chairpersons. Technical Committee members and Workshop Chairs and made this event possible. The support of INNS and EPRI is gratefully acknowledged. Special thanks are due to Ms. Melissa Bishop, Ms. Donna Buggs, Ms. Joice DeBolt and Ms. Lynetta Holbrook for their excellent efforts in Workshop organization and coordination. Finally, the help from Stanford University in hosting the Workshop is much appreciated.

Dejan J. Sobajic
Chairman
Technical Committee

A. Perspectives

LEARNING AND GENERALIZATION CHARACTERISTICS OF THE RANDOM VECTOR FUNCTIONAL-LINK NET

Yoh-Han Pao

and

Gwang-Hoon Park

Electrical Engineering and Applied Physics
Case Western Reserve University
Cleveland, OH 44106-7221

INTRODUCTION

Of all the capabilities ascribable to neural-net computing, there is none more noteworthy than that of supervised learning. The generalized delta rule (GDR) net is preeminent in this task domain and is both widely known and widely used for learning functional mappings.

This learning capability can be used for assessment of the security or stability of systems; for monitoring the consistency of sets of sensor data; for predicting the future of systems; for estimating control actions necessary for achieving certain specific objectives; and for a host of other tasks.

Stripped of the details of the various task environments, the generic idea is that such a net can synthesize a network computational model of a known function if that function does indeed exist. The learning procedure is based on a finite number of known instances of presumed functional mapping, and it is assumed and hoped that the computational network model so learned is not only capable of replicating the known instances of the mapping but is also valid for the infinitude of all the other points in the neighborhoods of the exemplars.

In practice many things can go awry. For example, there may be, in fact, no deterministic function and the exemplars might be merely a set of random pairing of points. In such a case the training set of mappings can be learned very well, and any test set will exhibit the random scatter inherent to the stochastic process used to generate the data.

If the net does not have enough available adjustable parameters or if the learning process is terminated prematurely, then both the training set and test set errors may be large.

If the net has too many adjustable parameters or if the net is 'over-trained', then the training set error may be deceptively low without comparable performance attainable for test set points.

In this paper we discuss these matter not primarily for the sake of understanding the behavior of the GDR net because the principal behavioral aspects of that net have been articulated, but rather for a variant of the GDR net which we call the Functional-Link net.

In the past we have advocated the use of the Functional-Link net[1-3] for supervised learning of mappings from one-pattern space to another. The advantages of that approach include the fact that supervised learning can be achieved very well using a "flat" net (one that has no hidden layer). In addition convergence in learning is fast, and the same system architecture can be used in supervised and unsupervised learning and in optimization.

The Functional-Link (FL) net can be implemented in many ways, one of which is the "random-vector" version described recently.[4,5]

This paper is concerned with the learning and generalization characteristics of the random-vector implementation of the FL net.

BRIEF REVIEW OF THE FUNCTIONAL-LINK APPROACH

The basic GDR concept is illustrated in Figure 1(a) where we show a mapping from an N-dimensional space to a one-dimensional space with no non-linearity at the output node. There is no loss of generality, and the same system can be replicated to cover mappings from R^N to R^M spaces with non-linear mappings at the output if so desired.

The FL network shown in Figure 1(b) performs a nonlinear transformation of the input pattern before it is fed to the input layer of the network. The essential action is, therefore, the generation of an enhanced pattern to be used in place of the original. Experience indicates that supervised learning can be achieved readily using a "flat" net (one that has no hidden layers), and that the delta rule can be used in place of the generalized delta rule (GDR) if this enhancement is done correctly.

In the 'random-vector' FL net, the functional enhancements are achieved in essentially the same manner as in the first layer of the GDR net shown in Figure 1(a). We see in the feedforward action of the GDR net the computing achieved at each hidden layer node consists of computing a scalar product $\underline{A}_j \underline{x} + b_j$ where A_j is the vector of the link-weights, b_j is a scalar node threshold, and $\underline{x}$ is the input vector. At the jth node, the net input is $\underline{A}_j \underline{x} + b_j$ and the output is $g(\underline{A}_j \underline{x} + b_j)$. For the GDR nets the vector $\{\underline{A}_j\}$ are iteratively adjusted together with the next layer weights $\{\beta_j\}$ until the output error is at a minimum.

In the GDR net, the hidden layer nodes transform the input vector successively into a series of new representations until we come to a suitable new space where a desired linear separation $\{\beta_j\}$ becomes possible.

From the work of Hornik, Stinchcombe, and White[6] and Funahashi[7], we know that a net such as that shown in Figure 1(a) can perform as a universal approximator of any (nonpathological) known mapping from R^N to R, as made evident by the training set instances.

The random-vector implementation of the FL net can be viewed as essentially the same as the GDR except for the fact that the hidden layer is moved down to serve as an enhancement of the input vector and the weights vectors $\{\underline{A}_j\}$ are not learned but are randomly generated.

The random-vector implementation can also be viewed as a type of encoding comparable to the coarse-coding of patterns with linguistic symbolic feature values.[8]

LEARNING AND GENERALIZATION CHARACTERISTICS

The work of Hornik et al.[6] and of others principally show that it is possible to use a GDR net with a single hidden layer to approximate the mapping indicated by the known training set. There is no guarantee that the mapping synthesized by the network approximates the actual function except at the known instances. However, the generalization characteristics of the net are of the utmost importance and practical interest.

In general it can be readily demonstrated that the random-vector FL net can be trained well and rapidly. As is in the case of the GDR net, the generalization characteristics of the FL net are also of great interest and importance.

In this paper we report on the learning and generalization characteristics of the FL net as manifested in a specific but representative learning task. To provide perspective, we compare those characteristics with those of a corresponding GDR net.

In our case the function to be learned is the curve $f_1(x)$ shown in Figure 2. However, the function $f_1(x)$ is known to us only through the points marked with bold dots. The function $f_2(x)$ is another function learned erroneously by the GDR net if training is terminated prematurely.

As shown in Figure 3, the FL net can achieve satisfactory learning of the training set points if 300 enhancements are used. A system error of 0.000025 is attained after 12,065 iterations. The net is a flat net, i.e. a linear net.

In generalization mode the performance of the FL net is not bad, as illustrated in Figure 4.

Comparable learning is achieved by a GDR net with 20 hidden layer nodes. The accuracy of the learning achieved for the training set points is illustrated in Figure 5. But this was achieved only after 1,161,496 iterations.

In generalization mode the performance of the GDR net is not bad either, as illustrated in Figure 6.

It is interesting to note that if GDR learning were stopped at 200,000 iterations, the function learned is $f_2(x)$ and not $f_1(x)$. This is shown for the training set points and for the generalization mode in Figure 7 and Figure 8 respectively.

The random vector version of the FL net is characterized by the need for a relatively large number of 'auto-enhancements', i.e. the additional enhancements nodes. This is not objectionable because the number needed does not seem to grow much with the dimension of the patterns involved.

Figure 9 shows that the number of iteration required to achieve a system error 0.000025 decreases rapidly with increase in the number of auto-enhancements.

The difference between the performances achievable with the training and test set is well illustrated by Figures 10 and 11.

From the curves shown in Figure 10, we see that after about 20,000 iterations there is no further significant improvement in the performance of the FL net in so far as generalization is concerned. In this case it is probably not quite accurate to say that such a net is 'overtrained' but additional improvement is marginal.

Similar behavior is shown in Figure 11 for the GDR net.

In this brief paper we endeavor to promote an understanding of the learning and generalization characteristics of the Functional-Link net. We discuss these for the case where the function to be learned is a deterministic function, without noise. Under such circumstances we do not have the phenomenon of 'overtraining' but only that of saturation of training. The situation is different for 'real-world' data where the functional mapping to be learned is stochastic. For those cases we can have 'overtraining' and actually worsen the accuracy of estimated values as we increase the degree of training. However, that occurs only in the case of the GDR net. In both cases the training procedure should be the same, namely do not proceed further with training if there is no corresponding improvement in test set results.

REFERENCES

1. Pao, Y.H., 1989. *Adaptive Pattern Recognition and Neural Networks.* Addison-Wesley Publishing Co., Reading, NJ.
2. Klassen, M. S., Pao, Y. H. and V. Chen, 1988. Characteristics of the functional-link net: A higher order delta rule net. Proc. of IEEE Annual International Conference on Neural Networks, San Diego, CA.
3. Pao, Y. H., 1989. Functional-link nets: Removing hidden layers. *AI Expert*, Vol. 4, No. 4, pp 60-68.
4. Pao Y.H. and Y. Takefuji, 1991. Functional-link net computing: Theory, system architecture, and functionalities. Special issue of *Computer* on Computer Architectures for Intelligent Machines, IEEE Computer Society Press, Vol. 3, pp. 76-79, also published in Readings in Computer Architectures for Intelligent Systems (expanded version), J. Herath (ed.), in press.
5. Pao, Y.H., S. Phillips and D.J. Sobajic, 1992. Neural-net computing and the intelligent control of systems. Special Intelligent Control issue of *International Journal of Control,* Vol. 56, No. 2, pp. 263-290, Taylor and Francis, Ltd., London, U.K.
6. Hornik, K., M. Stinchcombe, and H. White, 1989. Multilayer feedforward networks are universal approximations. *Neural Networks*, Vol. 2, pp. 359-366.
7. Funahashi, 1989. On the approximate realization of continuous mappings by neural networks. *Neural Networks*, Vol. 2, pp. 183-192.

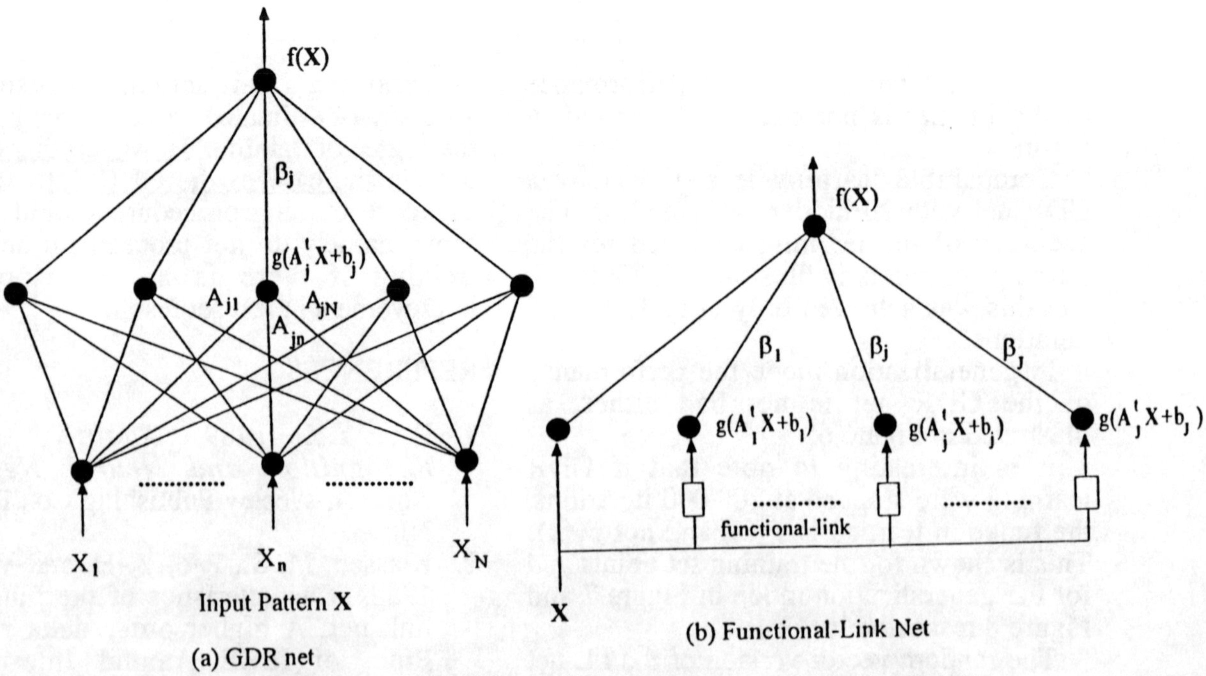

Figure 1 Schematic illustration of GDR and FL nets

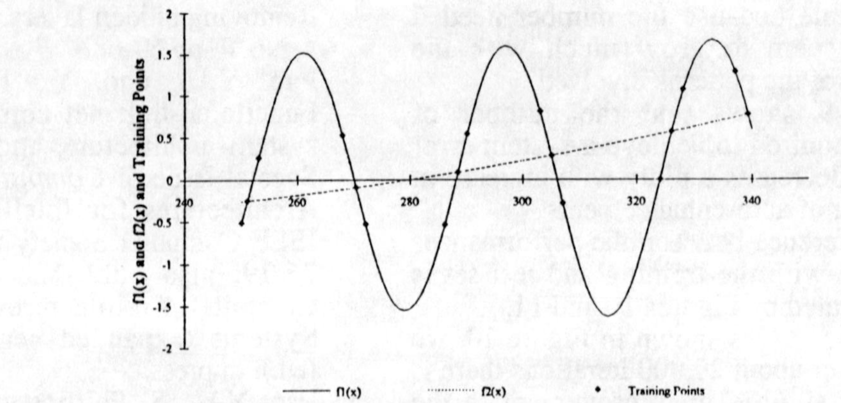

Figure 2 The function f1(x) to be learned, the training set points and the erroneous function f2(x)

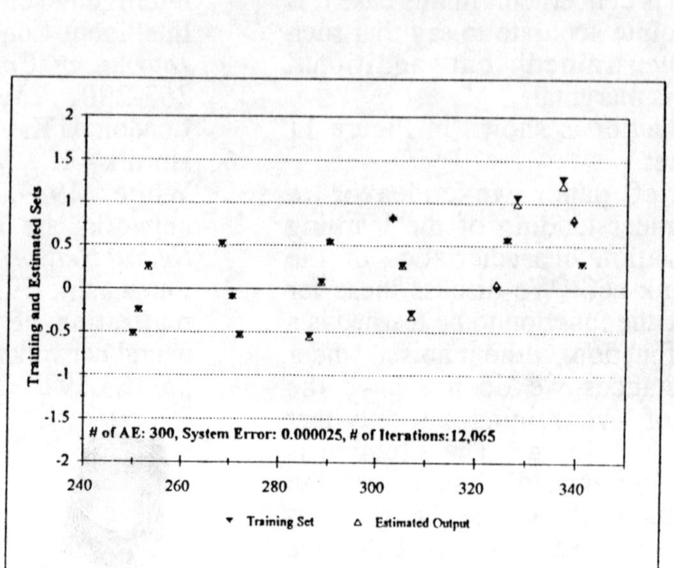

Figure 3 Learning achieved by FL net for training set points

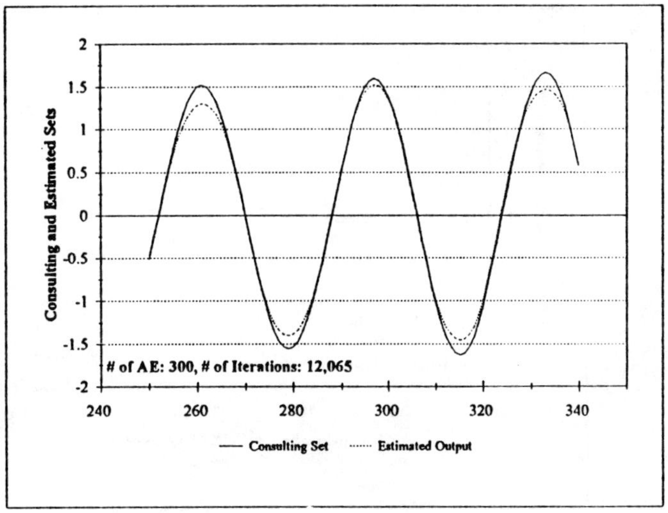

Figure 4 Comparison of actual and estimated function values for FL net in generalization mode

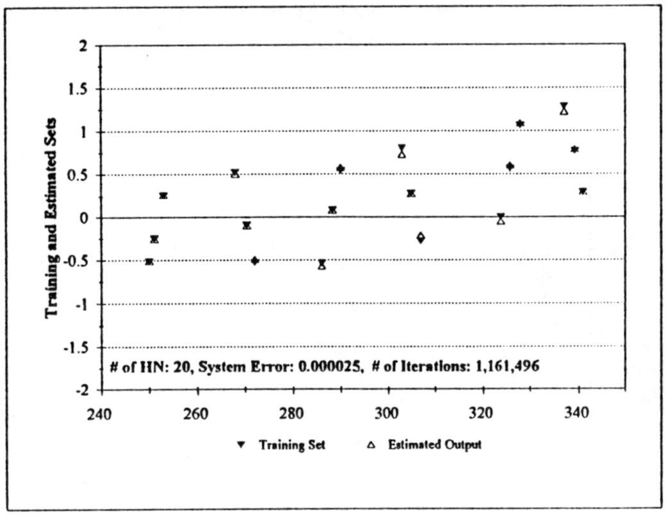

Figure 5 Learning achieved by GDR net for training set points

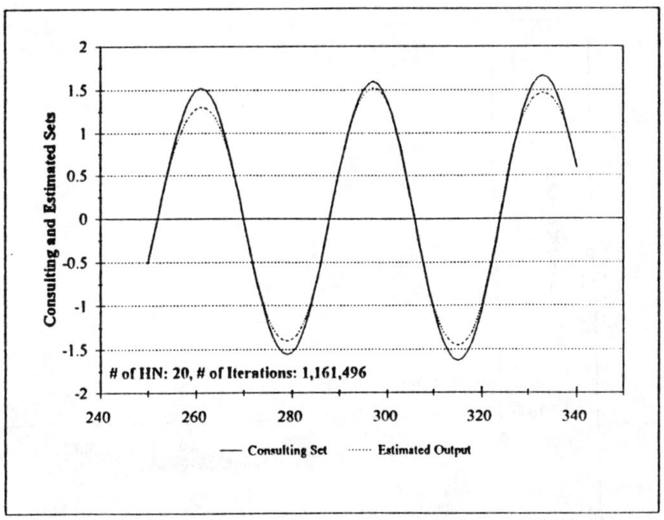

Figure 6 Comparison of actual and estimated function values for GDR net in consulting mode

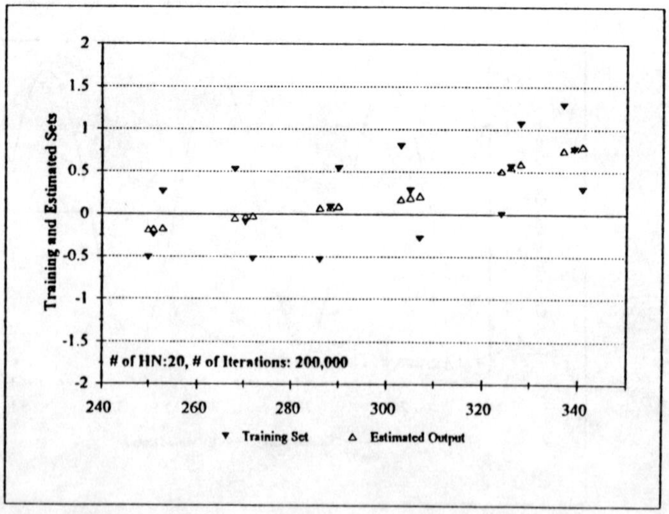

Figure 7 Failure of the GDR net to Learn the training set points if training is stopped at 200,000 iterations

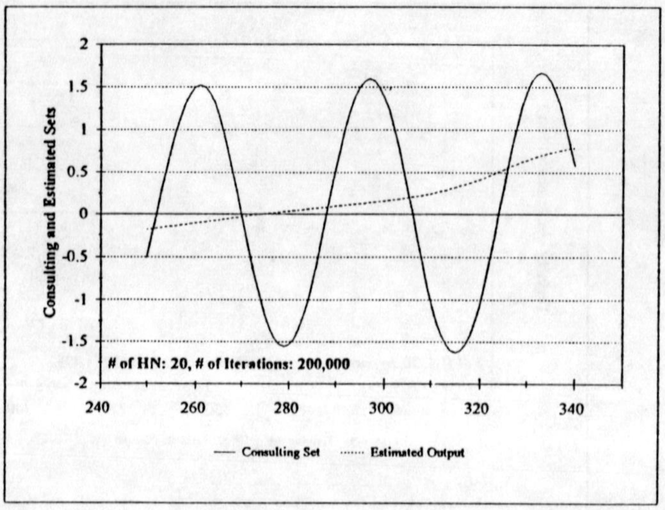

Figure 8 Failure of 'under trained' GDR net to generalize adequately in consulting mode

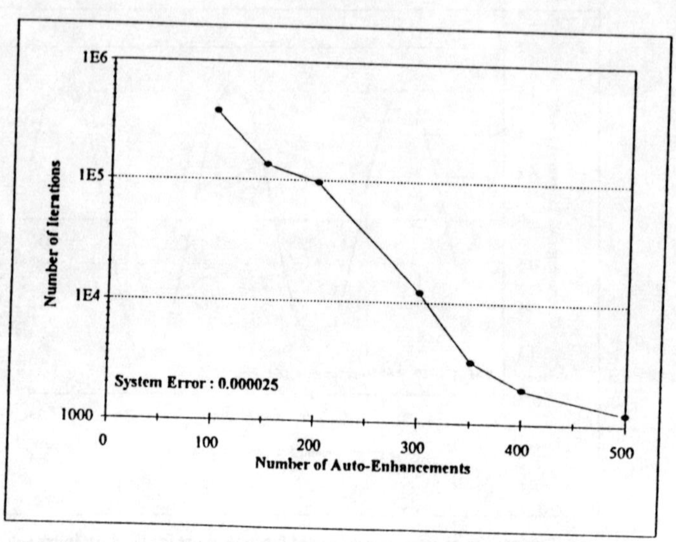

Figure 9 Variation of number of iterations needed for training with number of auto-enhancements (for FL net)

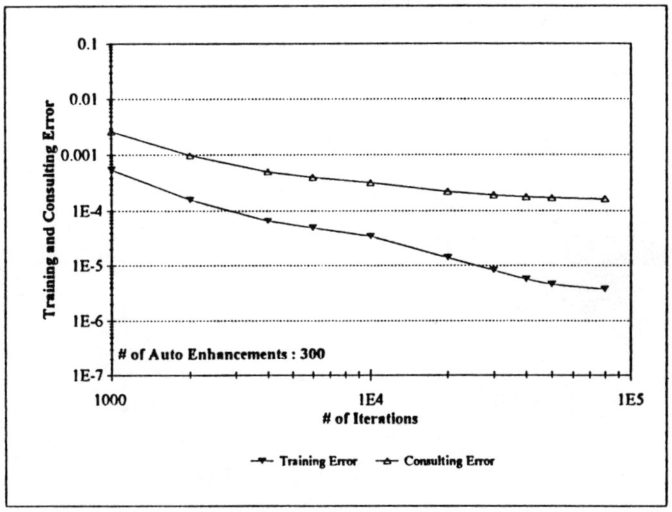

Figure 10 Saturation of training for the FL net

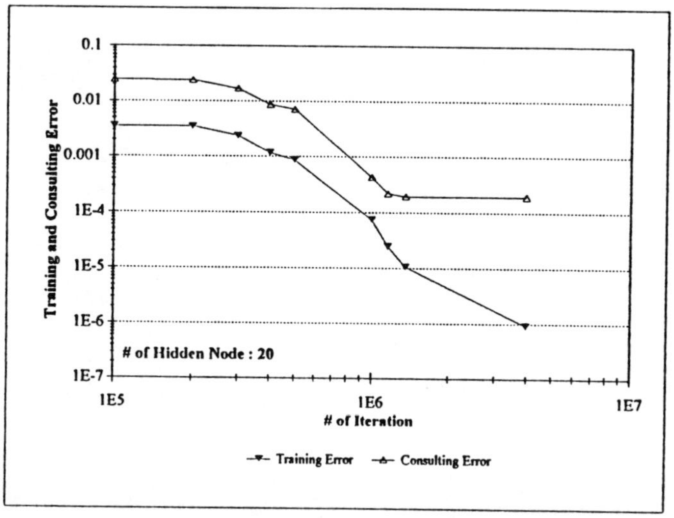

Figure 11 Saturation of training for the GDR net

Artificial Neural Networks and Expert Systems in the Power System Operations Environment

Mark J. Damborg Chen-Ching Liu

Electrical Engineering Department
University of Washington
Seattle, WA 98195

Abstract:

We suggest criteria to identify whether a power system operations task can be aided by an expert system or an artificial neural network or by neither. Examples are provided of expert system and neural network applications, from both the authors' experience and from the literature, which meet these criteria.

Proposition:

In this paper, we suggest that future power system operation will be aided by a combination of artificial neural network (ANN) and expert system (ES) tools which are incorporated into the communication, command and control system that is now evolving. We further suggest, that whether a task is suitable for ANN or ES support depends on specific properties of the task. Finally, we observe that ES technology is closer to being accepted in an EMS center than is ANN technology. Hence, the challenge is to develop ANN tools that respond to operations needs and that are compatible with the sensitivities of operations personnel.

First we must constrain our use of terms. An expert system will refer to a non-procedural computer program with an explicitly represented knowledge base which is executed through an inferencing mechanism. Our discussion of artificial neural networks will be restricted to feedforward, non-dynamic networks so that the ANN, however implemented, yields an input-output functional relationship.

Now for an attempt to identify when an ES or ANN approach is appropriate. We believe that the suitability of an ANN or ES approach to an operations task depends on at least three critical properties:

1. Does the information used in accomplishing the task result from a logical understanding of the task or from experiment?

2. Is this information best expressed as knowledge of task properties and constraints or as an input-output functional relationship?

3. Must the task be accomplished rapidly (on the order of the stable transient response time to a switching operation) or may it be carried out more leisurely in a response time associated with operator action?

Using these properties, we propose the following criteria for selecting an ES or ANN approach:

ES An ES may be appropriate if the information results from logical understanding, if this information can be expressed as properties and constraints and if rapid response is not required.

ANN An ANN is suggested if the information results in a functional representation which is identified through experiment.

By way of justifying our criteria, we recognize that at the heart of an ES is the knowledge base which contains information (properties and constraints) about the task. This knowledge base can take different forms (rules, frames etc.) but it is fundamentally unorganized. That is, the information units are not ordered, may be incomplete and may even be in conflict. Hence, an inferencing mechanism is required. This approach probably cannot be forced into a rapid response with present computation approaches.

On the other hand, the class of ANN considered here is fundamentally a functional representation with two features which make it especially interesting. First, the functional form need not be known but, rather, a standard structure can yield an arbitrarily good approximation. Second, this approximation can be developed from experience (pairs of required inputs and outputs) and can interpolate to respond to inputs not previously experienced. We expect an ANN to be able to execute quickly, using special purpose hardware if necessary, so response time is not a major concern.

We also suggest that other programming approaches are more appropriate when the above conditions fail. For example, if the information results from logical understanding but is best expressed as an input-output function, neither an ES or ANN is appropriate. No knowledge base or inference mechanism is required and, since the functional form is known, the parameters can be determined in other ways than with ANN training algorithms. Or if the information is from experiment but best represented as properties and constraints, an ANN is inappropriate and an ES can probably not be built due to the limited scope of experimental results. Other, more traditional approaches are also called for if the information results in procedures.

Finally, we add a comment on being sensitive to the needs of operations personnel. Whether an ES or ANN approach is attempted, a resulting "tool" will not be used unless the operator has confidence in it and control of it. Confidence probably is best achieved by involving operators in the development of the tool and by exhaustive testing, also involving the operators. One way to assure operators that they have control, is to make the tool advisory to them. That observation may be most appropriate for an ES in a human time frame. Another way to maintain operator control is to give them the ability to monitor the tool functionality.

ES Background:

The above proposition represents the experience of the authors. We will clarify our perspective by briefly describing three ES projects in power system operations that involved one or both of us. Each of these examples satisfies the criteria cited above. They are based on logic, use a knowledge base of properties and constraints and the response rates are on the order required by dispatchers or operations planners. Beyond these examples of our experience, the reader may wish to consult a recent special issue of the *IEEE Proceedings* on the subject [1].

CRAFT is a rule-based system that serves as a dispatcher's aid in an energy control center [2]. In the Puget Sound Power and Light system, automatic switches are installed on the 115 kV transmission lines which typically have service taps between the power circuit breakers at the substations. The automatic switches on any line, operating according to their design logic, are intended to isolate faults and restore service to at least some of the tap points on that line. For example, a switch may open automatically if a de-energized condition on a line section persists for x seconds. In other words, the operation of automatic devices is driven by faults and satisfies predefined design logic. Automatic switches and power circuit breakers on a transmission line make it possible to sectionalize the line so that the outage portion of a line can be reduced after a fault occurs. CRAFT uses rules, or logic, to locate the faulted section of the line. The expert system also determines the status of unmonitored automatic switches. The status of these devices can be determined by logic reasoning based on the design functions, timing and energized/de-energized status of the line sections on the transmission line. CRAFT is an interesting example of an expert system that performs reasoning based on the logic embedded in the design of a power line. It operates as an advisor to dispatchers for additional service restoration after all automatic actions have occurred.

The second example expert system is a system that includes empirical knowledge that may be difficult to define precisely. The system concerns the voltage security assessment problem [3]. To identify a potential threat to voltage security, the operational planner needs to examine the current operating point. The expert system selects the critical contingencies based on the structures of radials, loops and clusters. The definition of a structure may be somewhat imprecise. For example, a cluster is a collection of "several" buses interconnected by "low" impedance lines. Low impedance is defined as "less than the average impedance of all other lines at the same nominal voltage level." The idea here is to select only the critical contingency so that various contingencies related to the same cluster can be ignored. The contingencies are also weighted. One rule in this expert system is: "If a cluster has a boundary branch connected to the lowest voltage bus in the cluster and is carrying reactive power to the bus, then increase its weight based on the flow." In other words, this specific branch is considered more critical than some other branches inside the cluster. The amount of weight increase is based on empirical judgement.

Our third example illustrates a knowledge based system for a heuristic search. Power system restoration is a very complex problem involving a large number of generation, transmission, distribution and load devices. Mathematically, the problem formulation requires many variables (continuous and discrete) and many operating constraints (static and time varying). The objective of system restoration is to restore customers quickly without violating any equipment or operating constraints. The large-scale problem most likely does not have a unique solution, the number of options being very large. Hence, the search space is also large. Since the power system operating conditions will change as restoration progresses, the knowledge based system must track the operating condition and suggest a sequence of actions to the power system operators. Based on "heuristics," operators first start generating units with black start capabilities. The small amount of power available is then assigned to various non-black-start units. These units receive cranking power in order to re-start eventually providing more power to pick up the loads. This approach provides a good, but not necessarily optimal, solution to the large combinatorial problem [4].

Areas for ANN Contribution:

We cite three general areas where we think ANN approaches are likely to contribute to solving operations problems. Certainly there are others, many others. One need only scan the proceedings of a recent IEEE Power System Society conference or topical symposium on this subject to see a wide array of ideas for application. Any exclusion of a topic here is not intended to discriminate. Rather, our purpose is to illustrate that common examples satisfy our criteria.

1. Load Modeling

System loads are usually modeled simplistically, primarily due to our ignorance. Yet they are known to depend on voltage and frequency and, in fact, are dynamic systems. Transient stability analysis, used in operations planning, would obviously benefit from dynamic load models while static analysis, such as state estimation, would benefit from inclusion of load sensitivity to voltage and frequency. In both cases, the necessary models are functions whose form is not known but for which experimental data can be obtained and, hence, used to train an ANN [5].

2. Data Enhancement

Here we have in mind increasing the information available to dispatchers through additional monitoring of the power system and its components. ANNs are often proposed as monitors of the proper operation of individual equipment items such as switchgear to avoid failures or to schedule maintenance [6]. Larger scale efforts involve developing monitors for generating plants [7]. Such applications fit our criteria of being based on functions which are developed from experimental data. At present, most of these suggestions involve monitoring at the equipment location but we believe these kinds of monitors can also provide critical new data to system operators.

3. Security Monitoring

One system monitoring task which may be appropriate for ANNs is security monitoring. In fact, we would argue that this task, however performed, may well be the primary task of an EMS center: to determine when it is necessary to operate the system defensively against contingencies rather than according to economic criteria. The economic and performance impact of this decision, and of the defensive measures taken, can be enormous. The security problem is often separated into static problems (power flows, voltages and frequencies out of allowable ranges) and dynamic problems (loss of synchronism, voltage collapse). Attempts to build ANN security monitors have met with marginal success [8, 9]. The application successfully meets our criteria since it can be formulated as a functional representation to be determined through experiment. Further, the potential for rapid response of an ANN is important. However, the approach has only been successfully illustrated on very small systems and attempts to extend to larger systems have not yet proven feasible.

References:

1. C. Liu, Ed., "Special Issue on Knowledge-Based Systems in Electric Power Systems," *Proceedings of the IEEE*, Vol. 80, No. 5, May 1992.

2. C. Liu, M. Damborg, "CRAFT: On-Line Expert System for Customer Restoration and Fault Testing," EPRI Report EL-6880, Vol. 1-5, March 1990.

3. J. Cote, C. Liu, "Voltage Security Assessment Using Generalized Operational Planning Knowledge," *Proceedings, IEEE Power Engineering Society Winter Meeting*, 1992.

4. C. Liu, K. Liou, R. Chu, A. Holen, "Generation Capability Dispatch for Bulk Power System Restoration: A Knowledge-Based Approach," *Proceedings, IEEE Power Engineering Society Winter Meeting*, 1992.

5. R. Thomas, B.-Y. Ku, "Approximations of Power System Dynamic Load Characteristics by Artificial Neural Networks," *Proceedings, First International Forum on Applications of Neural Networks to Power Systems*, Seattle, WA, July 1991, pp. 178-182.

6. H. Ogi, H. Tanaka, Y. Akimoto, Y. Izui, H. Taoka, T. Sakaguchi, "Preventive Maintenance System for Gas Insulated Switchgear Using an Artificial Neural Network," *Proceedings, Third Symposium on Expert Systems Applications to Power Systems*, Tokyo/Kobe Japan, April 1991, pp. 627-633.

7. G. Zwingelstein, M. Masson, B. Dubuisson, M. Pavard, J. Mazalerat, "The Applications of Neural Networks for the Predictive Maintenance of Nuclear Power Plants," *Proceedings, Third Symposium on Expert Systems Applications to Power Systems*, Tokyo/Kobe Japan, April 1991, pp. 567-573.

8. M. Damborg, M. El-Sharkawi, M. Aggoune, R. Marks II, "Potential of Artificial Neural Networks in Power System Operation," *Proceedings, International Symposium on Circuits and Systems*, New Orleans, LA, May 1990, pp. 2933-2937.

9. S. Weerasooriya, M. El-Sharkawi, M. Damborg, R. Marks II, "Towards Static-Security Assessment of a Large-Scale Power System Using Neural Networks," *IEE Proceedings-C*, Vol. 139, No. 1, January 1992, pp. 64-70.

A Utility Perspective on Neural Networks, Fuzzy Logic, and Artificial Intelligence

Tennessee Valley Authority
Edward J. Bradley

The objective of this paper is to present the categories of problems that are plaguing the electric utility industry that may be solvable by the prudent application of neural networks, fuzzy logic, and other forms Artificial Intelligence. My viewpoint is different from the other participants in the neural network conference in that I am a utility engineer with little background in neural networks. My motivation for participation was to help initiate a dialogue between the power industry and the neural network community. The conference discussions were quite helpful in that it became quite apparent that even though there is little knowledge of the neural network community in the power industry there appears to be a good knowledge of the power industry and it's problems in the neural network community. Again this paper presents several categories of problems that plague the power industry and details several of these problems that could be solved through neural networks.

The problems fall into four general categories with a good bit of overlap. These are problems associated with Safety, Stability, Maintenance, and Economics. The biggest motivator being economics which overlap the other areas significantly. The reason that economics takes precedence over safety is not due to priorities but is due to the current regulatory environment. Presently the NRC has not accepted the use of microprocessor based control and instrumentation in safety systems without extensive design detail and historical operating experience. They also want proven algorithmic solutions that can be mathematically derived and fully exercised throughout their known operating range. Neural networks do not meet these criteria presently and therefore can only be used in the non-safety systems or as backups to proven systems in the safety systems. Many of the problems to be discussed have solutions in normal logic and analytic methods but by use of fuzzy logic (FL) and neural networks (NN) there may be simpler or faster methods of solving these problems.

Again the largest motivator for a utility to use fuzzy logic and neural networks will be on those changes that result in economic savings in plant generating capacity, operations costs, or engineering costs. The most likely systems are the balance of plant systems such as feedwater/main turbine control, heater drains and vents, reactor recirculation/power control. The balance of plant systems have not received the same level of attention as the safety systems and subsequently they are poorly maintained and consist of soon to be obsolete equipment. Each reactor trip due to process system failures or inability to respond to transient conditions result in very high costs due to the need to purchase the replacement power ($.5 M per day). Also for every 1/10% increase in power output result in half a million dollars per annum; therefore, the motivations are there to change.

The utility industry need solutions that address:

Instrumentation redundancy (either more signals, recognized failures, or pseudo signals based on diverse plant parameters, autoassociative networks),

Anticipated/imminent failure (recognize system/equipment past performance and recognize off normal conditions that may soon be failures),

Stable control response (quicker control action with less oscillations and overshoot),

Adaptive control (based on plant/system status, modify control strategy or tuning based on plant conditions, available mitigating systems etc),

Transient response controls that learn from past plant trips

Automated process/plant startup and shutdown,

Plant condition/status monitoring (recognize normal operating maneuvering versus abnormal transients, develop operating guidance or smart annunciation)

The above problems will drastically cut operating costs and reduce downtime that directly effect economics, but they also improve safety by having fewer challenges of the safety systems and by having controlled startups and shutdowns rather than the typical severe trip shutdown and it's resulting thermal transients.

Other problems can result in economic savings such as better flow measurement. The present steam flow measurement is dp across a venturi flow element. The venturi has a not well known fouling problem that results in reduced thermal power output from the plant as the flow uncertainty increases. The reduced thermal power is a direct result of using the flow measurement in the plant calorimetric calculations which are used to calibrate the neutron flux measurements daily.

As stated earlier there are many problems needing solutions but there is little data on these problems to use in training neural networks and fuzzy logic. Some of the problems can be more closely monitored to get the data such as the venturi fouling but others may not yield more data such as plant trip response enhancement or smart annunciation. Tools for developing more data or plant simulations would be very useful to a utility engineer who needs a quick and simple system simulation to solve controls problems. Tools similar to the Honeywell system categorized discussed in the conference appear to be quite useful. With such a system and a few transient parameter dumps may be all that is needed to develop an approximate n-th order mathematical model of the systems which can then be used to solve control problems or even to teach neural networks about the plant behavior where data is not available.

In conclusion I again recognize that there is a significant knowledge of the utility industry in the neural network community but I also note that the utility industry does not have much knowledge of the neural network community or of the potential for development of quick, simple, and implementable solutions to complex plant problems. The utility industry needs the expertise and tools to tackle problems with neural networks, fuzzy logic, and artificial intelligence and to accomplish this it needs guidance and involvement from the neural network community.

Typical Utility Problem Areas

SAFETY:	Sensor Redundancy (auto associative networks)
	More Stable Control Algorithms (fuzzy)
	Fault Tolerant Control System Design
	Smart Annunciator driver (plant condition recognition and alarm masking)
ECONOMY:	Venturi Fouling and Power Derating
	Plant Calorimetric via diverse parameters
	Cross Comparison/Calibration from diverse parameters
	More Stable Control Algorithms
STABILITY:	More Stable Control Algorithms
	Semi-Automated Process Startup/Shutdown
	Recognition of True Signals in Noisy Measurements
	Dynamic System/Equipment Simulations
MAINTENANCE:	Maintenance Recommender Based on Operating/Calibration History
	ASME Section 11 Pump Degradation Testing/Diagnosis Online
	Valve Leak Detection
	Motor Degradation (current trace monitoring)

and simple typical simulations to resolve [illegible] problems [illegible] [illegible] [illegible] [illegible] [illegible] [illegible] [illegible] [illegible] [illegible] [illegible] where data is [illegible] available.

[illegible] recognize that there is [illegible] [illegible] of the [illegible] in [illegible] neural networks [illegible] [illegible] [illegible] [illegible] development of [illegible] [illegible] [illegible] and [illegible] [illegible] [illegible] [illegible] utility [illegible] [illegible] [illegible] [illegible] [illegible] [illegible] and [illegible] [illegible] [illegible] [illegible] [illegible] [illegible] [illegible] [illegible] neural networks [illegible]

Nuclear Utility Priority Areas

[illegible]	[illegible] (also bridge/valve networks)
	[illegible] of Alarm, [illegible]
	Digital Control System Design
	Steam Generator Drum Level Condition Monitoring and [illegible]
[illegible]	Fuel Loading and Power Distribution
	Prediction of [illegible] (dP) [illegible]
[illegible]	Data Compression/Extraction from Diverse Parameters
	Plant Startup Control Algorithms
[illegible]	[illegible] Control Algorithms
	[illegible] Redundancy [illegible]
[illegible]	Reconciliation of [illegible] Flow Measurements
	Dynamic System Equipment Simulations
MAINTENANCE	Maintenance Recommender Based on Operating/Calibration History
	ASME Section XI [illegible] Inspection/Testing [illegible]
	[illegible]
	Motor Degradation [illegible] Current [illegible]

B. Neural Network Methodologies

Backpropagation and its Applications

Bernard Widrow Michael A. Lehr
Stanford University Department of Electrical Engineering, Stanford, CA 94305-4055

Backpropagation remains the most widely used neural network algorithm. We present a basic review of this method and its applications. The backpropagation algorithm's roots are also explored in a discussion of its predecessor, the LMS Algorithm.

Introduction

The field of neural networks has enjoyed major advances since 1960, a year which saw the introduction of two of the earliest feedforward neural network algorithms: the Perceptron rule [1] and the LMS algorithm (or Widrow-Hoff rule) [2]. Around 1961, Widrow and his students devised Madaline Rule I (MRI), the earliest learning rule for feedforward networks with multiple adaptive elements [3]. The major extension of the feedforward neural network beyond Madaline I took place in 1971 when Werbos developed a backpropagation algorithm for training multilayer networks. In 1974, this algorithm was incorporated in his doctoral dissertation [4]†. Werbos's work remained almost unknown in the scientific community. In 1982 Parker rediscovered the technique and in 1985, published a report on it at MIT [6]. Not long after Parker published his findings, Rumelhart, Hinton, and Williams [7] refined the technique and succeeded in making it widely known.

Early applications of LMS and MRI were developed by Widrow and his students in their studies of speech and pattern recognition, weather forecasting, and adaptive controls. After some success in these areas, work shifted in the mid-1960's to adaptive filtering and adaptive signal processing. This proved to be a fruitful avenue for research with applications including adaptive antennas, adaptive inverse controls, adaptive noise cancelling, and seismic signal processing. Outstanding work by R. W. Lucky and others at Bell Laboratories led to major commercial applications of adaptive filters and the LMS algorithm to adaptive equalization in high speed modems and to adaptive echo cancellers for long distance telephone and satellite circuits.

†We should note, however, that in the field of variational calculus the idea of error backpropagation through nonlinear systems existed centuries before Werbos first thought to apply the concept to neural networks. In the past 25 years, these methods have been used widely in the the field of optimal control, as discussed by Le Cun [5].

With the development of the backpropagation algorithm, it has now become possible to successfully attack problems requiring neural networks with high degrees of nonlinearity and high precision. Examples are shown in [8]. Backpropagation networks with fewer than 150 neural elements have been successfully applied in vehicular control simulations, speech generation, and undersea mine detection. Small networks have also been used successfully in airport explosive detection, expert systems, nonlinear system identification, and scores of other applications. Furthermore, efforts to develop parallel neural network hardware are advancing rapidly, and these systems are now becoming available for attacking more difficult problems like continuous speech recognition.

The networks used to solve the above applications varied widely in size and topology. A basic component of the neural networks used in all of these applications, however, is the adaptive linear combiner.

The Adaptive Linear Combiner

The adaptive linear combiner is diagrammed in Fig. 1. Its output is a linear combination of its inputs. In a digital implementation, this element receives at time k an input signal vector or input pattern vector $\mathbf{X}_k = [x_0, x_{1_k}, x_{2_k}, \ldots x_{n_k}]^T$, and a desired response d_k, a special input used to effect learning. The components of the input vector are weighted by a set of adaptive coefficients, the weight vector $\mathbf{W}_k = [w_{0_k}, w_{1_k}, w_{2_k}, \ldots w_{n_k}]^T$. The sum of the weighted inputs is then computed, producing a linear output, the inner product $s_k = \mathbf{X}_k^T \mathbf{W}_k$. The components of $\mathbf{X}_k$ may be either continuous analog values or binary values. The weights are essentially continuously variable, and can take on negative as well as positive values. By using a training algorithm to adapt its weights, the adaptive linear combiner has the ability to implement a wide range of responses to the patterns in a given training set.

When a nonlinearity is placed on the output of an adaptive linear combiner, the cascade is called an Adaline (ADAptive LINear Element). In the 1960's, the Adaline's nonlinearity was usually the sign or signum element. The Adaline was also known as a linear threshold element whose output was either $+1$ or -1. Modern neural networks usually use an "S"-shaped "sigmoid" function, a "soft" quantizer which varies smoothly from -1 to $+1$. An Adaline using a sigmoid nonlinearity is sometimes referred to as a Sigmoid Adaline. The Adaline usually includes a bias weight w_{0_k} which is connected to a constant input, $x_0 = +1$. This weight effectively

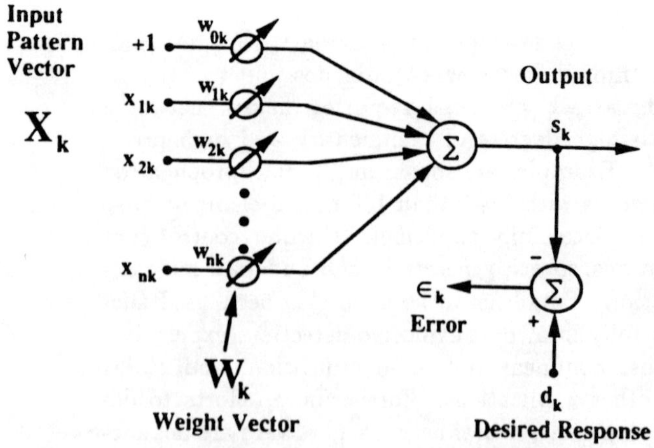

Figure 1: Adaptive linear combiner.

controls the threshold level of the quantizer or sigmoid.

Multilayer Networks

The Madaline networks of the 1960s had adaptive first layers and fixed threshold functions in the second (output) layers [8]. The feedforward neural networks of today often have many layers, and usually all layers are adaptive. The backpropagation networks of Rumelhart *et al.* [9] are perhaps the best-known examples of multilayer networks. A fully-connected three-layer feedforward adaptive network is illustrated in Fig. 2. In a fully-connected layered network, each Adaline receives inputs from every output in the preceding layer.

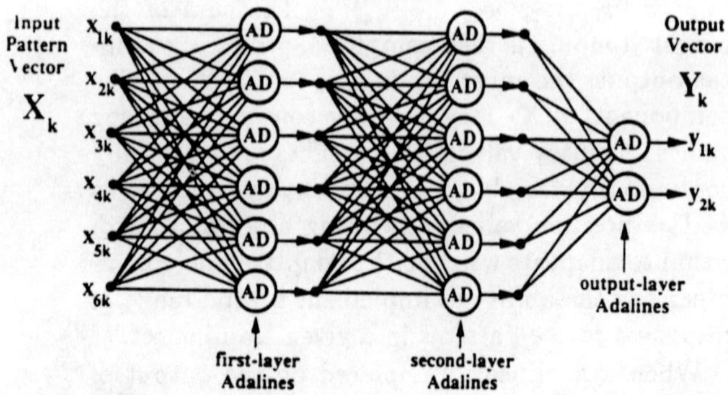

Figure 2: A three-layer adaptive neural network.

During training, the responses of the output elements in the network are compared with a corresponding set of desired responses. Error signals associated with the elements of the output layer are thus readily computed, so adaptation of the output layer is straightforward. The fundamental difficulty associated with adapting a layered network lies in obtaining "error signals" for hidden layer Adalines, that is, for Adalines in layers other than the output layer. The backpropagation algorithm provides a method for establishing these error signals.

Steepest-Descent Rules

Usually, the objective of adaptation for a feedforward neural network is to reduce the error between the desired response and the network's response averaged in some way over the training set. The most common error function is mean-square-error (MSE), although in some situations other error criteria may be more appropriate [8]. The most popular approaches to mean-square-error reduction in both single-element and multi-element networks are based upon the method of steepest descent. More sophisticated gradient approaches such as quasi-Newton and conjugate gradient techniques often have better convergence properties, but the conditions under which the additional complexity is warranted are not generally known. The discussion that follows is restricted to minimization of MSE by the method of steepest descent [10]. More sophisticated learning procedures usually require many of the same computations used in the basic steepest-descent procedure.

Adaptation of a network by steepest-descent starts with an arbitrary initial value $\mathbf{W}_0$ for the system's weight vector. The gradient of the mean-square-error function is measured and the weight vector is altered in the direction corresponding to the negative of the measured gradient. This procedure is repeated, causing the MSE to be successively reduced on average and causing the weight vector to approach a locally optimal value.

The method of steepest descent can be described by the relation

$$\mathbf{W}_{k+1} = \mathbf{W}_k + \mu(-\nabla_k), \tag{1}$$

where μ is a parameter that controls stability and rate of convergence, and ∇_k is the value of the gradient at a point on the MSE surface corresponding to $\mathbf{W} = \mathbf{W}_k$.

The LMS Algorithm The LMS algorithm works by performing approximate steepest descent on the mean-square-error surface in weight space. Because it is a quadratic function of the weights, this surface is convex and has a unique (global) minimum[‡]. An instantaneous gradient based upon the square of the

[‡]unless the autocorrelation matrix of the pattern vector set has m zero eigenvalues, in which case the minimum MSE solution will be an m dimensional subspace in weight space [11].

instantaneous error is

$$\hat{\nabla}_k = \frac{\partial \epsilon_k^2}{\partial \mathbf{W}_k} = \left\{ \begin{array}{c} \frac{\partial \epsilon_k^2}{\partial w_{0k}} \\ \vdots \\ \frac{\epsilon_k^2}{\partial w_{nk}} \end{array} \right\}. \qquad (2)$$

LMS works by using this crude gradient estimate in place of the true gradient ∇_k. Making this replacement into Eq. (1) yields

$$\mathbf{W}_{k+1} = \mathbf{W}_k + \mu\left(-\hat{\nabla}_k\right) = \mathbf{W}_k - \mu \frac{\partial \epsilon_k^2}{\partial \mathbf{W}_k} \qquad (3)$$

The instantaneous gradient is used because (a) it is an unbiased estimate of the true [11] gradient, and (b) it is easily computed from single data samples. The true gradient is generally difficult to obtain. Computing it would involve averaging the instantaneous gradients associated with all patterns in the training set. This is usually impractical and almost always inefficient.

The present error ϵ_k is defined to be the difference between the desired response d_k and the linear output $s_k = \mathbf{W}_k^T \mathbf{X}_k$ before adaptation:

$$\epsilon_k \triangleq d_k - \mathbf{W}_k^T \mathbf{X}_k. \qquad (4)$$

Performing the differentiation in Eq. (3) and replacing the error with (4) gives

$$\begin{aligned} \mathbf{W}_{k+1} &= \mathbf{W}_k - 2\mu\epsilon_k \frac{\partial \epsilon_k}{\partial \mathbf{W}_k} \\ &= \mathbf{W}_k - 2\mu\epsilon_k \frac{\partial \left(d_k - \mathbf{W}_k^T \mathbf{X}_k\right)}{\partial \mathbf{W}_k}. \qquad (5) \end{aligned}$$

Noting that d_k and $\mathbf{X}_k$ are independent of $\mathbf{W}_k$, we obtain

$$\mathbf{W}_{k+1} = \mathbf{W}_k + 2\mu\epsilon_k \mathbf{X}_k. \qquad (6)$$

This is the LMS algorithm. The learning constant μ determines stability and convergence rate. For input patterns independent over time, convergence of the mean and variance of the weight vector is ensured under fairly general conditions [12, 8] if

$$0 < \mu < \frac{1}{3 trace[\mathbf{R}]}, \qquad (7)$$

where $\mathbf{R}$ is the input correlation matrix $\mathrm{E}[\mathbf{X}_k \mathbf{X}_k^T]$ and where $trace[\mathbf{R}] = \sum$(diagonal elements of $\mathbf{R}$) is the sum of the signal powers of the components of the $\mathbf{X}$-vectors, i.e. $\mathrm{E}(\mathbf{X}^T \mathbf{X})$. With μ set within this range, the LMS algorithm converges in the mean to $\mathbf{W}^*$, the optimal Wiener solution [11].

In the LMS algorithm, and other iterative steepest-descent procedures, use of the instantaneous gradient is perfectly justified if the step size is small. For small μ, $\mathbf{W}$ will remain essentially constant over a relatively small number of training presentations, K. The total weight change during this period will be proportional to

$$\begin{aligned} -\sum_{\ell=0}^{K-1} \frac{\partial \epsilon_{k+\ell}^2}{\partial \mathbf{W}_{k+\ell}} &\simeq -K\left(\frac{1}{K}\sum_{\ell=0}^{K-1} \frac{\partial \epsilon_{k+\ell}^2}{\partial \mathbf{W}_k}\right) \\ &= -K\frac{\partial}{\partial \mathbf{W}_k}\left(\frac{1}{K}\sum_{\ell=0}^{K-1} \epsilon_{k+\ell}^2\right) \\ &\simeq -K\frac{\partial \xi}{\partial \mathbf{W}_k}, \qquad (8) \end{aligned}$$

where ξ denotes the mean-square-error function. Thus, on average the weights follow the true gradient.

Backpropagation for the Sigmoid Adaline

Fig. 3 shows a "Sigmoid Adaline" element which incorporates a sigmoidal nonlinearity. The input-output relation of the sigmoid can be denoted by $y_k = sgm(s_k)$. A typical sigmoid function is the hyperbolic tangent:

$$y_k = tanh(s_k) = \left(\frac{1 - e^{-2s_k}}{1 + e^{-2s_k}}\right). \qquad (9)$$

We shall adapt this Adaline with the objective of

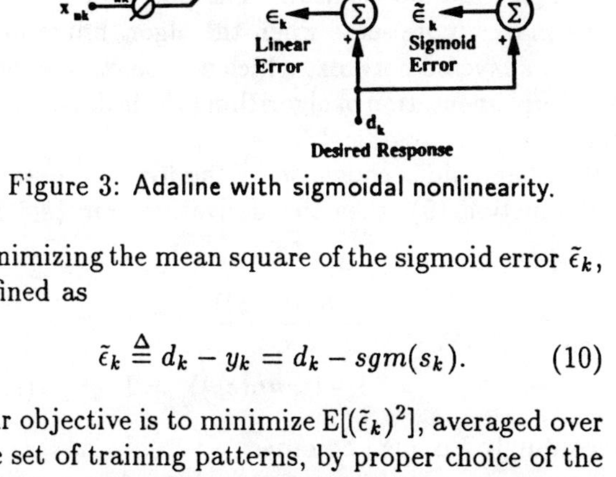

Figure 3: Adaline with sigmoidal nonlinearity.

minimizing the mean square of the sigmoid error $\tilde{\epsilon}_k$, defined as

$$\tilde{\epsilon}_k \triangleq d_k - y_k = d_k - sgm(s_k). \qquad (10)$$

Our objective is to minimize $\mathrm{E}[(\tilde{\epsilon}_k)^2]$, averaged over the set of training patterns, by proper choice of the weight vector. To accomplish this, we shall derive a backpropagation algorithm for the Sigmoid

Adaline element. An instantaneous gradient is obtained with each input vector presentation, and the method of steepest descent is used to minimize error as was done with the LMS algorithm of Eq. (6).

Referring to Fig. 3, the instantaneous gradient estimate obtained during presentation of the kth input vector $\mathbf{X}_k$ is given by

$$\hat{\nabla}_k = \frac{\partial(\tilde{\epsilon}_k)^2}{\partial \mathbf{W}_k} = 2\tilde{\epsilon}_k \frac{\partial \tilde{\epsilon}_k}{\partial \mathbf{W}_k}. \tag{11}$$

Differentiating Eq. (10) yields

$$\frac{\partial \tilde{\epsilon}_k}{\partial \mathbf{W}_k} = -\frac{\partial sgm(s_k)}{\partial \mathbf{W}_k} = -sgm'(s_k)\frac{\partial s_k}{\partial \mathbf{W}_k}. \tag{12}$$

We may note that

$$s_k = \mathbf{X}_k^T \mathbf{W}_k. \tag{13}$$

Therefore,

$$\frac{\partial s_k}{\partial \mathbf{W}_k} = \mathbf{X}_k. \tag{14}$$

Substituting into Eq. (12) gives

$$\frac{\partial \tilde{\epsilon}_k}{\partial \mathbf{W}_k} = -sgm'(s_k)\mathbf{X}_k. \tag{15}$$

Inserting this into Eq. (11) yields

$$\hat{\nabla}_k = -2\tilde{\epsilon}_k sgm'(s_k)\mathbf{X}_k. \tag{16}$$

Using this gradient estimate with the method of steepest descent provides a means for minimizing the mean-square-error even after the summed signal s_k goes through the nonlinear sigmoid. The algorithm is

$$\begin{aligned} \mathbf{W}_{k+1} &= \mathbf{W}_k + \mu(-\hat{\nabla}_k) &(17)\\ &= \mathbf{W}_k + 2\mu\tilde{\epsilon}_k sgm'(s_k)\mathbf{X}_k. &(18) \end{aligned}$$

Algorithm (18) is the *backpropagation* algorithm for the single Adaline element. The backpropagation name makes more sense when the algorithm is utilized in a layered network, which will be studied below. Implementation of algorithm (18) is illustrated in Fig. 4.

If the sigmoid is chosen to be the hyperbolic tangent function (9), then the derivative $sgm'(s_k)$ is given by

$$\begin{aligned} sgm'(s_k) &= \frac{\partial\,(tanh(s_k))}{\partial s_k} \\ &= 1 - (tanh(s_k))^2 = 1 - y_k^2. &(19) \end{aligned}$$

Accordingly Eq. (18) becomes

$$\mathbf{W}_{k+1} = \mathbf{W}_k + 2\mu\tilde{\epsilon}_k(1 - y_k^2)\mathbf{X}_k. \tag{20}$$

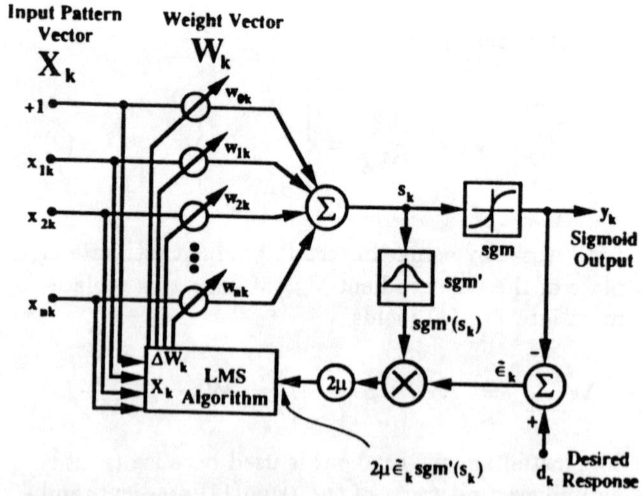

Figure 4: Implementation of backpropagation for the Sigmoid Adaline element.

Backpropagation for Networks The publication of the backpropagation technique by Rumelhart *et al.* [7] has unquestionably been the most influential development in the field of neural networks during the past decade. In retrospect, the technique seems simple. Nonetheless, largely because early neural network research dealt almost exclusively with hard-limiting nonlinearities, the idea eluded neural network researchers throughout the 1960's.

The basic concepts of backpropagation are easily grasped. Unfortunately, these simple ideas are often obscured by relatively intricate notation, so formal derivations of the backpropagation rule are often tedious. We present an informal derivation of the algorithm and illustrate how it works for the simple network shown in Fig. 5. A more formal derivation would require the use of ordered derivatives to precisely identify which quantities are treated as variables in each of the partial derivatives used below [13].

The backpropagation technique is a substantial generalization of the single Sigmoid Adaline case discussed in the previous section. When applied to multi-element networks, the backpropagation technique adjusts the weights in the direction opposite to the instantaneous error gradient:

$$\hat{\nabla}_k = \frac{\partial \varepsilon_k^2}{\partial \mathbf{W}_k} = \left\{ \begin{array}{c} \frac{\partial \varepsilon_k^2}{\partial w_{1k}} \\ \vdots \\ \frac{\partial \varepsilon_k^2}{\partial w_{nk}} \end{array} \right\}. \tag{21}$$

Now, however, $\mathbf{W}_k$ is a long n-component vector of all weights in the entire network. The instantaneous sum squared error ε_k^2 is the sum of the squares of the errors at each of the N_y outputs of the network.

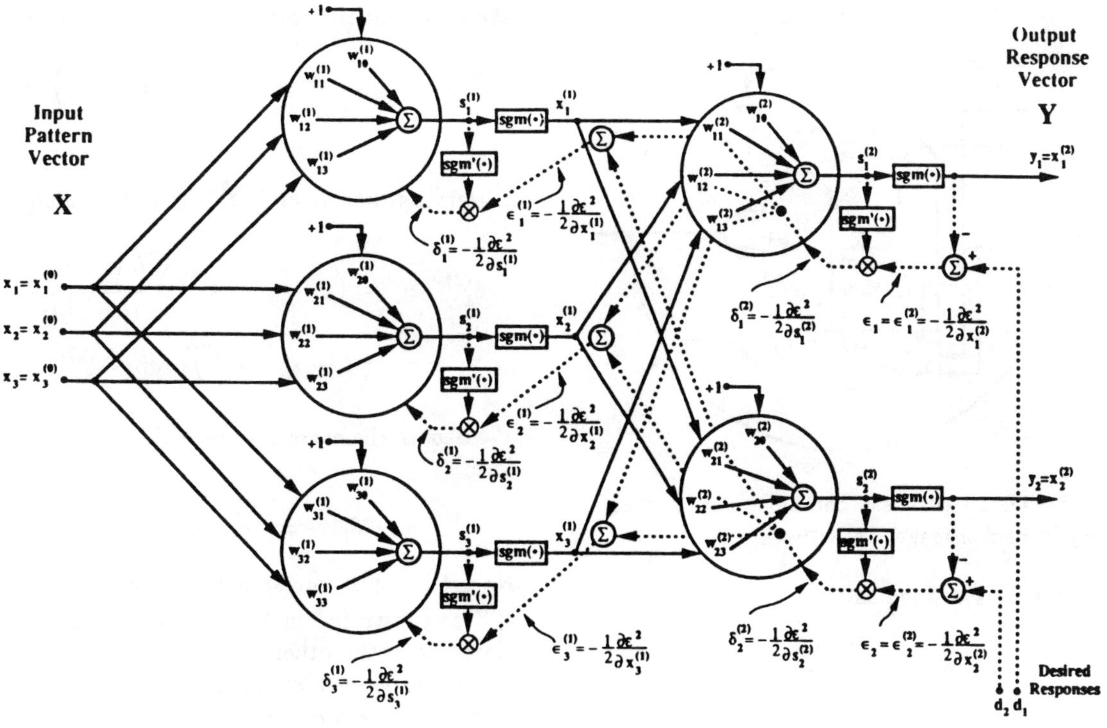

Figure 5: Example two-layer backpropagation network architecture.

Thus

$$\varepsilon_k^2 = \sum_{i=1}^{N_y} \epsilon_{ik}^2. \tag{22}$$

In the network example shown in Fig. 5, the sum square error is given by

$$\varepsilon^2 = (d_1 - y_1)^2 + (d_2 - y_2)^2, \tag{23}$$

where we now suppress the time index k for convenience.

In its simplest form, backpropagation training begins by presenting an input pattern vector $\mathbf{X}$ to the network, sweeping forward through the system to generate an output response vector $\mathbf{Y}$, and computing the errors at each output. The next step involves sweeping the effects of the errors backward through the network to associate a "square error derivative" δ with each Adaline, computing a gradient from each δ, and finally updating the weights of each Adaline based upon the corresponding gradient. A new pattern is then presented and the process is repeated. The initial weight values are normally set to small random numbers. The algorithm will not work properly with multilayer networks if the initial weights are either zero or poorly chosen nonzero values.

We can get some idea about what is involved in the calculations associated with the backpropagation algorithm by examining the network of Fig. 5. Each of the five large circles represents a linear combiner, as well as some associated signal paths for error backpropagation, and the corresponding adaptive machinery for updating the weights. This detail is shown in Fig. 6. The solid lines in these diagrams represent forward signal paths through the network, and the dotted lines represent the separate backward paths that are used in association with calculations of the square error derivatives δ. From Fig. 5, we see that the calculations associated with the backward sweep are of a complexity which is roughly equal to that represented by the forward pass through the network. The backward sweep requires the same number of function calculations as the forward sweep, but it requires no weight multiplications in the first layer.

As stated above, after a pattern has been presented to the network, and the response error of each output has been calculated, the next step of the backpropagation algorithm involves finding the instantaneous square error derivative δ associated with each summing junction in the network. The square error derivative associated with the jth Ada-

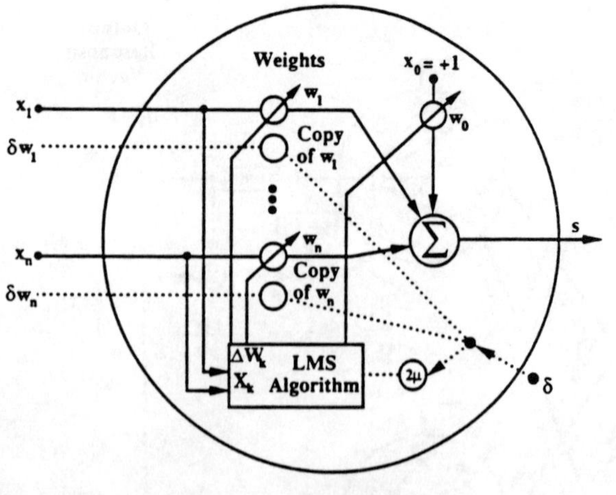

Figure 6: Detail of linear combiner and associated circuitry in backpropagation network.

line in layer ℓ is defined as[§]

$$\delta_j^{(\ell)} \triangleq -\frac{1}{2}\frac{\partial \varepsilon^2}{\partial s_j^{(\ell)}}. \tag{24}$$

Each of these derivatives in essence tells us how sensitive the sum square output error of the network is to changes in the linear output of the associated Adaline element.

The instantaneous square error derivatives are first computed for each element in the output layer. The calculation is simple. As an example, below we derive the required expression for $\delta_1^{(2)}$, the derivative associated with the top Adaline element in the output layer of Fig. 5. We begin with the definition of $\delta_1^{(2)}$ from Eq. (24),

$$\delta_1^{(2)} \triangleq -\frac{1}{2}\frac{\partial \varepsilon^2}{\partial s_1^{(2)}}. \tag{25}$$

Expanding the squared error term ε^2 by Eq. (23) yields

$$\begin{aligned}\delta_1^{(2)} &= -\frac{1}{2}\frac{\partial\left((d_1-y_1)^2+(d_2-y_2)^2\right)}{\partial s_1^{(2)}} \qquad (26)\\ &= -\frac{1}{2}\frac{\partial\left(d_1-sgm(s_1^{(2)})\right)^2}{\partial s_1^{(2)}}\\ &\quad -\frac{1}{2}\frac{\partial\left(d_2-sgm(s_2^{(2)})\right)^2}{\partial s_1^{(2)}}. \qquad (27)\end{aligned}$$

[§]In Fig. 5, all notation follows the convention that superscripts within parentheses indicate the layer number of the associated Adaline or input node, while subscripts identify the associated Adaline(s) within a layer.

We note that the second term is zero. Accordingly,

$$\delta_1^{(2)} = -\frac{1}{2}\frac{\partial\left(d_1-sgm(s_1^{(2)})\right)^2}{\partial s_1^{(2)}}. \tag{28}$$

Observing that d_1 and $s_1^{(2)}$ are independent yields

$$\begin{aligned}\delta_1^{(2)} &= -\left(d_1-sgm(s_1^{(2)})\right)\frac{\partial\left(-sgm(s_1^{(2)})\right)}{\partial s_1^{(2)}}\\ &= \left(d_1-sgm(s_1^{(2)})\right)sgm'(s_1^{(2)}). \qquad (29)\end{aligned}$$

We denote the error $d_1 - sgm(s_1^{(2)})$, by $\epsilon_1^{(2)}$. Therefore,

$$\delta_1^{(2)} = \epsilon_1^{(2)} sgm'(s_1^{(2)}). \tag{30}$$

Note that this corresponds to the computation of $\delta_1^{(2)}$ as illustrated in Fig. 5. The value of δ associated with the other output element in the figure can be expressed in an analogous fashion. Thus each square error derivative δ in the output layer is computed by multiplying the output error associated with that element by the derivative of the associated sigmoidal nonlinearity. Note from Eq. (19) that if the sigmoid function is the hyperbolic tangent, Eq. (30) becomes simply

$$\delta_1^{(2)} = \epsilon_1^{(2)}(1-(y_1)^2). \tag{31}$$

Developing expressions for the square error derivatives associated with hidden layers is not much more difficult (refer to Fig. 5). We need an expression for $\delta_1^{(1)}$, the square error derivative associated with the top element in the first layer of Fig. 5. The derivative $\delta_1^{(1)}$ is defined by

$$\delta_1^{(1)} \triangleq -\frac{1}{2}\frac{\partial \varepsilon^2}{\partial s_1^{(1)}}. \tag{32}$$

Expanding this by the chain rule, noting that ε^2 is determined entirely by the values of $s_1^{(2)}$ and $s_2^{(2)}$, yields

$$\delta_1^{(1)} = -\frac{1}{2}\left(\frac{\partial \varepsilon^2}{\partial s_1^{(2)}}\frac{\partial s_1^{(2)}}{\partial s_1^{(1)}} + \frac{\partial \varepsilon^2}{\partial s_2^{(2)}}\frac{\partial s_2^{(2)}}{\partial s_1^{(1)}}\right). \tag{33}$$

Using the definitions of $\delta_1^{(2)}$ and $\delta_2^{(2)}$, and then substituting expanded versions of Adaline linear outputs $s_1^{(2)}$ and $s_2^{(2)}$ gives

$$\delta_1^{(1)} = \delta_1^{(2)}\frac{\partial s_1^{(2)}}{\partial s_1^{(1)}} + \delta_2^{(2)}\frac{\partial s_2^{(2)}}{\partial s_1^{(1)}} \tag{34}$$

$$= \delta_1^{(2)} \frac{\partial}{\partial s_1^{(1)}} \left(w_{10}^{(2)} + \sum_{i=1}^{3} w_{1i}^{(2)} sgm(s_i^{(1)}) \right)$$
$$+ \delta_2^{(2)} \frac{\partial}{\partial s_1^{(1)}} \left(w_{20}^{(2)} + \sum_{i=1}^{3} w_{2i}^{(2)} sgm(s_i^{(1)}) \right).$$

Noting that $\partial[sgm(s_i^{(\ell)})]/\partial s_j^{(\ell)} = 0, i \neq j$, leaves

$$\begin{aligned} \delta_1^{(1)} &= \delta_1^{(2)} w_{11}^{(2)} sgm'(s_1^{(1)}) + \delta_2^{(2)} w_{21}^{(2)} sgm'(s_1^{(1)}) \\ &= \left[\delta_1^{(2)} w_{11}^{(2)} + \delta_2^{(2)} w_{21}^{(2)} \right] sgm'(s_1^{(1)}). \end{aligned} \quad (35)$$

Now, we make the following definition:

$$\epsilon_1^{(1)} \triangleq \delta_1^{(2)} w_{11}^{(2)} + \delta_2^{(2)} w_{21}^{(2)}. \quad (36)$$

Accordingly,

$$\delta_1^{(1)} = \epsilon_1^{(1)} sgm'(s_1^{(1)}). \quad (37)$$

Referring to Fig. 5, we can trace through the circuit to verify that $\delta_1^{(1)}$ is computed in accord with Eqs. (36) and (37). The easiest way to find values of δ for all the Adaline elements in the network is to follow the schematic diagram of Fig. 5.

Thus, the procedure for finding $\delta^{(\ell)}$, the square error derivative associated with a given Adaline in hidden layer ℓ, involves respectively multiplying each derivative $\delta^{(\ell+1)}$ associated with each element in the layer immediately downstream from a given Adaline by the weight which connects it to the given Adaline. These weighted square error derivatives are then added together, producing an error term $\epsilon^{(\ell)}$, which, in turn, is multiplied by $sgm'(s^{(\ell)})$, the derivative of the given Adaline's sigmoid function at its current operating point. If a network has more than two layers, this process of backpropagating the instantaneous square error derivatives from one layer to the immediately preceding layer is successively repeated until a square error derivative δ is computed for each Adaline in the network. This is easily shown at each layer by repeating the chain rule argument associated with Eq. (33).

We now have a general method for finding a derivative δ for each Adaline element in the network. The next step is to use these δ's to obtain the corresponding gradients. Consider an Adaline somewhere in the network which, during presentation k, has a weight vector $\mathbf{W}_k$, an input vector $\mathbf{X}_k$, and a linear output $s_k = \mathbf{W}_k^T \mathbf{X}_k$.
The instantaneous gradient for this Adaline element is

$$\hat{\nabla}_k = \frac{\partial \varepsilon_k^2}{\partial \mathbf{W}_k}. \quad (38)$$

This can be written as

$$\hat{\nabla}_k = \frac{\partial \varepsilon_k^2}{\partial \mathbf{W}_k} = \frac{\partial \varepsilon_k^2}{\partial s_k} \frac{\partial s_k}{\partial \mathbf{W}_k}. \quad (39)$$

Note that $\mathbf{W}_k$ and $\mathbf{X}_k$ are independent so

$$\frac{\partial s_k}{\partial \mathbf{W}_k} = \frac{\partial \mathbf{W}_k^T \mathbf{X}_k}{\partial \mathbf{W}_k} = \mathbf{X}_k. \quad (40)$$

Therefore,

$$\hat{\nabla}_k = \frac{\partial \varepsilon_k^2}{\partial s_k} \mathbf{X}_k. \quad (41)$$

For this element,

$$\delta_k = -\frac{1}{2} \frac{\partial \varepsilon_k^2}{\partial s_k}. \quad (42)$$

Accordingly,

$$\hat{\nabla}_k = -2\delta_k \mathbf{X}_k. \quad (43)$$

Updating the weights of the Adaline element using the method of steepest descent with the instantaneous gradient is a process represented by

$$\mathbf{W}_{k+1} = \mathbf{W}_k + \mu(-\hat{\nabla}_k) = \mathbf{W}_k + 2\mu\delta_k \mathbf{X}_k. \quad (44)$$

Thus, after backpropagating all square error derivatives, we complete a backpropagation iteration by adding to each weight vector the corresponding input vector scaled by the associated square error derivative. Eq. (44) and the means for finding δ_k comprise the general weight update rule of the backpropagation algorithm.

There is a great similarity between Eq. (44) and the LMS algorithm (6), but one should view this similarity with caution. The quantity δ_k, defined as a squared error derivative, might appear to play the same role in backpropagation as that played by the error in the LMS algorithm. However, δ_k is not an error. Adaptation of the given Adaline is effected to reduce the squared output error ε_k^2, not δ_k of the given Adaline or of any other Adaline in the network. The objective is not to reduce the δ_k's of the network, but to reduce ε_k^2 at the network output.

It is interesting to examine the weight updates that backpropagation imposes on the Adaline elements in the output layer. Substituting Eq. (29) into Eq. (44) reveals that the Adaline which provides output y_1 in Fig. 5 is updated by the rule

$$\mathbf{W}_{k+1} = \mathbf{W}_k + 2\mu\epsilon_1^{(2)} sgm'(s_1^{(2)}) \mathbf{X}_k. \quad (45)$$

This rule turns out to be identical to the single Adaline version (18) of the backpropagation rule. This is not surprising since the output Adaline is provided with both input signals and desired responses, so its training circumstance is the same as that experienced by an Adaline trained in isolation.

In our experience, weight initialization and input normalization can strongly affect the training speed of neural networks. In recent years, we have relied primarily on a heuristic designed to ensure that none of the sigmoids will be saturated during the early stages of training. This saturation can impede learning since the derivative of a saturated sigmoid is near zero, and this derivative is proportional to the gradient vector of the corresponding Adaline. In the input layer, we avoid this problem by normalizing the input patterns so each component is zero-mean with a variance of roughly 0.2¶. The weights are then randomly initialized to have a distribution which is zero-mean i.i.d. Gaussian with a variance $1.5/N$, where N is the number of inputs to the Adaline being initialized. If the standard $\tanh(\cdot)$ sigmoid is used, this causes the signal level at the input of each sigmoid in the network to be distributed roughly zero-mean Gaussian with a variance of 0.3 when training commences. The output of each sigmoid will then be distributed with a variance of roughly 0.2, just like the network inputs. Due to the linearity of the sigmoid at small signal levels, the use of relatively low-variance sigmoid inputs helps keep the size of the derivatives backpropagated through the network relatively uniform, if the number of Adalines per layer does not vary drastically.

Another weight initialization approach with which we have had considerable success in many instances is one devised by Nguyen [14]. This approach involves choosing the initial weights of each hidden layer in a quasi-random manner which ensures that at each position in a layer's input space the outputs of all but a few of its Adalines will be saturated, while ensuring that each Adaline in the layer is unsaturated in some region of its input space.

One of the most promising new areas of neural network research involves backpropagation variants for training various recurrent (signal feedback) networks. Recently, backpropagation rules have been devised for training recurrent networks to learn static associations [15, 16]. More interesting is the on-line technique of Williams and Zipser [17] which allows a wide class of recurrent networks to learn dynamic associations and trajectories. A more general and computationally viable variant of this technique has been advanced by Narendra and Parthasarathy [18]. These on-line methods are generalizations of a well-known steepest-descent algorithm for training linear IIR filters [19, 11].

¶To do this we merely compute a mean and sample-variance for each network input over the training pattern set. Then, for each input pattern vector $\mathbf{X}$ in both the training and generalization sets, we subtract the appropriate mean from each component, divide by the corresponding standard deviation, and multiply the result by the square root of 0.2.

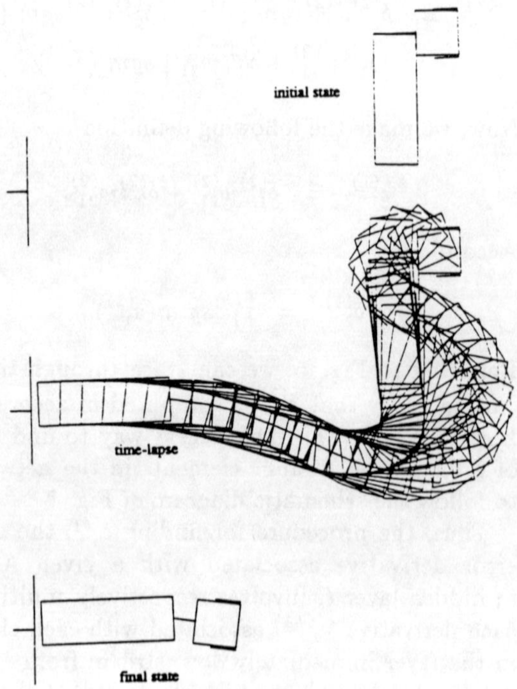

Figure 7: Example Truck Backup Sequence.

An equivalent technique which is usually far less computationally intensive also exists [4, 7, 20]. This approach, called "backpropagation-through-time," has been used by Nguyen and Widrow [21] to enable a neural network to learn without a teacher how to back up a computer-simulated trailer truck to a loading dock (Fig. 7). The approach used is described in detail in another paper in these proceedings [22]. It involved two networks: a neural emulator trained by backpropagation to model the kinematics of the truck, and a neural controller trained by backpropagation-through-time to generate the required steering signals. The process of backing up a truck is a complicated and highly nonlinear steering task. Nevertheless, the controller network was able to learn of its own accord to solve this problem. Once trained, the controller network could successfully back up the truck from any initial position and orientation in front of the loading dock.

Summary

The backpropagation algorithm can be employed to solve a wide range of problems. There is an enormous pool of applications to which it has has been successfully applied—only a few are mentioned in this paper—and there are scores more that will become realities in the next few years. Just as neural networks can be used to emulate the kinematics of a trailer truck and control its steering, they can also be used to model and control processes in the power industry. Likely applications may include magnetic bearing control to increase generator efficiency and reduce bearing wear; nuclear reactor monitoring and modeling for safety; and control of fuel metering to improve voltage regulation and efficiency. In the coming years, we expect to see a growing amount of research on backpropagation applications in the power industry. In the long run, we hope and believe the industry will benefit greatly from the fruits of this activity.

Acknowledgments

This work was sponsored by SDIO Innovative Science and Technology Office and managed by ONR under contract #N00014-86-K-0718, by the Department of the Army Belvoir R D & E Center under contract #DAAK70-92-K-0001, by a grant from the Lockheed Missiles and Space Company, by NASA under contract #NCA2-389, and by Rome Air Development Center under contract #F30602-88-D-0025, subcontract E-21-T22-S1.

References

[1] F. Rosenblatt. On the convergence of reinforcement procedures in simple perceptrons. *Cornell Aeronautical Laboratory Report VG-1196-G-4*, Buffalo, New York, February 1960.

[2] B. Widrow and M. E. Hoff, Jr. Adaptive switching circuits. In *1960 IRE Western Electric Show and Convention Record, Part 4*, pages 96–104, August 23 1960.

[3] B. Widrow. Generalization and information storage in networks of adaline "neurons". In M. Yovitz, G. Jacobi, and G. Goldstein, editors, *Self-Organizing Systems 1962*, pages 435–461. Spartan Books, Washington, DC, 1962.

[4] P. Werbos. *Beyond Regression: New Tools for Prediction and Analysis in the Behavioral Sciences*. PhD thesis, Harvard University, Cambridge, MA, August 1974.

[5] Y. le Cun. A theoretical framework for backpropagation. In D. Touretzky, G. Hinton, and T. Sejnowski, editors, *Proceedings of the 1988 Connectionist Models Summer School*, pages 21–28, San Mateo, CA, June 17-26 1988. Morgan Kaufmann.

[6] D. Parker. Learning-logic. Technical Report TR-47, Center for Computational Research in Economics and Management Science, MIT, April 1985.

[7] D. E. Rumelhart, G. E. Hinton, and R. J. Williams. Learning internal representations by error propagation. In D. E. Rumelhart and J. L. McClelland, editors, *Parallel Distributed Processing*, volume 1, chapter 8. The MIT Press, Cambridge, MA, 1986.

[8] B. Widrow and M. A. Lehr. 30 years of adaptive neural networks: Perceptron, madaline, and backpropagation. *Proc. IEEE*, pages 1415–1442, September 1990.

[9] D. E. Rumelhart and J. L. McClelland, editors. *Parallel Distributed Processing*, volume 1 and 2. The MIT Press, Cambridge, MA, 1986.

[10] R. V. Southwell. *Relaxation Methods in Engineering Science*. Oxford, New York, 1940.

[11] B. Widrow and S. D. Stearns. *Adaptive Signal Processing*. Prentice-Hall, Englewood Cliffs, NJ, 1985.

[12] L. L. Horowitz and K. D. Senne. Performance advantage of complex lms for controlling narrow-band adaptive arrays. *IEEE Trans. Circuits Systems*, CAS-28(6):562–576, June 1981.

[13] P. Werbos. Generalization of backpropagation with application to a recurrent gas market model. *Neural Networks*, 1:339–356, 1988.

[14] D. Nguyen and B. Widrow. Improving the learning speed of 2-layer neural networks by choosing initial values of the adaptive weights. In *Proceedings of the International Joint Conference on Neural Networks*, San Diego, CA, June 1990.

[15] F. J. Pineda. Generalization of backpropagation to recurrent neural networks. *Physical Review Letters*, 18(59):2229–2232, 1987.

[16] L. B. Almeida. A learning rule for asynchronous perceptrons with feedback in a combinatorial environment. In *Proceedings of the IEEE First International Conference on Neural Networks*, volume II, pages 609–618, San Diego, CA, June 1987.

[17] R. J. Williams and D. Zipser. A learning algorithm for continually running fully recurrent neural networks. ICS Report 8805, Institute for Cognitive Science, University of California at San Diego, La Jolla, CA 92093, October 1988.

[18] K. S. Narendra and K. Parthasarathy. Identification and control of dynamical systems using neural networks. *IEEE Transactions on Neural Networks*, 1(1):4–27, March 1990.

[19] S. A. White. An adaptive recursive digital filter. In *Proc. 9th Asilomar Conf. Circuits Syst. Comput.*, page 21, November 1975.

[20] B. Pearlmutter. Learning state space trajectories in recurrent neural networks. In D. Touretzky, G. Hinton, and T. Sejnowski, editors, *Proceedings of the 1988 Connectionist Models Summer School*, pages 113–117, San Mateo, CA, June 17-26 1988. Morgan Kaufmann.

[21] D. Nguyen and B. Widrow. The truck backer-upper: An example of self-learning in neural networks. In *Proceedings of the International Joint Conference on Neural Networks*, volume II, pages 357–363, Washington, DC, June 1989.

[22] B. Widrow and F. Beaufays. Neural control systems. In *Proceedings of the EPRI/INNS Workshop on Neural Network Computing for the Electric Power Industry*, Stanford, Ca, August 17-19 1992.

Using Flow Graph Interreciprocity to Relate Recurrent-Backpropagation and Backpropagation-Through-Time.

Françoise Beaufays and Eric A. Wan *

Abstract

In this paper, we show that two popular algorithms used to adapt dynamic neural networks, recurrent-backpropagation and backpropagation-through-time, can be easily related when represented by signal flow graphs. Starting with the flow graph for recurrent-backpropagation, we use a simple transposition to produce a second graph. The new graph is shown to be interreciprocal with the original and to correspond to the backpropagation-through-time algorithm. Interreciprocity provides a theoretical argument to verify that both flow graphs implement the same overall weight update.

Introduction

Two adaptive algorithms, recurrent-backpropagation (RBP) and backpropagation-through-time (BPTT), are currently used to train multilayer neural networks with output feedback connections. RBP, first introduced for single layer fully recurrent networks by Williams and Zipser [1], is well suited for on-line adaptation of dynamic networks where a desired response is specified at each time step. BPTT [2, 3, 4], typically used with feedforward networks embedded in dynamic structures, does not allow on-line adaptation, but has been shown to be computationally less expensive than RBP. Both algorithms are generally derived independently and then compared in terms of performances and domain of application. In this letter, we use flow graph theory as a common support for comparing the two algorithms.

We begin by deriving a general flow graph diagram for the weight updates associated with RBP. A second flow graph is obtained by transposing the original, i.e., by reversing the arrows that link the graph nodes, and by interchanging the source and sink nodes. Flow graph theory shows that transposed flow graphs are interreciprocal, and for single input single output (SISO) systems, have identical transfer functions. This basic property, which was first presented in the context of electrical circuits analysis [5], finds applications in a wide variety of engineering disciplines, such as the reciprocity of emitting and receiving antennas in electromagnetism [6], the relationship between controller and observer canonical forms in control theory [7], and the duality between decimation in time and decimation in frequency formulations of the FFT algorithm in signal processing [8].

Network Equations

A neural network with output recurrence is shown in Figure 1. Let $\mathbf{r}(k-1)$ denote the vector of external reference inputs to the network and $\mathbf{x}(k-1)$ the recurrent inputs. The output vector $\mathbf{x}(k)$ is a function of the recurrent and external inputs, and of the adaptive weights $\mathbf{w}$ of the network:

$$\mathbf{x}(k) = \mathcal{N}(\mathbf{x}(k-1), \mathbf{r}(k-1), \mathbf{w}). \tag{1}$$

The neural network $\mathcal{N}$ is most generally a feedforward multilayer architecture [2]. If $\mathcal{N}$ has only a single layer of neurons, the structure of Figure 1 represents a completely recurrent network [1, 9].

Adapting the neural network amounts to finding the set of weights $\mathbf{w}$ that minimizes the cost function $J = \frac{1}{2} E\left[\sum_{k=1}^{K} \mathbf{e}(k)^T \mathbf{e}(k)\right] = \frac{1}{2}\sum_{k=1}^{K} E\left[\mathbf{e}(k)^T\mathbf{e}(k)\right]$, where the expectation $E[\cdot]$ is taken over the external reference inputs $\mathbf{r}(k)$ and over the initial values of the recurrent inputs $\mathbf{x}(0)$. The error $\mathbf{e}(k)$ is defined

*The authors are with the department of Electrical Engineering, Stanford University, Stanford, CA 94305-4055. This work was sponsored by EPRI under contract RP8010-13, and by NSF under grant NSF IRI 91-12531.

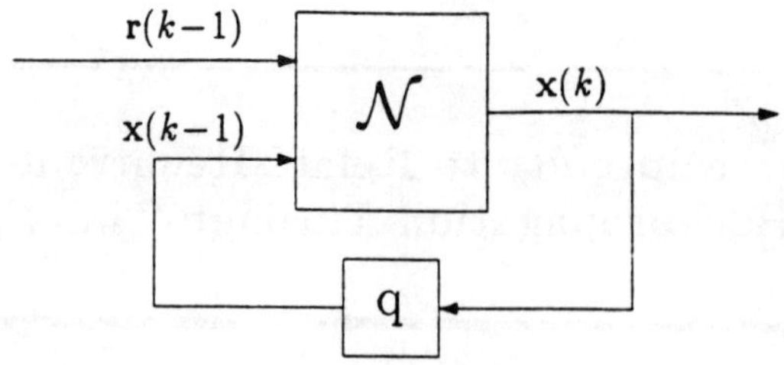

Figure 1: Recurrent neural network (q represents a unit delay operator).

at each time step as the difference between the desired state $\mathbf{d}(k)$ and the recurrent state $\mathbf{x}(k)$, whenever the desired vector $\mathbf{d}(k)$ is defined, and is otherwise set to zero:

$$\mathbf{e}(k) = \begin{cases} \mathbf{d}(k) - \mathbf{x}(k) & \text{if } \exists\ \mathbf{d}(k) \\ 0 & \text{otherwise} \end{cases} \tag{2}$$

For such problems as terminal control [10, 3] a desired response may be given only at the final time $k = K$, while for other problems such as system identification [11, 12] it is more common to have a desired response vector for all k.

In both RBP and BPTT, a gradient descent approach is used to adapt the weights of the network. At each time step, the contribution to the weight update is given by

$$\Delta\mathbf{w}(k)^T = -\frac{\mu}{2}\frac{d\left[\mathbf{e}(k)^T\mathbf{e}(k)\right]}{d\mathbf{w}} = \mu\,\mathbf{e}(k)^T \cdot \frac{d\mathbf{x}(k)}{d\mathbf{w}}, \tag{3}$$

where μ is the learning rate[1]. The accumulation of weight updates over $k = 1...K$ is given by $\Delta\mathbf{w} = \sum_{k=1}^{K}\Delta\mathbf{w}(k)$. For consistency, we assume that in both algorithms the weights are actually adjusted using the agregate $\Delta\mathbf{w}$ instead of being updated at each time step[2].

Flow Graph Representation of the Adaptive Algorithms

RBP was originally derived for fully recurrent single layer networks. A more general algorithm is obtained by using equation 1 to directly evaluate the state gradient $d\mathbf{x}(k)\,/\,d\mathbf{w}$ in the above weight update formula. Applying the chain rule, we get:

$$\frac{d\mathbf{x}(k)}{d\mathbf{w}} = \frac{\partial\mathbf{x}(k)}{\partial\mathbf{x}(k-1)}\cdot\frac{d\mathbf{x}(k-1)}{d\mathbf{w}} + \frac{\partial\mathbf{x}(k)}{\partial\mathbf{r}(k-1)}\cdot\frac{d\mathbf{r}(k-1)}{d\mathbf{w}} + \frac{\partial\mathbf{x}(k)}{\partial\mathbf{w}}\cdot\frac{d\mathbf{w}}{d\mathbf{w}}, \tag{4}$$

in which $d\mathbf{r}(k-1)\,/\,d\mathbf{w} = 0$ since the external inputs do not depend on the network weights, and $d\mathbf{w}\,/\,d\mathbf{w} = I$, where I is the identity matrix. With these simplifications, equation 4 reduces to:

$$\frac{d\mathbf{x}(k)}{d\mathbf{w}} = \frac{\partial\mathbf{x}(k)}{\partial\mathbf{x}(k-1)}\cdot\frac{d\mathbf{x}(k-1)}{d\mathbf{w}} + \frac{\partial\mathbf{x}(k)}{\partial\mathbf{w}}. \tag{5}$$

Equation 5 is then applied recursively, from $k = 1$ to $k = K$, with initial conditions $d\mathbf{x}(0)\,/\,d\mathbf{w} = 0$. For sake of clarity, let $\mathcal{X}^{rec}(k) = d\mathbf{x}(k)\,/\,d\mathbf{w}$, $\mathcal{X}(k) = \partial\mathbf{x}(k)\,/\,\partial\mathbf{w}$, and $\mathbf{J}(k) = \partial\mathbf{x}(k)\,/\,\partial\mathbf{x}(k-1)$. With this new notation, equation 5 can be rewritten as:

$$\mathcal{X}^{rec}(k) = \mathbf{J}(k)\cdot\mathcal{X}^{rec}(k-1) + \mathcal{X}(k) \qquad \forall k = 1...K \tag{6}$$

with initial condition $\mathcal{X}^{rec}(0) = 0$. The weight update at each time step is given by:

$$\Delta\mathbf{w}^T(k) = \mu\,\mathbf{e}(k)^T\cdot\mathcal{X}^{rec}(k). \tag{7}$$

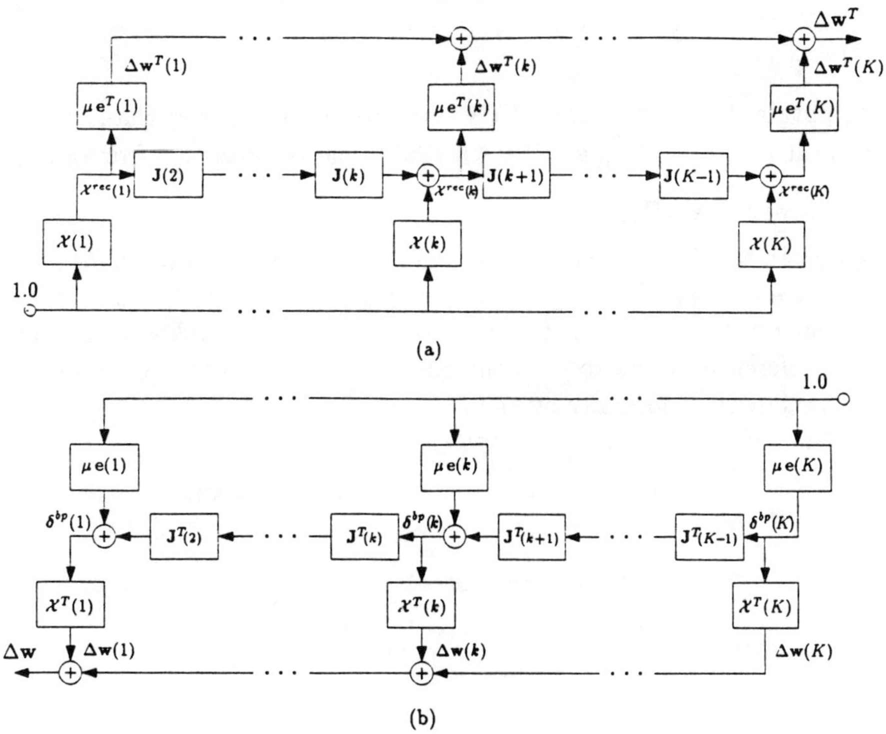

Figure 2: (a)Flow graph associated with the recurrent-backpropagation algorithm. (b) Transposed flow graph: a representation of the backpropagation-through-time algorithm.

Equations 6 and 7 can be illustrated by a flow graph (see Figure 2a). The input to the flow graph, or source node variable, is set to 1.0 and propagated along the lower horizontal branch of the graph. The center horizontal branch computes the state derivatives $\mathcal{X}^{rec}(k)$, and the upper horizontal branch accumulates the weight changes $\Delta \mathbf{w}(k)^T$. The total weight change, $\Delta \mathbf{w}^T$, is readily available at the output (sink node). RBP is completely defined with this flow graph.

Let us now build a new flow graph by transposing the flow graph of Figure 2a. Transposing the original flow graph is accomplished by reversing the branch directions, transposing the branch gains, replacing summing junctions by branching points and vice versa, and interchanging source and sink nodes. The new flow graph is represented in Figure 2b. From the work by Tellegen [14] and Bordewijk [15], we know that transposed flow graphs are a particular case of interreciprocal graphs. This means, in a SISO case, that the sink value obtained in one graph, when exciting the source with a given input, is the same as the sink value of the transposed graph, when exciting its source by the same input. Thus, if an input of 1.0 is distributed along the upper horizontal branch of the transposed graph, the output, which is now accumulated on the lower horizontal branch, will be equal to $\Delta \mathbf{w}$. This $\Delta \mathbf{w}$ is identical to the output of our original flow graph[3].

With the notation introduced before, and calling $\boldsymbol{\delta}^{bp}(k)$ the signal transmitted along the center horizontal branch, we can write down the equations describing the new flow graph:

$$\boldsymbol{\delta}^{bp}(k) \quad = \quad \mathbf{J}^T(k+1) \cdot \boldsymbol{\delta}^{bp}(k+1) + \mu\, \mathbf{e}(k) \qquad \forall k = K...1 \tag{8}$$

with initial condition $\boldsymbol{\delta}^{bp}(K+1) = 0$. The weight update at each time step is given by:

$$\Delta \mathbf{w}(k) \quad = \quad \mathcal{X}^T(k) \cdot \boldsymbol{\delta}^{bp}(k). \tag{9}$$

[1]We define the derivative of a vector $\mathbf{a} \in \Re^n$ with respect to another vector $\mathbf{b} \in \Re^m$ as the matrix $d\mathbf{a}/d\mathbf{b} \in \Re^{n \times m}$ whose $(i,j)^{th}$ element is da_i/db_j. Similarly, the partial of a vector $\mathbf{a} \in \Re^n$ with respect to another vector $\mathbf{b} \in \Re^m$ is the matrix $\partial\mathbf{a}/\partial\mathbf{b} \in \Re^{n \times m}$ whose $(i,j)^{th}$ element is $\partial a_i/\partial b_j$.

[2]RBP also allows for on-line adaptation, i.e., modifying the weights at each time k by $\Delta \mathbf{w}(k)$. This alternative adaptation scheme has slightly different properties and will not be considered in this paper.

[3]The flow graphs introduced here are in fact single input multi output (SIMO). The arguments of interreciprocity may be applied by considering the SIMO graph to be a stack of SISO graphs, each of which can be independently transposed.

Equations 8 and 9, obtained from the new flow graph, are nothing other than the description of BPTT: $\delta^{bp}(k)$ is the error gradient $-\mu\, d\left[\sum_{n=1}^{K} \mathbf{e}(n)^T \mathbf{e}(n)\right] / d\mathbf{x}(k)$ backpropagated from $k = K...1$.

Extension to Controller-Plant Structures

Flow graph theory can also be applied to more complicated network arrangements, such as the dynamic controller-plant structure illustrated in Figure 3. A discrete-time dynamic plant $\mathcal{P}$ described by its state-space equations is controlled by a neural network controller $\mathcal{C}$. Let $\mathbf{x}(k-1)$ be the state of the plant, $\mathbf{r}(k-1)$ the external reference inputs to the controller, and $\mathbf{u}(k-1)$ the control signal used to drive the plant. Figure 3 can be described formally by the following equations:

$$\mathbf{x}(k) = \mathcal{P}(\mathbf{x}(k-1), \mathbf{u}(k-1)) \tag{10}$$

$$\mathbf{u}(k-1) = \mathcal{C}(\mathbf{x}(k-1), \mathbf{r}(k-1), \mathbf{w}). \tag{11}$$

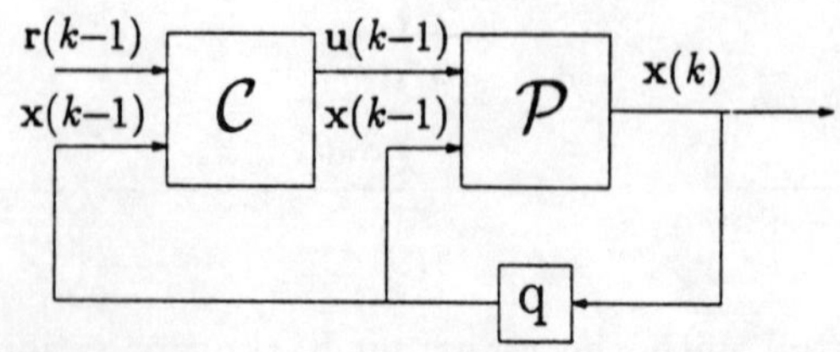

Figure 3: Controller-Plant Structure

As before, the error vector $\mathbf{e}_k$ is defined as the difference between the desired state $\mathbf{d}(k)$ and the actual state $\mathbf{x}(k)$ when there exists a desired state, and zero otherwise. Using RBP to adapt the weights of the controller requires the evaluation of the derivatives of the state with respect to the weights. Applying the chain rule to equations 10 and 11, we get:

$$\frac{d\mathbf{x}(k)}{d\mathbf{w}} = \frac{\partial \mathbf{x}(k)}{\partial \mathbf{x}(k-1)} \frac{d\mathbf{x}(k-1)}{d\mathbf{w}} + \frac{\partial \mathbf{x}(k)}{\partial \mathbf{u}(k-1)} \frac{d\mathbf{u}(k-1)}{d\mathbf{w}} \tag{12}$$

$$\frac{d\mathbf{u}(k-1)}{d\mathbf{w}} = \frac{\partial \mathbf{u}(k-1)}{\partial \mathbf{x}(k-1)} \frac{d\mathbf{x}(k-1)}{d\mathbf{w}} + \frac{\partial \mathbf{u}(k-1)}{\partial \mathbf{w}}. \tag{13}$$

Equations 12 and 13 can then be represented by a flow graph (see Figure 4a). Transposing this flow graph, we get a new graph, which corresponds to the BPTT algorithm for the controller-plant structure (see Figure 4b). Again, the argument of interreciprocity immediately shows the equivalence of the weight updates performed by the two algorithms.

Conclusion

We have shown that recurrent-backpropagation and backpropagation-through-time are easily related when represented by signal flow graphs. In particular, the flow graphs corresponding to the two algorithms are the exact transpose of one another. As a consequence, flow graph theory could be applied to verify that the weight updates performed by the algorithms are identical. These principles were then extented to a controller-plant structure to illustrate how flow graph techniques can be applied to a variety of adaptive dynamic systems. A more comprehensive paper which also addresses the network specific implementation of the flow graphs and contrast the computational requirements of the two algorithms can be found in [16].

References

[1] R.J. Williams and D. Zipser. A learning algorithm for continually running fully recurrent neural networks. *Neural Computation*, 1(2):270–280, Summer 1989.

[2] David E. Rumelhart and James L. McClelland, editors. *Parallel Distributed Processing.* The MIT Press, Cambridge, MA, 1986.

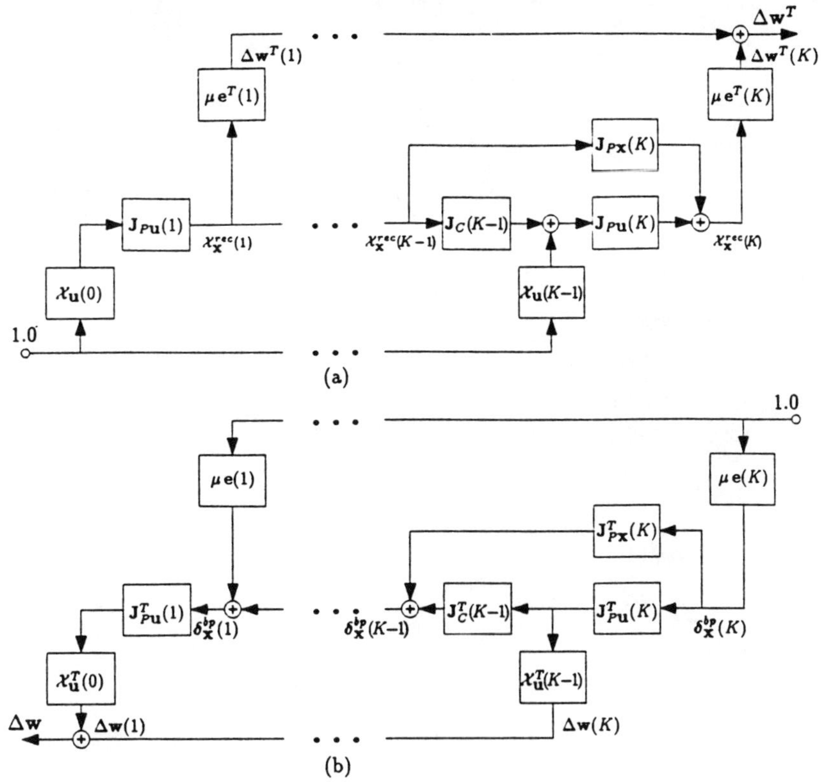

Figure 4: (a) Flow graph corresponding to RBP. (b) Transposed flow graph: a representation of BPTT. Notation: $\mathbf{J}_C(k) \doteq \partial \mathbf{u}(k)/\partial \mathbf{x}(k), \mathbf{J}_{P\mathbf{x}}(k) \doteq \partial \mathbf{x}(k)/\partial \mathbf{x}(k-1), \mathbf{J}_{P\mathbf{u}}(k) \doteq \partial \mathbf{x}(k)/\partial \mathbf{u}(k-1), \mathcal{X}_{\mathbf{u}}(k) \doteq \partial \mathbf{u}(k)/\partial \mathbf{w}, \mathcal{X}_{\mathbf{x}}^{rec}(k) \doteq d\mathbf{x}(k)/d\mathbf{w}, \mathcal{X}_{\mathbf{u}}^{rec}(k) \doteq d\mathbf{u}(k)/d\mathbf{w}, \delta_{\mathbf{x}}^{bp}(k) \doteq -\mu\, d[\sum_{k=1}^{K} \mathbf{e}^T(k)\mathbf{e}(k)]/d\mathbf{x}(k)$.

[3] Derrick Nguyen and Bernard Widrow. Neural networks for self-learning control systems. *IEEE Control Systems Magazine*, April 1990.

[4] Paul J. Werbos. Backpropagation through time: What it does and how to do it. *Proc. IEEE, special issue on neural networks*, 2:1550–1560, October 1990.

[5] P. Penfield, R. Spence, and S. Duiker. *Tellegen's theorem and electrical networks.* MIT Press, Cambridge, Mass., 1970.

[6] S. Ramo, J. R. Whinnery, and T. Van Duzer. *Fields and waves in communication electronics.* John Wiley & Sons, second edition, 1984.

[7] T. Kailath. *Linear Systems.* Prentice-Hall, Englewood Cliffs, NJ, 1980.

[8] A. V. Oppenheim and R. W. Schafer. *Digital Signal Processing.* Prentice-Hall, Englewood Cliffs, NJ, second edition, 1989.

[9] F.J. Pineda. Generalization of back-propagation to recurrent neural networks. *Physical Review Letters*, 59:2229–2232, 1987.

[10] A. E. Bryson, Jr. and Y. Ho. *Applied Optimal Control.* Blaisdell Publishing Co., New York, 1969.

[11] Ljung and Lenart. *System Identification: theory for the user.* Prentice-Hall, Englewood Cliffs, NJ, 1987.

[12] Kumpati S. Narendra and Kannan Parthasarathy. Identification and control of dynamical systems using neural networks. *IEEE Transactions on Neural Networks*, 1(1):4–27, March 1990.

[13] S. A. White. An adaptive recursive digital filter. In *Proc. 9th Asilomar Conf. Circuits Syst. Comput.*, page 21, November 1975.

[14] B. D. H. Tellegen. A general network theorem, with applications. *Philips Res. Rep.*, 7:259–269, 1952.

[15] J.L. Bordewijk. Inter-reciprocity applied to electrical networks. *Appl. Sci. Res.*, 6B:1–74, 1956.

[16] F. Beaufays and E. Wan. Relating recurrent-backpropagation and backpropagation-through-time: An application of flow graph interreciprocity. *Submitted to Neural Computation*, 1992.

Neural Network Based Inferential Sensing and Instrumentation

Aloke Guha
Honeywell Sensor and System Development Center
3660 Technology Drive
Minneapolis, MN 55418.

ABSTRACT

We present an approach for building neural network-based "inferential" sensor and instrumentation systems which infer the values of complex process variables by integrating information from multiple sensors. Such a sensing approach is especially useful in processes where direct instrumentation for monitoring the process is not possible, precluding closed-loop control. Inferential sensors can provide estimates of process variables that are usually measured off-line or through analytical laboratory instruments such as composition of fluids in a mixture. A special class of inferential sensors, intangible sensors, can be constructed when subjective properties of a process or material are involved, especially in quality control. Examples of such properties include smoothness, toughness, softness, and flavor. Attractive features of inferential sensors include noise immunity, fault tolerance and diagnostics through the incorporation of analytical redundancy in the selection of sensors. Furthermore, the neural network underlying inferential sensors embodies a process model on which sensitivity analysis can be executed. The results of the sensitivity analysis can be used to derive the control laws and the means of optimizing the process.

1. INTRODUCTION: Neural Networks for Sensing

The difficulty in controlling many complex processes begins with the problems in sensing the process. In many instances, the lack of appropriate sensors to monitor the process precludes closed-loop control. Frequently, the instrumentation to provide information on the state of a process or material where a specific sensor technology is required can be either unavailable or very expensive. An alternate approach to solving this problem is to infer the state from a multiplicity of other related sensor information. Examples of such inferential sensing include the on-line determination of flavor of a food product.

Because of their properties of universal approximation and generalization, fast execution, and fault tolerance, neural networks are well suited for building such inferential sensors. Inferential sensors based on neural networks are especially applicable in the controls of material processes (Fig. 1) as we describe next.

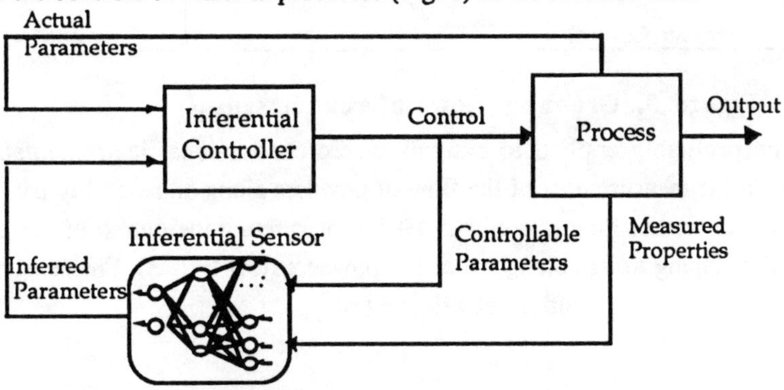

Figure 1. Inferential sensing and control scheme

2. Inferential Sensors

The basic function of inferential sensors is to map multiple similar or dissimilar sensors onto some inferred value. For example, in materials processing, inferential sensors can be used to indicate the level of resin cure from color measurements or the state of a carbonization of a composite from its acoustic signature [Guha 92a]. We describe two specific instances.of neural network-based inferential sensors to illustrate the concept.

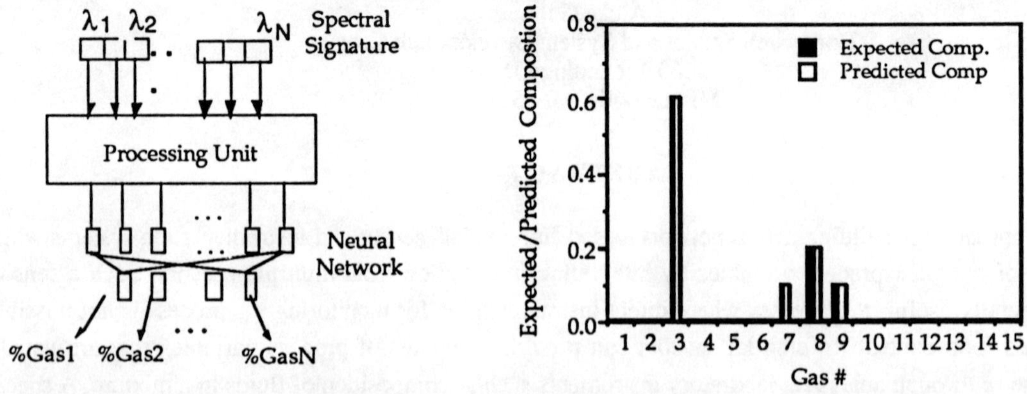

Figure 2. Gas composition sensor and its performance on a four-gas mixture

One example of an inferential sensor is that of an on-line gas composition sensor (GCS) [Guha 92b]. In many semiconductor fabrication, composite synthesis, food processing, and combustion processes, an on-line non-invasive gas composition detector is very useful. In the GCS, a neural network processes an optical spectral signature to produce a vector indicating the relative concentrations of gases with negligible error (Fig. 2). The role of the neural network is to provide a general deconvolution function, as opposed to pure linear deconvolution possible with digital signal processing. In addition to the GCS, there have been other reports on detecting gases using multiple non-spectral sensor data [Sundgren 91]. Though not as versatile, such non-spectral techniques still offer advantages of on-line analysis of some gas mixtures.

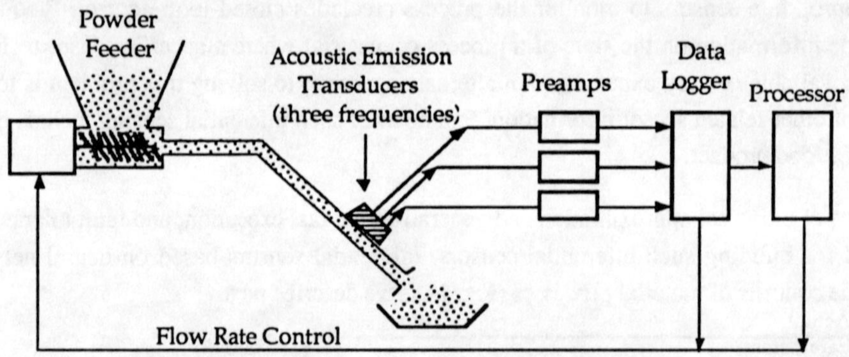

Figure 3. Granular flow inferential sensor

Inferential sensors can also be profitably applied to systems where conventional instrumentation has proved unreliable or uneconomical, as in the measurement of the flow of powders along an assembly path. An inferential sensor developed at Honeywell can predict flow rate within 4% by interpreting the output of three tuned acoustic transducers placed at the end of a sloping trough along which the powders flow (Fig. 3). The sensor is also capable of operation over a wide range of granular compositions and flow rates.

3. Intangible Sensors

An interesting subclass of inferential sensors is that of *intangible sensors* [Guha 91], so called because they model subjective judgements or evaluation. Examples of intangible sensors include those that measure the consistency or smoothness of mixtures, the softness or toughness of materials, and the flavors of food or beverage products. While these properties may ultimately depend on the chemistry of the substances involved, no single sensor is available to measure the desired property. For example, functional definitions of creaminess rely on the judgements of process or product experts and are not simple functions of variables that can be easily measured. The neural network underlying the intangible sensor is trained to estimate the desired subjective measure. The resulting sensor can then be used as part of an on-line control strategy. Honeywell has noted excellent results from experiments on intangible sensing on consumer goods production processes.

Much of the work in intangible sensing requires modeling a complex subjective quality, deduced either from off-line measurements or from on-line sensing. Our empirical results show that neural network models are very effective in predicting flavor without any *a priori* assumptions on the sensors-flavor relationship. We have also verified the feasibility of on-line intangible sensing of quality on a food manufacturing process using on-line raw material and process parameters. Despite availability of limited sample data on 15 process parameters, an intangible sensor could predict a multivalued process quality with good accuracy in 86% of the test cases.

While the degree of generalization achieved by the intangible sensor depends on the subset of sample data used to train the network, partial generalization is possible even when only a small number of samples is available. As more data is collected, an improved intangible sensor can be developed, allowing continuous improvement and increased confidence in prediction.

4. Fault Detection and Noise Immunity

Besides exploiting the fault tolerance of neural networks, inferential sensors can also be designed for sensor fault detection and immunity from noise in measurements. The detection of sensor failures and the tolerance to noise in sensor readings can be achieved by incorporating analytical redundancy into the sensor set read by the neural network.

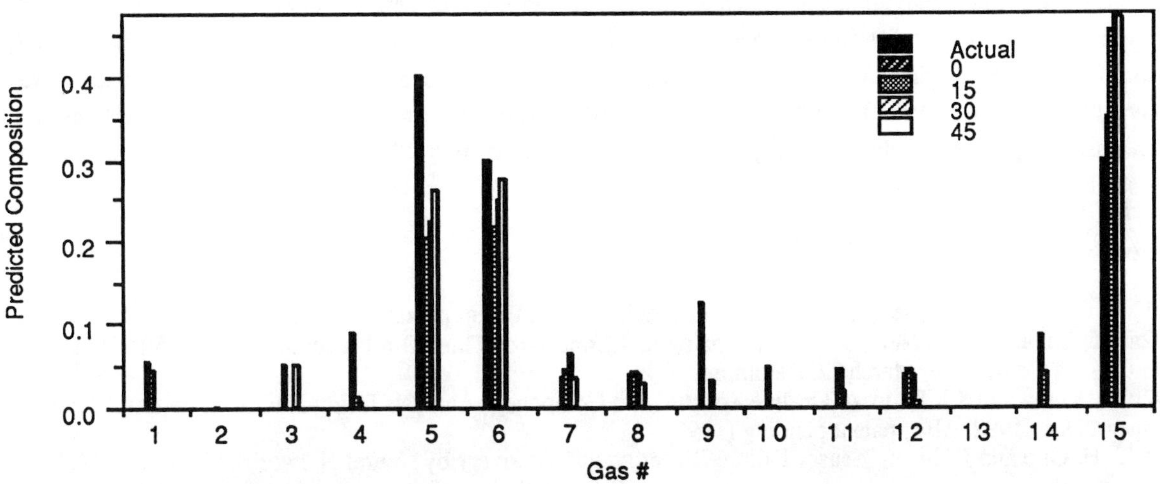

Figure 4. Predicted composition by GCS on noisy data

Sensor failure detection is extremely useful in process control where unchecked failure can be expensive, and occasionally dangerous, when processes tread the extreme ranges of variables [Guo 1991]. As an added benefit, the inferential sensor can be augmented to provide health or status data on the sensors. Because the sensors can be selected by design to have some correlation, neural networks can be trained to detect the failure of a sensor based on the combined activities of the sensor array. The network may also be trained to produce corrected estimates of the quantity previously measured by the faulty sensor. Related work [Naidu 90] has shown that a neural network can discern critical from noncritical sensor problems more accurately than a more conventional approach based on finite integral squared error diagnostics.

One example of noise immunity that we have examined is that of the GCS. The GCS was trained with on noisy data generated randomly for each component signature. Fig. 4 depicts the relative performance of the GCS in discriminating noise from the signal. Results indicate that the errors in predicting composition is less than 1%, 2% and 5% for signal to noise ratios of 45 dB, 30 dB, and 15 dB, respectively [Guha 92b].

5. Inferential Controllers

The inferential sensor described uses a neural network to develop a "process model" that relates measurable properties of the material and process to inferred state or property of the process. Since many neural networks employ differentiable functions, sensitivity analysis can be conducted to derive the critical control parameters of the process [Hashem 92]. Thus, the neural network underlying the inferential sensor can be used to derive the key control variables for the process without assumptions on the nature of the process. The sensitivity analysis will also indicate the rate of change of the output with respect to changes in the critical variable. A control law can therefore be derived for controlling the properties measured inferentially. This form of closed-loop neurocontroller development can be classified as direct forward modeling [Samad 91] where a process model is used to compute the controller output. Although sensing systems are typically static, both inferential sensors and controllers may have dynamic behavior in which case dynamic or recurrent neural networks may be used to model the processes.

6. Conclusions

We have described the use of neural networks as a basis for a new class of sensors that may provide significant advantages in monitoring and instrumenting complex materials and processes. Such sensors are especially well suited in processes where there is a paucity of process knowledge and where the specific sensor technology necessary for instrumentation is either unavailable or uneconomical. The inferential sensor relies on the modeling and generalization property of neural networks since they embody a uniform process model. This model can be realized on-line to infer complex sensing information. In addition to synthesizing a sensor system, inferential sensors have added advantages of noise immunity and fault detection abilities. Furthermore, the process model underlying the inferential sensor can be used to derive the key control variables and the control law, which in turn can be realized in a closed loop controller.

References

[Guha 91] A. Guha, Intangible sensor and Method of Making Same. Patent pending.

[Guha 92a] A. Guha, "Neural Network Based Sensing and Control for Thin Film Processes," Proc. 35th Society of Vaccuum Coaters Conference, March 92, Baltimore, MD.

[Guha 92b] A. Guha and S.J. Bedros, "On-line Spectral and Component Analysis Using Neural Networks," ANNIE '92, to appear, Nov. 1992. Also, patent pending (1992).

[Guo 91]. T.-H. Guo and J. Nurre, "Sensor Failure Detection and Recovery by Neural Networks," Proc. IJCNN, July 1991.

[Hashem 92] S. Hashem, "Sensitivity Analysis for Feedforward Artificial Neural Networks with Differentiable Activation Functions," Proc. IJCNN, June 1992, Baltimore, Maryland.

[Naidu 90] S.R. Naidu, et al., "Use of Neural Networks for Sensor Failure Detection in a Control System," IEEE Control Systems Magazine, April 1990.

[Samad 91] T. Samad, "Neural Networks for Control–An Overview," ISA/91 International Conference and Exhibition, Oct. 1991.

[Sundgren 91] H. Sundgren, F. Winquist, I. Lukkar, and I. Lundstrom, "Artificial Neural Networks and Gas Sensor Arrays: Quantification of Individual Components in a Gas Mixture," Measurements Science Technology 2 (1991), 464-469, United Kingdom.

OPTIMIZING NEURAL NETWORKS WITH GENETIC ALGORITHMS

STEVEN A. HARP & TARIQ SAMAD

Honeywell SSDC, 3660 Technology Drive, Minneapolis, MN 55418

ABSTRACT

We describe an approach to application-specific neural network design using genetic algorithms. A genetic algorithm is a robust optimization method particularly well suited for search spaces that are high-dimensional, discontinuous and noisy—features that typify the neural network design problem. Our approach is relevant to virtually all neural network applications: it is network-model independent and it permits optimization for arbitrary, user-defined criteria. We have developed an experimental system, NeuroGENESYS, and have conducted several experiments on small-scale problems. Performance improvements over manual designs have been observed, the interplay between performance criteria and network design aspects has been demonstrated, and general design principles have been uncovered.[1]

INTRODUCTION

With the growing interest in the practical use of neural networks, addressing the problem of customizing networks for specific applications is becoming increasingly critical. It has repeatedly been observed that different network structures and learning parameters can substantially affect performance. Such important aspects of neural network applications as generalization, learning speed, connectivity and tolerance to network damage are strongly related to the choice of network architecture. Yet there are no analytical or empirical approaches that neural network researchers and application developers can use for reliably assisting them in designing networks.

In a research effort over the last three years, we have been investigating the use of genetic algorithms (Holland, 1975; Goldberg, 1989) for synthesizing neural network architectures (Harp, Samad & Guha, 1989, 1990; Harp & Samad, 1991a, 1991b). The genetic algorithm is an optimization method particularly appropriate for this problem: it has been distinguished by its relative immunity to relevant characteristics such as high dimensionality, local minima and noise. In addition, we have been able to suggest tentative generalizations based on our experiments. These preliminary results give us some grounds for believing that both of the objectives of this research program can be realized: First, to develop a next-generation development tool for artificial neural networks; such a tool would allow an application developer to describe the problem or class of problems to be solved and would automatically search for an optimal (by the user's criterion) network design. A second objective is to uncover further evidence to help build a theory of neural network design. The GA can help us answer questions about what kinds of architectures are appropriate under a given set of conditions.

Several other researchs have looked at applying genetic algorithms to neural networks. In much of this work, however, a genetic algorithm is used as a learning rule—the network structure is predetermined and weight values are optimized by genetic search (Whitley and Hanson, 1989). Miller, Todd and Hegde (1989) used a genetic algorithm to determine unit connectivity at a low-level: the network size is essentially fixed.

THE NEURAL NETWORK DESIGN PROBLEM

The artificial neural network model that is by far the most popular choice for practical applications is the multi-layered feedforward network trained with backpropagation. In its full generality, this is a very large class of networks. Any feedforward structure—an arbitrary acyclic directed graph—can be used, and different connections can have different learning rate values and different initial conditions. The vast majority of actual applications use one of two network structures (Figure 1). The only choice left to the designer is the number of units in a hidden layer. In addition, typically one learning rate is used for all connections in the network. Thus the designer is left with two or three degrees of freedom (a few other parameters—convergence criteria, initial weight ranges—are also sometimes considered). The resulting complexity is certainly manageable, but the constraints on network design it imposes imply that we are not extracting the full potential of neural network technology. Nonstandard designs can dramatically improve performance, see e.g. Lang and Witbrock (1989) and Fahlman and Lebiere (1990).

---

[1] A longer version of this paper appears in the proceedings of the 1992 American Power Conference.

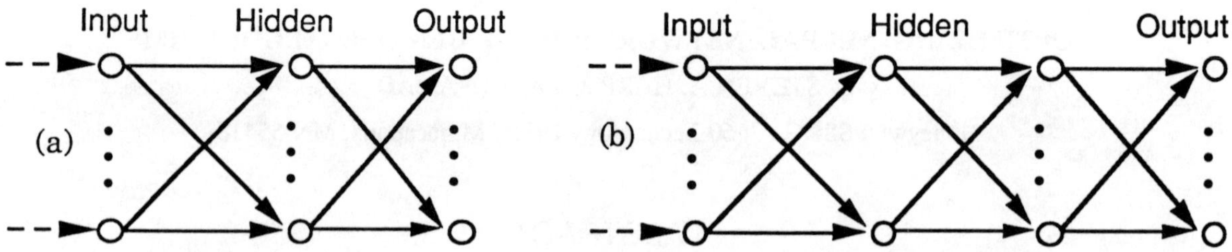

Figure 1. Two network architectures commonly used with backpropagation learning. A single hidden layer is present in the network (a), while the network (b) employs two. All units are fully connected between layers.

AN EVOLUTIONARY APPROACH TO DESIGN

We have employed the genetic algorithm, an optimization technique loosely modeled after biological evolution, for determining appropriate network architectures. An important aspect of our work is that networks can be tailored to task-specific requirements: any combination of cost and performance considerations such as network size, generalization, learning speed, tolerance to damage, and suitability for hardware implementation, can be included as optimization criteria. This work is embodied in a software system called NeuroGENESYS.

Our overall scheme is depicted in Figure 2. A network is represented by a *blueprint*—a data-structure that encodes a number of characteristics of the network including structural properties and learning parameter values. A blueprint is *instantiated* into an actual network, typically with random initial weights. We then train the network using a learning algorithm (parameter values for which are also obtained from the blueprint) and given training data. The trained network can then be tested in various ways—e.g., on non-training inputs, after disabling some units, or after perturbing learned weight values. Which testing scenarios are chosen depends on the requirements of the application. Finally, the network is evaluated and a *fitness* estimate is computed for it. The fitness estimate can be an arbitrarily complex function. In our current system, it is a weighted linear sum of a large number of relevant criteria such as the numbers of units and weights in the network, accuracy, learning speed, efficiency (this is useful for unsupervised clustering networks), average and maximum fanouts (i.e., the number of outgoing weights from a unit—a quantity that is important for VLSI implementation), and the various test scores.

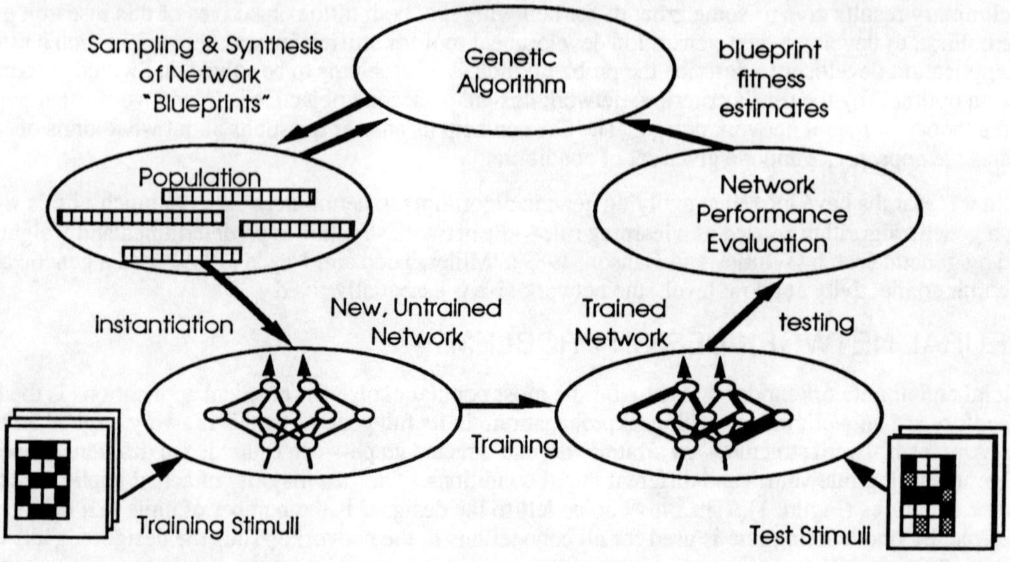

Figure 2. An overview of network architecture optimization in NeuroGENESYS.

This process of instantiation, training, testing and evaluation is performed for each member of the population. After the entire population is evaluated, the next *generation* of blueprints is produced. This process is mediated by a number of *genetic operators*, the most prominent of these being *crossover*, in which two parent blueprints are spliced together to produce a child blueprint. The higher the fitness of a blueprint, the greater the probability of it being selected as a parent for the next generation. Characteristics that are found useful will thereby tend to be emphasized in the next generation, whereas harmful ones will tend to be suppressed.

The details of a representation scheme are specific to the class of network models, and may differ dramatically between, for example, backpropagation networks, and self-organizing feature maps (Harp & Samad, 1991a and 1991b). Details aside, several features of the approach in the abstract deserve note:

- It is model-independent: it can be applied to any existing or future neural network model, including models without a training component. We have implemented and successfully demonstrated the approach on backpropagation networks and Kohonen feature maps.

- The definition of network performance depends on the application. The fitness of a network can be an arbitrary function of several distinct performance and cost criteria, some or all of which can thereby be optimized simultaneously.

- The genetic algorithm is an inherently parallel optimization procedure. The different networks present in a generation can all be trained simultaneously on different processors, and the NeuroGENESYS software takes advantage of this feature.

- It is generally compatible with other methods of network design. For example, the initial population can consist of manually optimized networks. An application developer could design one or more networks that could be the starting point for optimization by the genetic algorithm.

EXPERIMENTAL RESULTS

In several small-scale experiments, our approach has produced high-performing, and often surprising, architectures. We highlight some of these results below. Details are available in earlier publications (Harp, Samad & Guha, 1989, 1990; Harp & Samad, 1991a, 1991b):

- Our first experiment was for a simple digit recognition application. The training set consisted of 10 8×4 binary pixel images of the digits 0-9. The optimized structures NeuroGENESYS produced had no hidden units: we thereby discovered that the problem was linearly separable. Even for this simple problem, we were able to demonstrate the benefits of application-specific optimization. When optimized for average fanout in conjunction with accuracy, the best networks produced learned perfectly (but slowly) and had an average fanout of three connections per unit—the network was sparsely connected. With learning speed as the criterion, the networks produced learned substantially faster and had an average fanout almost an order of magnitude higher—there were several connections between each pair of input and output units. Recently, Izui and Pentland (1990) have shown theoretically that multiple connections between units can accelerate learning.

- With the exclusive OR problem NeuroGENESYS produced fast learning networks that consistently contained "bypass" connections from the input directly to the output layer (in addition to connections to and from hidden layers). Recent empirical work (Whitley, Starkweather & Bogart, 1990) provides further support for the hypothesis that bypass connections are useful for faster learning. With an appropriate choice of optimization criterion, NeuroGENESYS has also been able to produce known minimal networks for this problem.

- In experiments on a simplified system identification problem—process delay estimation—NeuroGENESYS produced a network that was 40% more accurate than our best manual design.

- Our recent work with Kohonen feature maps has produced further preliminary validations of our approach. Analyses of several experiments using the feature map version of NeuroGENESYS have produced novel and interesting generalizations relating parameter settings and performance criteria.

CONCLUSIONS

Successes achieved to date in applications of artificial neural networks are a testament to the versatility of the technology—virtually never is a systematic attempt made at optimizing network designs. With an effective approach to application-specific design, neural network solutions of higher quality yet can be anticipated. We believe genetic algorithms provide a powerful approach to solving the neural network design problem. The prototype system we have developed, NeuroGENESYS, has produced exciting results in several small-scale experiments, and we hope to conduct realistic evaluations of our approach in the near future.

REFERENCES

Fahlman, S.E. & C. Lebiere (1990). *The Cascade-Correlation Learning Architecture.* Technical Report CMU-CS-90-100, School of Computer Science, Carnegie Mellon University.

Goldberg, D. (1989). *Genetic Algorithms in Search Optimization and Machine Learning*. Addison-Wesley.

Harp, S., T. Samad & A. Guha (1989). Towards the genetic synthesis of neural networks. *Proceedings of the Third International Conference on Genetic Algorithms.* San Mateo, CA: Morgan Kaufmann.

Harp, S., T. Samad &A. Guha (1990). Designing application-specific neural networks using the genetic algorithm. In *Advances in Neural Information Processing Systems 2*, D.S. Touretzky, ed., San Mateo, CA: Morgan Kaufmann.

Harp, S. A. & T. Samad (1991a). Genetic optimization of self-organizing feature maps. *Proc. of the International Joint Conference on Neural Networks*, July, Seattle.

Harp, S. A. & T. Samad (1991b). Genetic synthesis of neural network architecture. In *Handbook of Genetic Algorithms*, L. Davis, ed.. New York: Van Nostrand Reinhold.

Holland, J. (1975). *Adaptation in Natural and Artificial Systems.* Ann Arbor, MI: University of Michigan Press.

Izui, Y. & A. Pentland (1990). Analysis of neural networks with redundancy. *Neural Computation, 2*, pp. 226-238.

Lang, K.J. & M.J. Witbrock (1989). Learning to tell two spirals apart. In *Proceedings of the 1988 Connectionist Models Summer School*, D. Touretzky, G. Hinton & T. Sejnowski, eds.. San Mateo, CA: Morgan Kaufmann.

Schaffer, J.D., R. Caruana & L. Eshelman (1989). Using genetic search to exploit the emergent behavior of neural networks. Philips Laboratories, 345 Scarborough Rd., Briar Cliff Manor, NY 10510.

Whitley, D. & T. Hanson (1989). Optimizing neural networks using faster more accurate search. *Proc. of the Third International Conference on Genetic Algorithms.* San Mateo, CA: Morgan Kaufmann.

C. Nuclear Power Plants

POTENTIAL USE OF NEURAL NETWORKS IN NUCLEAR POWER PLANTS

Robert E. Uhrig
306 Pasqua Engineering Building, University of Tennessee, Knoxville, TN 37996-2300

ABSTRACT

The application of neural networks, a rapidly evolving technology used extensively in defense applications to nuclear power plants has the potential to enhance the safety, reliability and operability of these systems. The work described here deals with modeling of nuclear power plant systems or parts of these systems that can be isolated. Monitoring of components involves the collection of data from these components and a detailed analysis to determine whether an actual or incipient fault exists. This paper describes the modeling of these systems or components using output from one or many sensors. Special emphasis is placed on the modeling of the interrelationship between two or more sensor outputs at a time when the component or system is known to be performing satisfactorily. The neural network is then used to predict one or more of the sensor signals using the other sensor signals as inputs. If there is a significant difference (beyond normal statistical variations) when the predicted and actual outputs are compared,, then the interrelationship between the sensor signals has changed, indicating that something in the component, system, or instrumentation has changed.

INTRODUCTION

Historically, utilities and other operators of complex systems, such as chemical processing plants and nuclear power plants, have relied on human operators to monitor the plants and to diagnose any problems which occur. With a few notable exceptions, this approach appears to have worked reasonably well. However, there is clear evidence linking accidents and a number of troublesome operational incidents over the years to "operator error." One possible solution is to automate the plants and take the operator out of the loop. [Uhrig 1989] For a variety of reasons (legal, regulatory, and other), such a solution is not practical. The alternative approach of "backstopping" the operators by providing them with the results of automated surveillance of the overall plant was considered the most practical way of addressing this problem. The overall objective is to provide the operator with additional information about the power plant in a way that is useful, timely and non-intrusive. Special emphasis was given to the early detection of abnormalities and deviations from normal operation, with the intent that the operator could take corrective or compensating action, if appropriate, to avoid shutting down the plant unnecessarily.

The work described here deals with modeling of nuclear power plant systems, or parts of such systems that can be isolated from the total system. [Uhrig et al, 1992] Typically, the measured variables from the systems are analog variables that must be sampled and normalized to expected peak values before they are introduced into neural networks. Often data must be preprocessed to put it into a form more acceptable to the neural network (e.g., a fast Fourier transformation of the time-series data to produce a spectral plot of the data). The neural networks are usually simulated on modern high-speed personal computers and workstations that carry out the calculations serially. However, it is possible to implement neural networks using specially designed microchips where the network calculations are truly carried out in parallel, thereby providing virtually instantaneous outputs (microsecond response times) for each set of inputs. Speed-ups of 100 to 1000 over simulation systems are possible using appropriate microchips to implement the neural network.

APPLICATIONS OF THE TECHNOLOGY

Specific applications under study include a) virtually instantaneous identification of transients or malfunctions, b) sensor validation, c) plant-wide monitoring, d) check valve monitoring, and e) analysis of vibration. These individual projects will be briefly described as examples of the application of the technology that has been developed. [Uhrig et al, 1992]

Generally, neural network technology developed involves three specific tasks that are undertaken using a variety of methods. These are 1) diagnostics based on pattern recognition in processed time-records of variables, 2) feature detection based on recognition of patterns in data, and 3) modeling of phenomena and systems with interpretation of input-output relationships. Many projects involve both pattern recognition and modelling. In most cases, comparison of predicted results (based on models developed from data taken when the system was working properly) or patterns (learned by neural network models from data presented to it) with actual results or patterns is involved. Typically, measured variables from the systems are analog variables that must be sampled and normalized to expected peak values before they can be utilized. Often, data must be processed to put it into an acceptable form (e.g., a fast Fourier transformation of the time-series to produce a spectral plot of the data). Five specific applications will be briefly described within the context of the technology developed.

Transient Diagnostics. When a complex system is operating normally, the readings of the hundreds, or even thousands, of instruments in a typical control room form a pattern (or unique set) of readings that represent a normal state of the plant or system. When a disturbance occurs, the instrument readings undergo a transition to a different pattern, representing a different state that may be normal or abnormal, depending upon the nature of the disturbance. The fact that the pattern of instrument readings undergo a transition to a new state that is different is sufficient to provide a basis for identifying the transient or the change of state of the system. Such identification requires a rapid (real-time), efficient method of "pattern recognition", such as neural networks, to implement a diagnostic tool based on this phenomenon.

Experience to date using data from the modern Singer-Link training simulator at TVA's Watts Bar Nuclear Power Plant to identify seven transients and steady state operation has been very successful, even with the inclusion of noise in the simulator signals. [Bartlett and Uhrig 1992] Some 27 variables were used to identify seven different transients (loss of coolant in the hot leg of the reactor coolant system (RCS), loss of coolant in the cold leg of the RCS, main steam-line break in containment, main feedwater line break in containment, total loss of off-site power, control rod ejection, and steam generator tube leak). Time records for the 27 variables constituted the input vectors to the neural network for training. In all cases, the network was able to immediately identify the transients. Even in the presence of 5% noise, the network was able to identify most transients before the automatic safety system shut down the plant. [Guo and Uhrig, 1992] This work has been extended by using both sensitivity analysis to identify the most important input variables associated with each specific transient or abnormality. [Guo and Uhrig, 1992] This led to the use of many simple "modular" neural networks, each capable of detecting a single different abnormality or transient, using only a few of the most important variables for that transient or abnormality as inputs, selected either by sensitivity analysis or genetic algorithm optimization. These modular neural networks have the advantage of simplicity, ease of training, and can be trained to detect degrees of abnormality of that specific type.

Sensor Validation. A neural network-based process for automated sensor validation during both steady-state and transient operations has been developed. [Upadhyaya, Eryurek, and Mathai, 1991] The use of neural networks for signal estimation has the advantage that the functional form relating a set of process variables is defined by the neural network model of the process and is implicitly nonlinear. Once the network is properly trained, predictions of the output variable can be made in real time. Furthermore, the state estimation using a neural network is less sensitive to measurement noise compared to direct model-based techniques.

Tests with data from the Experimental Breeder Reactor-II (EBR-II), a liquid metal fast breeder reactor at the Idaho National Engineering Laboratory and a commercial four-loop PWR power plant have shown that this technique can be used to validate data from sensors placed in nuclear power plants [Upadhyaya, Eryurek, and Mathai, 1991]. Networks for the validation of readings of hot leg temperature, reactor power and control rod position for a 4-loop PWR demonstrated the validity of this approach. For the hot leg temperature reading, the input signals were reactor power, pressurizer level and cold leg temperature.

Plant-Wide Monitoring Using Neural Networks. The technique for sensor validation has been expanded into a plant-wide monitoring system [Upadhyaya, and Eryurek, 1992] by using an autoassociative neural network where the inputs and the outputs are identically the same variables. The neural network is trained over the range of operation using the same data for the input vector and the desired output vector. Backpropagation using a sigmoidal function with an adjustable coefficient was used to train the network when the system is operating properly. Under these conditions, the neural network outputs represent estimates of the instantaneous values of the output variables, and all of these estimates are virtually identical to the actual outputs. When a sensor begins to drift or an error is introduced into a data channel, the predicted value and the actual value differ. Hence, monitoring the differences between the estimates predicted by the neural network and the actual values from the system provides a method of identifying drift or instrumentation system (or sensor) failure. An alternative interpretation of these differences might be that the input--output relationship of the system from which the signals come may be changing due to system failure or changes of some sort.

This technique was applied to data from eighteen signals from the EBR-II during ascension in power from 45% to 100%. Errors between estimates predicted by the neural network and the actual values were usually less than 0.5%. When one of the sensors fails or an error is introduced into one of the sensor outputs (the inputs to the network) , the corresponding output (the predicted value of the variable) of the network changed only slightly. Hence, a difference between the network output and the actual variable (the network input) identifies that data channel or related system as the one with a problem.

Check Valve Monitoring. Multiple check valve failures at San Onofre Unit 1 found in November 1985 led to the NRC (Nuclear Regulatory Commission) issuing NUREG-1190 that discussed the reasons for the failures of the five check valves and how the check valve failures induced water hammers to the feedwater system. Although there are many possible failure mechanisms for check valves, the most common problems associated with check valve failures are due to system flow oscillations or system piping vibrations. These vibrations and oscillations induce check valve component wear and thus component failure. Many check valve components emit sounds that can be detected by monitoring acoustical emissions from the valve body.

A technique involving the use of a neural network for the analysis of acoustical data from check valves to evaluate their status has been developed under support of the Electric Power Research Institute [Ikonomopoulos, Uhrig, and Tsoukalas, 1992] It involves training a neural network to model the internal behavior of the check valve. In this case, the power spectral density (PSD) of the sampled time-series at one point on the check valve is used as the input to the neural network, and the PSD of the sampled time-series at another point is the desired output of the neural network. The network is trained while the flow varies over the normal range of operation when the valve is known to be operating properly. The neural network is then used in a monitoring mode to predict the output sensor PSD from the input PSD, and a comparison is made between the predicted and actual output PSDs. Deviations indicate that the inter-relationship between the input and output signals has changed due to a change (failure) of the valve.

Analysis of Vibrations. The flow of fluids (water, air or steam) as well as rotating machinery can induce vibration and shock in most complex systems. Vibration per se is not necessarily bad if its amplitudes and the associated forces are within acceptable limits. Indeed, vibration in machinery can be the source of much information about the various systems involved. Analysis of vibration data can identify the operational problems of rotating machinery. Such analyses have proven very effective in identifying potential failures before they occur. Because it is difficult for neural networks to deal with power spectral densities that have narrow peaks (which are typical of PSDs taken on undamped systems such as rotating machinery) the PSDs were transformed using a recirculation neural network to reduce their dimensionality. This effectively spread the information contained in the peaks over the whole frequency spectrum and put the data in a form more appropriate for analysis using backpropagation training and recall. [Alguindigue, Loskiewicz-Buczak, and Uhrig, 1991] The result was that the accuracy in fault detection was very high (97 to 100%).

The two-sensor technique used for check valves was also used to analyze the progressive failure of a large motor/pump bearing in a nuclear power plant. [Alguindigue, Loskiewicz-Buczak, and Uhrig, 1992] A series of measurements of horizontal and vertical components of acceleration were taken periodically as the bearing began to fail. The power spectra of the horizontal and vertical components of acceleration on the bearing during the first five sets of measurements (when the bearing was operating properly) were the input and desired output respectively of a neural network while it was being trained. For the next seven sets of measurements, while the bearing progressed towards failure, the predicted value of the vertical component of acceleration (obtained from the neural network with the horizontal component as the input) was compared with the actual value of the vertical component. The mean square error (the sum of the squares of the differences of the individual spectral values of the actual and predicted values of vertical component of acceleration) grew as the bearing progressed towards failure.

CONCLUSION

In the light of the experience described above, it is clear that neural networks offer an interesting, challenging, and productive means of addressing many of the problems that occur in complex systems. Although much of the work described has been to demonstrate feasibility of the specific approach, the results are encouraging and indicate that the neural network techniques have the potential to enhance the performance and safety of complex systems in a cost effective way.

REFERENCES

I. Alguindigue, A. Loskiewicz-Buczak, and R. E. Uhrig, "Neural Networks for the Monitoring of Rotating Machinery, Proceedings of the 8th Power Plant Dynamics, Control & Testing Symposium, Knoxville, TN, May 27-29, 1992.

E. B. Bartlett and R. E. Uhrig, "Nuclear Power Plant Status Diagnostics Using An Artificial Neural Network," Nuclear Technology, Vol. 97, pp 272-281, March 1992.

Z. Guo and R. E. Uhrig, "Accident Scenario Diagnostics with Neural Networks", Proceedings of the 8th Power Plant Dynamics, Control & Testing Symposium, Knoxville, TN, May 27-29, 1992.

A. Ikonomopoulos, L. H. Tsoukalas, and R. E. Uhrig, "Use of Neural Networks to Monitor Power Plant Components," Proceedings of the American Power Conference, Chicago, IL, April 13-15, 1992.

R. E. Uhrig, "Opportunities for Automation and Control of Next Generation Nuclear Power Plants", Nuclear Technology, Vol. 88, pp 157-165, November 1989.

R. E. Uhrig, A. J. Maren, L. F. Miller, L. H. Tsoukalas, and B. R. Upadhyaya, "Enhancing Nuclear Power Plant Performance through the Use of Artificial Intelligence," Third Annual Report, Department of Energy Contract #DE-FG07-88ER12824, June 1, 1990-March 31, 1992. Issued April 1992.

B. R. Upadhyaya, G. Mathai, and E. Eryurek, "Development and Application of Neural Network Algorithms for Process Diagnostics," Proceedings of the 29th IEEE Conference on Decision and Control, Honolulu, Hawaii, December, 1990.

Belle R. Upadhyaya and Evren Eryurek, "Application of Neural Networks for Sensor Validation and Plant Monitoring," Nuclear Technology, Vol. 97, pp 170-176, February 1992.

Sensor Validation in Power Plants Using Neural Networks

Mostafa Khadem
Ali Ipakchi

ABB System Control, Inc.
2550 Walsh Ave.
Santa Clara, CA 95051

Frank J. Alexandro

University of Washington
Seattle, WA 98195

Robert W. Colley

Electric Power Research Institute
Palo Alto, CA 94303

Abstract

This paper presents a method for verifying the accuracy of the measurement of a process variable in power plants. The method is based on the use of a set of process variables dynamically related to the target process variable. The measurements of this set of variables are used as inputs to a neural network to obtain an estimate of the target variable reading. Agreement, within a set tolerance, of the target instrument reading, and the neural network output(s) establishes that the target instrument is operating properly.

The method was applied to the feedwater flow meters in the two feedwater flow loops of the TMI-1 nuclear power plant. These meters are susceptible to frequent fouling that degrades their measurement accuracy. A neural network whose inputs were the readings of a set of reference instruments, was designed to predict both feedwater flow rates simultaneously. A multi-layer feedforward neural network employing the Backpropagation algorithm was used. A normalization procedure based on the mean and standard deviation of each input vector was used to produce a faster training process.

Training and testing were done on data generated from a plant simulation computer program and on TMI-1 operating data from two different operating cycles. The results show that by using proper training data, the neural network can predict the correct flow rates with an absolute relative error of 2%.

Introduction

A typical power plant employs many instruments to measure critical parameters. While the accuracy of some of these measurements may be verified by using redundant instruments, this is not always feasible. The absence of redundant measurements for a critical instrument may require frequent manual calibration checks. Such checks are expensive and time consuming. One approach to the solution of this problem is to use analytical redundancy techniques. Analytical redundancy refers to the use of information from instruments measuring a set of process variables to obtain the correct measurement for a different process variable. The relationship between these variables is expressed by a set of differential equations that describe the system dynamics. To make use of the data from a set of reference instruments to obtain information about a target instrument, some method of digital processing is usually required. Fortunately, a typical modern power plant contains sophisticated computer and data acquisition systems that sample and store the output of practically all the plant instrumentation. A modern power plant computer system can easily make these data available to other applications through software and networking techniques. This has opened new opportunities for developing on-line or off-line software techniques for reducing plant instrument calibration efforts while improving the overall reliability of plant instrumentation.

In this project a methodology was developed to use neural networks to verify the measurement readings of power plant sensors, using the readings of other sensors measuring different process variables. The methodology was applied to the feedwater flow meters in the TMI-1 nuclear power plant. An inaccurate flow rate measurement affects the maximum permissible power level.

Statistical and neural network-based studies of instrument calibration and sensor validation methods have been reported on by other investigators [1-3].

Methodology

Using first principles of physics, a set of algebraic and differential equations describing the TMI-1 plant had previously been obtained on another EPRI sponsored project. These equations form the basis of a computer simulation of the plant that has the capability of determining a linear approximation of the system about different operating points. Using this linear model, observability computations were made to establish the degree to which a set of process variables are related, through the system dynamics, to the target instrument. Variables strongly related to a target instrument were selected for the reference set. The information from the reference set is processed by an artificial neural network. The inputs to the network are the data from the reference set, and the outputs are estimates of the feedwater flow rates in the two loops. The network is trained by applying the Backpropagation algorithm to a set of training data. After the training period is complete, the output of the neural network is compared to the measurement readings of the feedwater flow meters. Re-calibration is required if the two values differ by more than a set tolerance.

TMI-1 Tests and Results

Tests were made using both simulation data and plant operating data from TMI-1 plant. The simulation tests were based on a dynamic simulation of the power plant. The observability analysis for the linearized model led to the selection of reactor power, main feedwater pump speed (feedwater chain was modeled with only one pump), and steam pressure and temperature of loop A and B, as input parameters to the neural network. The neural network output parameters were feedwater flow rates of loops A and B. The neural network used had seven input nodes including the bias node, one hidden layer with four nodes, and two output nodes. The training data was made using an asymmetrical flow condition in the feedwater flow loops. the simulation provided data for 200 seconds of steady state at 83% power level, followed by a ramp of 400 seconds duration to 63% power level, followed by a final phase of 200 seconds steady state at 63%. Training data was prepared at intervals of one second for a total of 800 training pairs. The learning rate was 0.005 and the momentum rate was 0.25. It took 300 iterations on the DEC VAX 8700 computer for acceptable convergence. After training, the neural network was tested using data obtained from another simulation test. There were 1252 testing data at intervals of one second. Figure 1 shows the simulated and neural network predicted feedwater flow rates for both loops. Over the course of the test run, the absolute relative errors in loop A and Loop B remained below 2% and 1% respectively.

Tests were also made using TMI-1 Cycle 7 and Cycle 8 operating plant data. For these tests feedwater temperature entering each of the steam generators A and B, steam temperature and throttle pressure of both loops, and core power were used as input parameters to the neural network. Steam flow rate is not included as an input parameter, since it is not a measurable parameter in the plant. The output parameters of the neural network were the two feedwater flow rates. The network had one hidden layer with four nodes. The Cycle 7 data shows operation at steady state with very little operational changes. The training data for this cycle was for the period of 08/20/88 through 11/18/88 taken at eight hours intervals. There were 232 data points considered as good training points; however a number of points in the above training period were obviously bad and were removed. The learning rate for this test was 0.005 and the momentum rate was 0.5. It took 1000 iterations on DEC VAX 8700 computer for acceptable convergence. The network was then tested using 1093 data points taken from the rest of the cycle (11/19/88 to 12/18/89). Figures 2 and 3 shows the measured and neural network predicted feedwater flow rates for loop A and loop B respectively. The large spikes are points with bad data in one or more of the input parameters. More than 90% of the testing points for both loops have relative absolute error of less than 2%.

The Cycle 8 data shows operation essentially at steady state with more operational changes than Cycle 7. The training data for this cycle was for the period of 03/06/90 through 10/31/90 taken at eight hours intervals. There were 619 data points considered as good training points; however as in Cycle 7, a number of points in the above training period were obviously bad and were removed. The learning rate for this test was 0.005 and the momentum rate was 0.5. It took 1000 iterations on DEC VAX 8700 computer for acceptable convergence. The network was then tested using 703 data points taken from the rest of the cycle (11/01/90 to 06/30/91). Figures 4 and 5 show the measured and neural network predicted feedwater flow rates for loop A and loop B. Figure 6 shows the relative error for both loops. The large spikes are again points with bad data in one or more of the input parameters. Again as in Cycle 7, more than 90% of the testing points for both loops have relative absolute error of less than 2%.

Conclusions

This paper demonstrates a method for using a neural network to validate sensor accuracy. By computing the observability relationships for the plant process variables, sensors related to a target sensor(s) were identified and used as input parameters to a neural network. After using the Backpropagation algorithm to train the network, the output of the network which is an estimate of the correct reading of the target sensor is obtained. Applying the method to simulated and plant operating data, it was shown that the correct readings of the feedwater flow meters can be obtained with an absolute relative error of less that 2% when proper selection of the training data is made. Further investigations are in progress to determine the robustness of a neural network-based sensor validation system.

References

[1] R. J. Lord & J. E. Meyer, "Instrument Calibration Reduction Program", EPRI-NP-7207-CCML, March 1991.

[2] R. E. Uhrig, "Potential Application of Neural Networks to Nuclear Power Plants", American Power Conference, 1991.

[3] B. R. Upadhyaya & E. Eryurek, "Application of Neural Networks for Sensor Validation and Plant Monitoring", Nuclear Technology, Vol. 97, Feb. 1992.

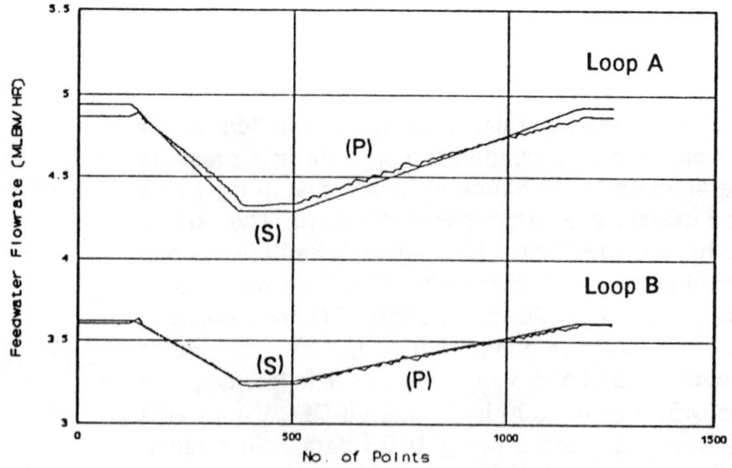

Figure 1: Simulated and Predicted FW Flow Rates

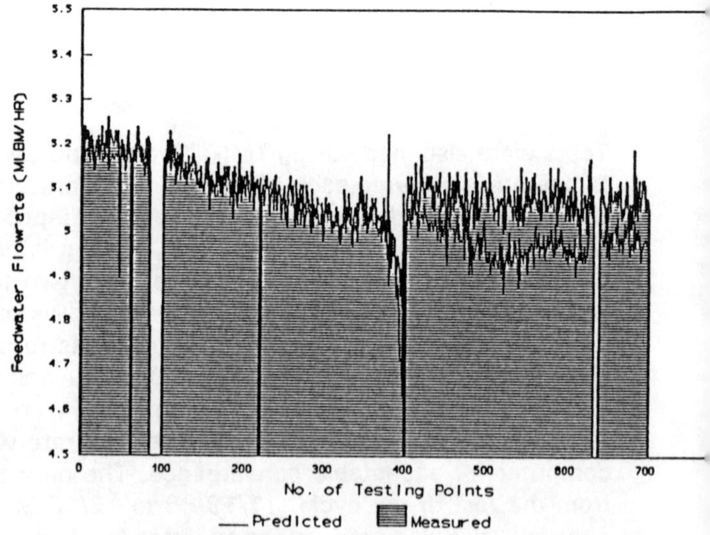

Figure 4: Measured and Predicted FW Flow Rate (loop A) Cycle 8

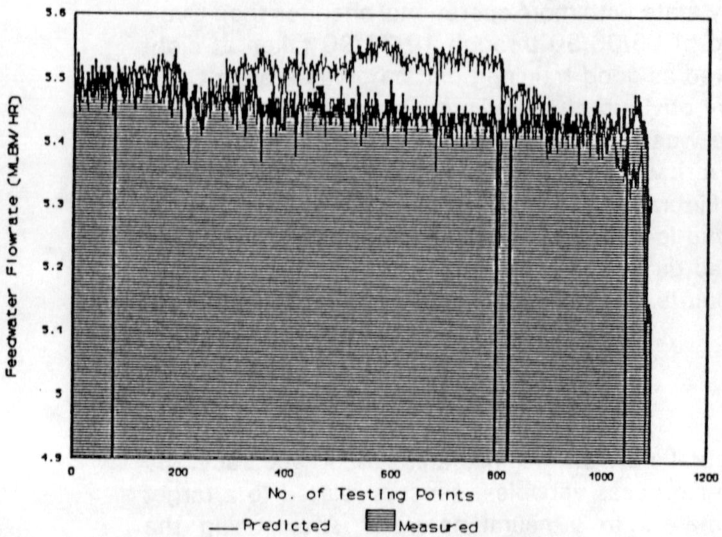

Figure 2: Measured and Predicted FW Flow Rate (loop A) Cycle 7

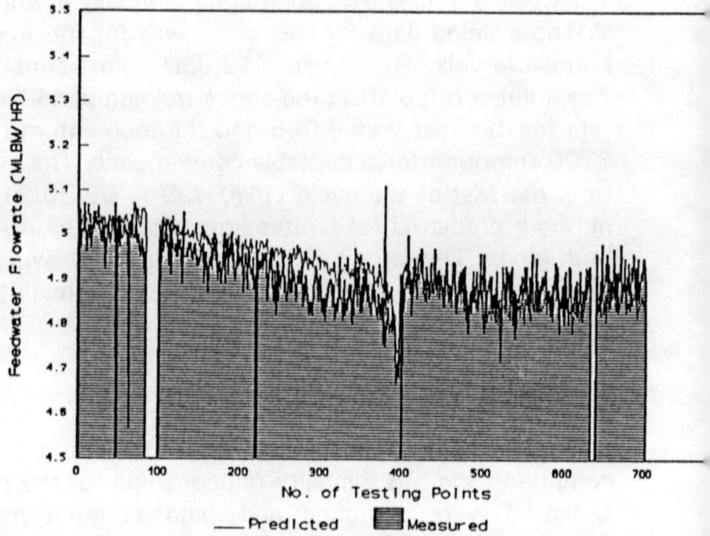

Figure 5: Measured and Predicted FW Flow Rate (loop B) Cycle 8

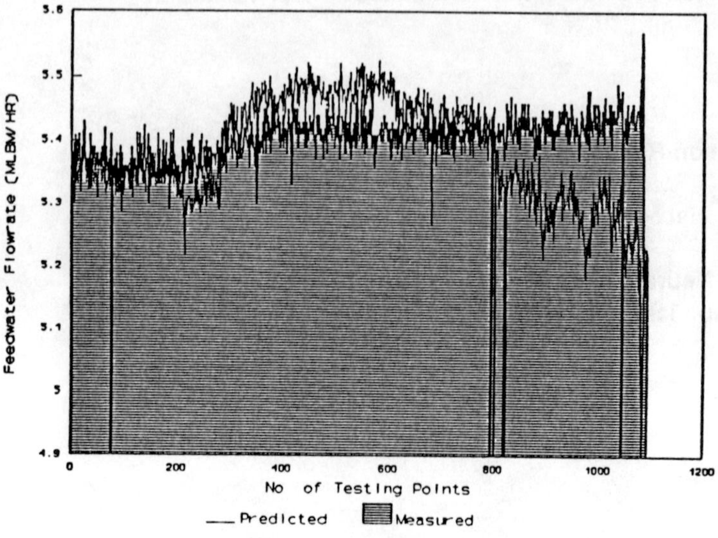

Figure 3: Measured and Predicted FW Flow Rate (loop B) Cycle 7

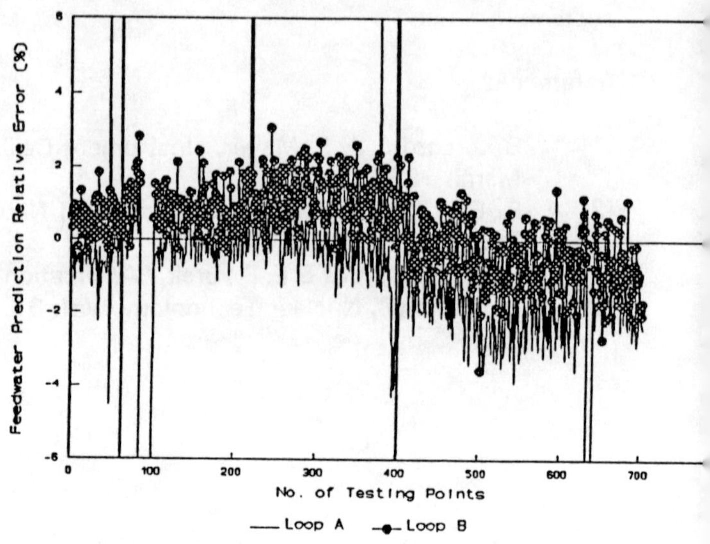

Figure 6: Prediction Relative Error for Cycle 8

MEASURING FUZZY VARIABLES IN A NUCLEAR REACTOR USING ARTIFICIAL NEURAL NETWORKS

A. IKONOMOPOULOS, L. H. TSOUKALAS, R. E. UHRIG
The University of Tennessee and Oak Ridge National Laboratory
Department of Nuclear Engineering
Knoxville, TN, 37996-2300

ABSTRACT

A novel approach is presented that couples Artificial Neural Networks (ANNs) with rule-based fuzzy expert systems for the purpose of monitoring nuclear reactor systems. A model-referenced approach is utilized which provides timely concise and task specific information about the state of the system. In this approach, the rule-based system performs the basic interpretation and processing of the model data, and pre-trained ANN's provide the model. Having access to a set of neural networks that typify general categories of system behavior, the rule-based system performs identification functions. The output of the neural networks is membership functions which condense information about the state of a system in a form convenient for a rule-based system. This allows the identification function to be performed in a time frame faster or at least comparable to the temporal evolution of the system. The methodology is demonstrated using actual data from an experimental nuclear reactor exhibiting excellent robustness to noisy as well as faulty signals.

INTRODUCTION

Monitoring the performance of equipment in a nuclear facility requires a set of procedures for recognizing whether the values of various parameters are within desirable or undesirable ranges. The monitored parameters are typically specific of a particular system, often the output of sensors and meters. Generally, the notions of *performance*, *desirable*, *normal*, are quantified in technical specifications and operating procedures, in terms of set-points and ranges. Yet, in the course of operations they are imputed with meanings that vary not only with the history of a particular equipment but also with different operators and the state of the plant as a whole [1].

There is a number of variables in a complex system which cannot be measured directly, or the failure of a sensor requires that a variable be inferred from other system parameters. In order to perform accurate and on-time evaluation of non-directly measurable system parameters, a *virtual measuring device* is proposed for measuring user-defined dynamic variables with operational significance. A promising feature of virtual instruments is that their function may be modified by changing their software, not hardware. Generally, measurement involves a mapping of complex input patterns to simple output patterns. Neural networks are employed for mapping spatiotemporal input patterns, pertained to directly measured system parameters, to a simplified set of membership functions [2] representing the values of a fuzzy variable [3].

A major obstacle encountered in the utilization of fuzzy reasoning for industrial applications is the difficulty in generating membership functions [4]. In the proposed approach neural networks are used to map a set of signals representing the state of a nuclear system to a set of membership functions that describe the values of a fuzzy variable, called in this study **VALVE_POSITION**. A set of pre-trained neural networks receive and filter several on-line time series corresponding to vital nuclear system parameters and compute the value of another system parameter in the form of a membership function. Each membership function corresponds to a different value of the monitored variable and has a unique shape.

MODEL DESCRIPTION

Fuzzy logic and rule-based systems are convenient tools for dealing with imprecision in an attempt to model human behavior [5]. In the process control area, the rule-based approach differs from analytical methodologies in the sense that the former uses qualitative information where the later is based on mathematical system description. Rule-based systems offer the advantage of describing a system whose behavior can be articulated in "if-then" type statements, called fuzzy or production rules [5]. In fuzzy controllers, fuzzy rules describe the relation between state variables and action or control variables, e.g.,

if pressure is *medium* and **temperature** is *high* then **valve_position** is *partially_open*

where, **pressure** and **temperature** are fuzzy variables describing the state of the system, *medium* and *high* are two of their fuzzy values, **valve_position** is a fuzzy variable describing an action and *partially_open* is one of its fuzzy values. Every one of the variables is termed as universe of discourse with individual labels identifying classes (or subsets) of the universe. Each fuzzy value of the fuzzy variable is a subset of the corresponding universe and is characterized by a membership function. The membership function maps the universe of discourse to the real interval [0,1]. The degree to which a measured (crisp) value belongs to a particular class is its grade of membership.

The proposed methodology is based on a set of pretrained neural networks which receive several on-line time series and compute another system parameter in the form of membership functions. Each membership function describes a different value of the monitored variable and has a unique shape. The methodology is demonstrated utilizing actual data obtained during a start-up period of the High Flux Isotope Reactor (HFIR), three-loop pressurized water reactor, operated at the Oak Ridge National Laboratory. The monitored parameter is the position of the Secondary Flow Control Valve. This particular valve controls the water flow at the secondary side of the system and is considered as a vital system component. The data used is normalized in the interval 0.1 to 0.9, sampled every 16 seconds, with a total of 1240 samples.

Five parameters have been chosen for describing the Secondary Flow Control Valve Position: **neutron flux, primary flow pressure variation (DP), core inlet temperature, core outlet temperature** and **secondary flow**. The time series of these five parameters are used to train a set of five neural networks, where each one them is responsible for recognizing a specific position of the valve-disc. The output of every neural network is a membership function uniquely labeling a particular position of the Secondary Flow Control Valve. The behavior of the Secondary Flow Control Valve is represented in the universe of discourse of human linguistics with the fuzzy variable **VALVE_POSITION** that takes five fuzzy values, namely: *closed*, *partially_closed*, *medium*, *partially_open* and *open*. A membership function, namely: μ_{closed}, $\mu_{partially\_closed}$, μ_{medium}, $\mu_{partially\_open}$ and μ_{open} is defined for every fuzzy value, respectively. These five membership functions compose a library of prototypes and are considered sufficient for describing the position of the valve at every instant during a start-up period. A schematic representation of the universe of discourse of the fuzzy variable **VALVE_POSITION**, is given in Figure 1.

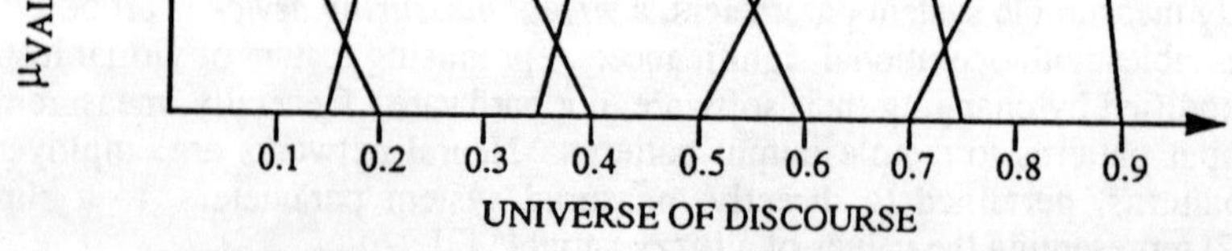

Figure 1: Membership function labeling of possible valve positions.

The area occupied by every membership function in the universe of discourse depicts the uncertainty associated with this particular class [6]. The fuzzy set (or class) *open* in the universe of discourse is characterized by the membership function μ_{open} (Figure 1) at this point. In this manner, the grade of membership (or level of presumption) of this point in the class *open*, is represented. This framework provides a natural way of dealing with problems where the source of imprecision is the absence of sharply defined boundaries of class membership, rather than the presence of random variables [6]. The membership functions developed for this particular study have trapezoidal shape, which is very useful for performing computations in the fuzzy control area.

The neural networks contemplated in this research are three-layer feed-forward [7]. The training algorithm used is back-propagation with generalized delta rule and momentum term, as it is supplied by the Plexi neural environment. The input layer of the network consists of five nodes, each one receiving information from a particular system parameter. The output layer consists of ten nodes corresponding to the coordinates of ten points of the trapezoidal membership function, describing the position of the valve at the universe of discourse. Typically, 7,000 training cycles will produce a sum square error of 0.02 when 30 nodes are used in the hidden layer. One of the major advantages of ANNs is their unique ability to be trained in a *user-defined environment*. Funahashi [8] has proved that any continuous mapping can be approximately realized by Rumelhart-Hinton-Williams' [9] neural networks with at least one hidden layer having sigmoid output function. Lapedes and Farber [10] have shown experimentally that a neural network trained with the back-propagation algorithm is able to make predictions of arbitrarily complex mappings with very good accuracy. Taking advantage from this property, the set of neural networks developed to compute the states of the **VALVE_POSITION**, has been trained to predict the position of the valve-disc at the near future. Thus, the ANNs receive information concerning the state of the system at instant t in time and predict what the valve position may be at instant $t+\Delta t$. The time interval Δt has been set equal to one sampling interval (i.e., 16 seconds).

Generally, the output of the neural networks will be somewhat different than the prototype membership functions. Moreover, one or two (since overlap of membership functions has been allowed) will represent correct values, while the rest need to be ignored. In order to identify the correct output one could consider the membership functions as *fuzzy numbers*, and use the *dissemblance index* to estimate the "distance" between the prototype membership function and the membership function predicted from the neural network [11].

It is in the nature of neural computing to associate new patterns with existing ones, in a manner that implicitly constructs the variable *time*. On the other hand, the notion of measurement is associated with the variation of system parameters in time. This makes imperative the construction of a measuring device that will be inherently dynamic. At time t, there is an indication of the possible valve-position at the next time step along with knowledge of the present, as well as, previous states of the valve. By using multi-variate regressive models, the optimum number of previous network responses required from the rule-based system to validate the present network response, has been found to be four. The rule-based system operates in a dynamic fashion since it automatically accepts or discards responses at a certain time and considers the new values when it draws conclusions during future evaluations.

The overlap of adjacent membership functions creates an environment where the system can be estimated at two separate states at the same instant in time. In order to address this issue, the characteristic function of the set of network responses which have been accepted by the rule-based system (called the set of *satisfactory_responses, (SR))* is considered continuous, rendering the set *SR* fuzzy [6]. The network predicted membership functions are treated as fuzzy numbers, and their product with the level of presumption of their own satisfactory responses, is calculated [11]. The resultant membership functions are added in the universe of discourse creating a new fuzzy number. The membership function computed from the summation of independent membership functions describes uniquely and unambiguously the state of the valve.

MODEL TESTING

The performance of the proposed methodology has been tested by substituting each one of the five time series with a signal corresponding to 100% random noise. This random signal along with the remaining four original signals compose the test vectors supplied to the measuring device during the recall process.

The results presented here (Figures 2 and 3) depict the substitution of the average flux signal by a totally random one. In Figure 2, the responses of the virtual measuring device are plotted with a frequency of appearance equal to one every ten time steps for a time span covering the first 200 time steps of the start-up duration. The measuring device has successfully recorded the two abrupt openings of the valve at around 60 and 120 time steps after the start-up initiation, where the valve-position has been identified as **medium**. The results obtained throughout the start-up period are shown in Figure 3, plotting one membership function every 100 time steps. Around 700 time steps after the transient initiation the valve-position is characterized as **open**, and the same characterization is repeated thereafter. Furthermore, during the time interval between time steps 200 and 700, the position of the valve-disc is initially defined as **medium** and later as **partially_open**, following the transition of the valve from one state to another.

CONCLUSIONS

A new approach has been presented for measuring linguistic variables. The methodology is based on the utilization of ANNs as membership function generators. This novel implementation offers the advantage of describing the state of a system in a condensed form suitable for control applications. The membership functions computed from the *virtual measuring device* can be used as input to a fuzzy controller without the prerequisite of fuzzifying the crisp input values obtained from measurements.

The methodology presented here has been subjected to numerous tests exhibiting robustness to noisy inputs and faulty signals. The virtual measuring device has been proven capable of identifying the correct system state even in the case where 20% of the required input information is eliminated. The optimization algorithm developed considers the responses of the independently firing networks and constructs the membership function that most appropriately describes the state of the system in the universe of discourse.

REFERENCES

1. L. H. Tsoukalas and A. Ikonomopoulos, "Modeling Complex Systems with Artificial Neural Networks," in *Intelligent Engineering Systems Through Artificial Neural Networks*, edited by C. Dagli, S. Kumara and Y. Shin (ASME Press, New York, 581-587 1991).
2. A. Ikonomopoulos, L. H. Tsoukalas, J. A. Mullens and R. E. Uhrig, "Monitoring Nuclear Reactor Systems Using Neural Networks and Fuzzy Logic," *Proc. of the 1992 ANS Topical Meeting on Advances in Reactor Physics*, Vol. 2, Charleston, S. C., 140-151 (1992).
3. L. A. Zadeh, "Outline of a New Approach to the Analysis of Complex Systems and Decision Processes," *IEEE Trans. on Systems Man and Cybernetics*, SMC-3, 28-44 (1973).
4. L. A. Zadeh, "The Calculus of Fuzzy if/then Rules," *AI Expert*, 23-27 (March 1992).
5. L. A. Zadeh, "Fuzzy Logic," *IEEE Computer*, 83-93 (April 1988).
6. L. A. Zadeh, "Fuzzy Sets," *Information and Control*, 8, 338-353 (1965).
7. J. Hertz, A. Krogh and R. Palmer, *Introduction to the Theory of Neural Computation*, (Addison-Wesley, Redwood City, 1991).
8. K. Funahashi, "On the Approximate Realization of Continuous Mappings by Neural Networks," *Neural Networks*, Vol. 2, 183-192 (1989).
9. D. E. Rumelhart, G. E. Hinton and R. J. Williams, "Learning Representations by Back-Propagating Errors," *Nature*, 323, 533-536 (1986).
10. A. Lapedes and R. Farber, "Nonlinear Signal Processing Using Neural Networks: Predictions and System Modeling," *Los Alamos Nat'l. Lab. report LA-UR-87-2662* (1987).
11. A. Kaufmann and M. M. Gupta, *Introduction to Fuzzy Arithmetic*, (Van Nostrand Reinhold, New York, 1991).

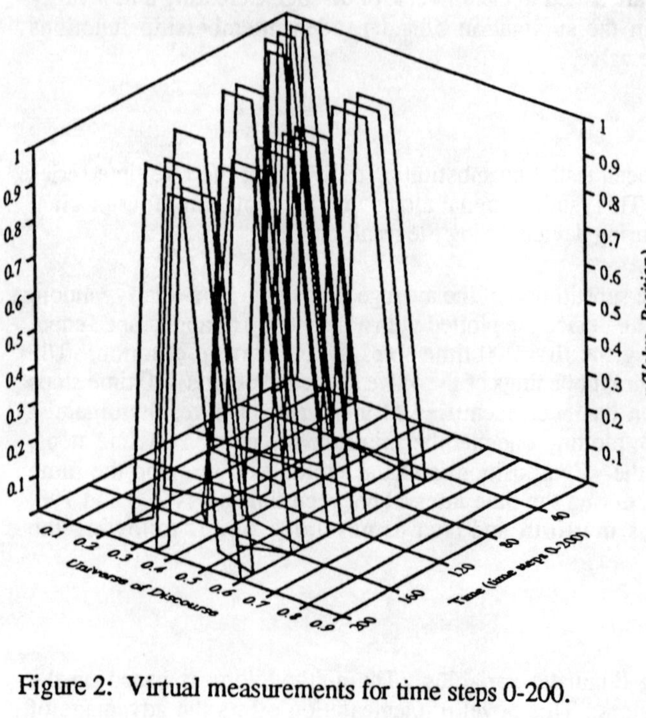

Figure 2: Virtual measurements for time steps 0-200.

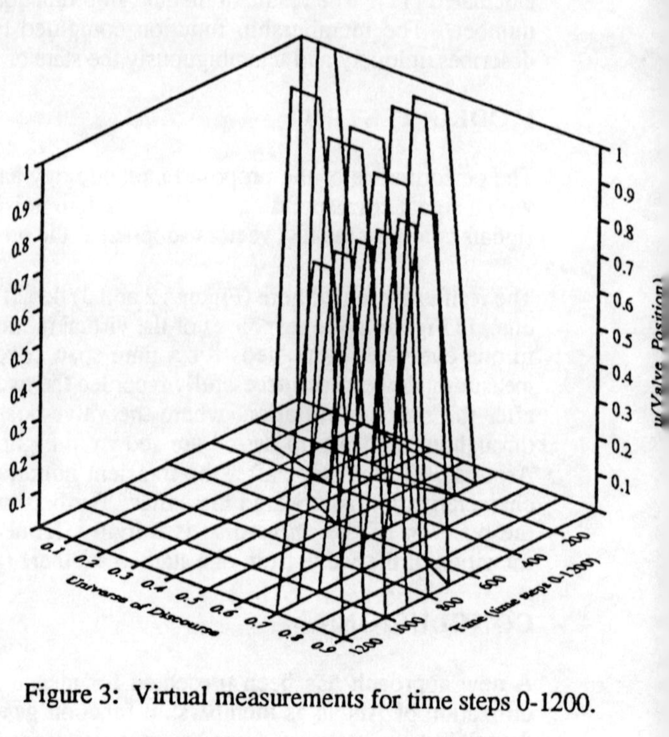

Figure 3: Virtual measurements for time steps 0-1200.

Application of a Real Time Artificial Neural Network for Classifying Nuclear Power Plant Transient Events

Y. D. Lukic, C. R. Stevens
Arizona Public Service
P. O. Box 52034
Phoenix, AZ 85072-2034

J. Si
Department of Electrical Engineering
Arizona State University
Tempe, AZ 85287-5706

Abstract - In this paper we demonstrate the application of ANN methodology in the recognition of transient events in a Nuclear Power Plant. The end product is a feed forward, back propagation network capable of classifying the transient event categories outlined in this report. The network is able to classify all transient events trained on after 5 minutes following event initiation. The network achieves this reliability using test data with up to 5% sensor input noise.

Keywords - Artificial Neural Networks, Back propagation, Nuclear Power Plant, Transient Events, Severe Accident Management.

1. Introduction

Modern day Nuclear Powered Generating Stations are very complex. During abnormal operations and accidents, nuclear power plants are heavily reliant on the skill and training of operators to stabilize the plant and restore it to a safe condition. Since operators rely on plant process inputs to diagnose and respond to these events, this lends itself to an artificial neural network (ANN) application. In addition, the capability of ANN's to generalize should greatly improve on the operators ability to diagnose events. Plant procedures for categorizing events attempt to cover all anticipated accidents, however this may not always be possible, especially for cases of noisy or unreliable plant process data. This project is an effort to build a neural network that is capable of categorizing these events from the time signatures of key plant process data. This paper demonstrates the capability of an ANN to classify 11 transients/accident events and steady state operation for a Combustion Engineering System 80 pressurized water reactor (PWR).

This project investigated several key areas: (a) evaluation on the effect of varying the number of processing elements (PEs) in the hidden layer and number of hidden layers as it relates to the memory capacity of the ANN; (b) evaluation of techniques to improve ANN noise tolerance; and, (c) evaluation of the relative importance of individual plant process parameters in the recognition of transient events.

A search of related technical literature resulted in identifying several recent attempts at an AI application to categorize nuclear power plant transient events. Bartlett and Uhrig[1] teamed up to develop a single ANN that can recognize 8 nuclear power plant transient events. The trained ANN is able to correctly classify the 8 events with up to 2% noise. Ikonomopoulos, Uhrig, and Tsoukalas[2] have performed a feasibility study for categorizing two simple, but similar in nature, transients for a Westinghouse four-loop Pressurized Water Reactor. The results are reportedly very promising in that the system not only could easily distinguish between the two transients (with 99% accuracy) but could do so with up to 20% noise in all five input parameters. Finally, Silverman, Howard, and Harpster[3] have developed a rule based system for classifying Loss of Coolant Accidents (LOCAs). Once the rule based system classifies the LOCA break size it directs it to one of 15 ANNs which predict the time to core uncovery, lower support plate failure, and containment building failure in terms of operating equipment.

2. ANN Architecture

In this project a single three layer ANN was developed to classify all events. The net has 24 input and 4 output PEs. The input layer PEs correspond to 24 individual plant process parameters, while the 4 output PEs correspond to the binary classification of 12 separate events that the net was trained on. As shown in Table 1 these include 11 transient events plus normal operations at power. The single hidden layer size was optimized as discussed in Sections 5 and 6.

3. Transient Events

Simulations with the Combustion Engineering Plant Analysis Code[4] (CEPAC) were used to generate plant process parameters as a function of time for selected Palo Verde Nuclear Generating Station (PVNGS) transients. The number of benchmark transient events available to teach the network was limited to cases that CEPAC could simulate; events which require 2-phase flow modeling, i.e. large and medium size pipe breaks in the primary and secondary loops, and rapid reactivity excursions at power could hence not be modeled. Using the CEPAC simulated plant process parameters (see Table 2) as input to the ANN, the net was trained to classify transient events described in Table 1. Several cases (e.g. transients with different break sizes) were modeled for selected transient categories in order to ensure the

ANN was trained to recognize important variations in the progression of these transients.

Table 1: Transient Events Modeled

Event No.	Transient Event Category	No. of Cases	Output Code
1	Normal Steady State Conditions at Power	1	0000
2	Hot and Cold Leg Breaks	4	0001
3	Pressurizer Relief Valve Leaks	4	0010
4	Steam Generator #1 Tube Ruptures	2	0011
6	Steam Generator #1 ADV Stuck Open	1	0100
8	Steam Generator #1 Steam Line Break	1	0101
10	Loss of Main Feedwater Train A	1	0110
5	Steam Generator #2 Tube Ruptures	2	1011
7	Steam Generator #2 ADV Stuck Open	1	1100
9	Steam Generator #2 Steam Line Break	1	1101
11	Loss of Main Feedwater Train B	1	1110
12	Normal Trip from Power	1	1111

4. ANN Input Process Parameters

The CEPAC computer program source code was modified for this project to output the 24 chosen process parameters in the training input format necessary for NeuralWorks[5], the commercial neural network software package used in this project. An initial list of process parameters was developed at the start of the project; the list was based on a composite of plant process parameters which the operators use in transient event diagnostic, and are provided by CEPAC. Table 2 shows a list of the 24 plant process parameters that were used to train the ANN in this project.

Table 2: ANN Input List

Input No.	Plant Process Parameter	Units	Importance Rank
1	Pressurizer Pressure	psia	2
2	Pressurizer Level	feet	8
3,4	T_{cold} Train A and B	deg F	1
5,6	T_{hot} Train A and B	deg F	9
7	Sub-cooled Temperature Margin	deg F	4
8	High Pressure Safety Injection Flow	lbm/sec	15
9	Charging Pump Flow	gpm	3
10	Letdown Flow	gpm	10
11,12	Steam Generator No. 1 and 2 Level	feet	7
13,14	Steam Generator No. 1 and 2 Pressure	psia	6
15,16	Main Feedwater Flow Train A and B	lbm/sec	5
17,18	Auxiliary Feedwater Flow Train A and B	lbm/sec	12
19,20	ADV Steam Flow Train SG No. 1 and 2	lbm/sec	14
21,22	Main Steam Flow SG No. 1 and 2	lbm/sec	11
23,24	I-131 Activity SG No. 1 and 2	μcuries/cm^3	13

5. Network Training and Testing

Training and test data were generated from the same CEPAC code transient event simulation runs. For each simulated event, the training data was collected starting at 20 sec. of the simulation initiation and every 20 sec. thereafter, up to a simulation time of 580 sec. The test data was collected starting at 30 sec. after simulation start, and at every 20 sec. interval thereafter, up to a simulation time of 590 sec. All the training and test data sets from individual simulation runs were grouped into a separate training and test file.

The NeuralWorks computer program was used to train the network using the back propagation training algorithm with the hyperbolic tangent squashing function. Prior to training the net, a NeuralWorks[5] program feature which permits the normalization of the input training data between the limits of -1.0 and 1.0, and output training data between the limits of -0.8 and 0.8 was activated. The learning and momentum coefficient as a function of training cycles were set for both the hidden and output layers, with larger more aggressive coefficients initially. For example, the starting learning coefficient for the hidden layer was 0.3, and for the output layer 0.15; the starting momentum for both was 0.4. The learning and momentum coefficient are changed later to achieve a more relaxed learning rate after the character of hidden layer has been established.

The net was trained for incremental number of cycles. After each incremental training period the network was presented with the test data file and the RMS error of the test data output was calculated. It was found that the nets converged to an overall asymptotic RMS error within 250 to 450 cycles per training pattern. A plot of the asymptotic RMS error as a function of the number of PEs in the hidden layer is shown in Figure 1. As can be observed in this figure, the RMS error consistently decreases between 3 and 9 hidden layer PEs and does not significantly improve thereafter. No evidence of over-training the nets was observed.

Figure 2 shows a typical result of the network output (with 12 hidden PEs) when presented with test data. The figure shows the time evolution of the output bits for the SG Tube Rupture event (SG #1, 1 ruptured tube) with the ANN weights trained and tested with 0% input noise. For the purposes of this project the threshold values are defined as follows: a PE output value of <0.25 is taken to be equivalent to be a bit output value of 0; similarly a PE output value >0.75 is taken to be equivalent to be a bit output value of 1. Output PE values between 0.25 and 0.75 are considered as undeveloped, i.e. the event may not be classified yet at that time during the event.

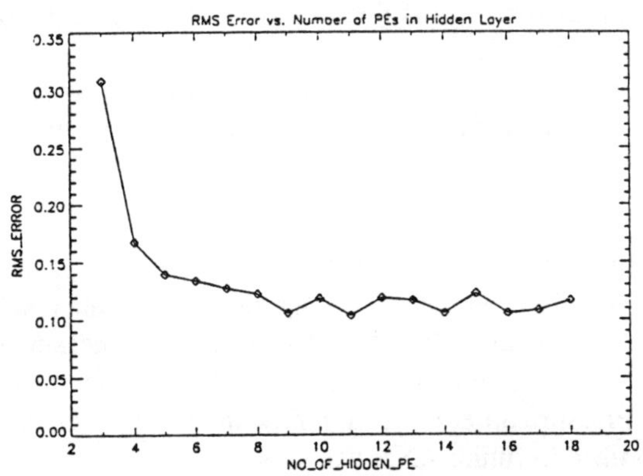

Figure 1 RMS error of the ANN output as function of the hidden layer size

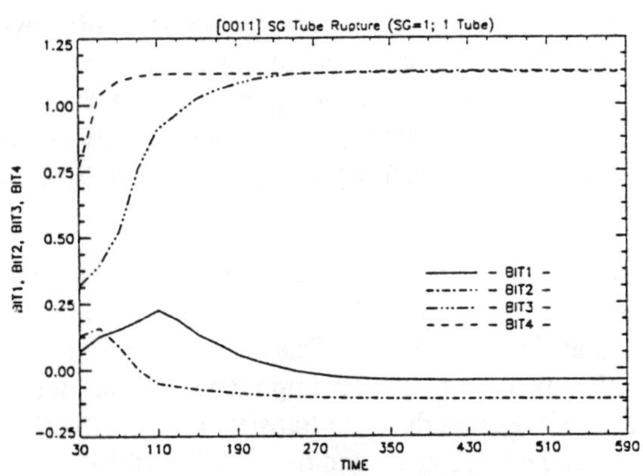

Figure 2 Time evolution of the ANN output bits for SG Tube Rupture (SG No. 1; 1 tube)

6. Discussion of Key Results

Minimal number of hidden layer PEs and of hidden layers

Based on Figure 1 results, we verified empirically for our data that the minimal number of hidden processing elements is 9. Given that the number of distinct training patterns is 638, the relationship $P \approx 2^H$ between the number of distinct training patterns **P** and the number of hidden nodes **H,** predicts quite accurately the number of minimal hidden nodes at 9.

Experimentation on the number of hidden layers was conducted for a two hidden layer configuration containing 12 and 6 PEs respectively. The results do not appear to improve noticeably over those derived with a single hidden layer ANN having 12 PEs.

Evaluation of input noise impact

The network is reasonably noise tolerant for noise input up to approximately 5%. The net was trained with 0% and 10% input noise and then each was tested with 0-20% input noise. An important finding as a result of this training is that a net trained with input noise appears to become less sensitive to input noise in the test data. The general tolerance of the nets to expected levels of noise in the plant process parameters seems to reduce the need to develop separate networks trained for each transient category. However there may be other practical reasons why multiple nets may be desirable: the limited analysis conducted thus far for failed and degraded inputs, implies that the net is not robust enough to withstand large variations in input process parameter values. Several options exist to be explored in the future for this project which could involve: (1) adding more input plant process parameters; (2) adding an input preprocessor possibly a self-associative net; (3) adding a reinforced learning feature to the existing net; or, (4) multiple nets.

Event Diagnostic Time

Comparison of the desired and calculated output vectors as a function of simulation time reveals that some transient events take longer to develop than others, e.g. Steam Generator Tube Ruptures which take approximately 300 seconds to correctly classify the event vs. Main Feed water Pump Trip which is classified within 20 seconds of the event initiation. This is qualitatively consistent with the results in reference 1.

We have also evaluated the qualitative impact of noise on the ANN as it relates to event diagnostic time. The results indicate that with increasing input noise it generally takes longer to reach the threshold for positive classification for selected events; this finding is also consistent with reference 1. For large input noise levels the net may not be able to correctly classify the event at all.

The net developed for this project is expected to only give valid results between 300 and 600 seconds of event initiation. This net has not been tested for the ability to classify events beyond 600 seconds, hence its usefulness beyond this time as currently trained, is questionable. Some nuclear power plant transients may continue for hours and may be compounded by combinations of events resulting from additional equipment failures, e.g., any transient events resulting in a temporary loss of secondary system heat removal will result in an increase in primary system pressure. This could lead to a Pressurizer Relief Valve opening and the valve failing to re-close. Hence it may be difficult to train the net for long periods of event evolution for two reasons: (1) as the plant is finally stabilized following most events, the process parameters all go to the same stable states, this may ultimately confuse the net; and, (2) compounding events as a result of equipment failures or operator error as discussed above. For item (1) further research will be needed. For item (2) the number of likely compounding events is fortunately limited and therefore it may be tractable to handle this using a larger net, or multiple nets.

Pruning of Inputs

Minimizing the number of unnecessary inputs is important for reducing the amount of data that must be manipulated. An analysis was conducted by zero-ing out all ANN inputs, one at the time, and by observing the effect on the resulting output signal RMS error. This analysis permitted us to establish a first cut of the ANN input ranks, as a measure of each plant process parameter importance in transient event classification. The ranks as shown in Table 1, are a composite result accross different transient events; note that the High Pressure Safety Injection flow input shows the least impact on the RMS error, and hence will be the first input to be pruned in future ANN versions.

7. Conclusions and Future Work

The objectives of this project have been met by demonstrating the ability to train ANNs to identify transient events from plant process parameters. We intend to continue this project at Arizona Public Service, possibly leading to development of on-line control-room diagnostic tools useful in severe accident management. Further work is contemplated in the area of training the ANNs with simulations of additional transient events and applicable combinations of these. Following the initiative of Reference 3, we also intend to explore the ANN applications leading to forecast of core uncovery time, time to vessel failure and time to containment failure. Finally, we have great interest in pursuing additional work which could help the operators validate individual plant process variable signals in terms of other process variable signals.

8. References

1. E. B. Bartlett, R. E. Uhrig, "Nuclear Power Plant Status Diagnostics Using Artificial Neural Networks," Proceedings of the American Nuclear Society, Jackson Lake WY Meeting, September 1991
2. A. Ikonomopoulos, R. E. Uhrig, L. Tsoukalas, "A Hybrid Neural Network - Fuzzy Logic Approach to Nuclear Power Plant Transient Identification," Proceedings of the American Nuclear Society, Jackson Lake WY Meeting, September 1991
3. E. B. Silverman, J. H. Howard, J. L. Harpster, ARD Corporation, Personal Communication, July 1992
4. Combustion Engineering Plant Analysis Code, User's Manual, Report No. CE-CES-61, Combustion Engineering, April 1986
5. NeuralWorks Professional II/Plus, Reference Guide, NeuralWare Inc., Pittsburgh, PA, 1991

Control Rod Wear Recognition Using Neural Nets

James A. Boshers, Duke Power, P.O. Box 1006, Charlotte, NC 28201
Charles Saylor, Shyam Kamadolli, Kaman Sciences, 258 Genesee St., Utica, NY 13502
Richard Wood, Can Işık, Syracuse University, 121 Link Hall, Syracuse, NY 13244

Abstract

Cladding wear of power control rods of nuclear reactors causes rod clusters to be replaced prematurely. To mitigate wear by repositioning rods, it is necessary to identify the location of each wear scar and track its depth, by using inspection data. The purpose of this study is to investigate methods to automatically perform such analysis, which is currently performed by human operators. The paper describes a method which processes efficiently a large volume of data, identifies the wear scars and finds certain quantitative properties of the wear. The method combines algorithmic, neural network and rule-based components. The paper also summarizes the results of the feasibility study of using feed forward neural networks for the identification of wear marks.

Background

Wear occurs in the cladding of nuclear reactor control rods, at locations adjacent to guide cards and a continuous guide. Guide cards are support pieces located at roughly 1 foot intervals over the 12 foot length of the control rods, the continuous guide is a longer support located towards the end of the rod. Each guide card contacts the control rod over an axial length of approximately 0.75 inch. The wear is believed to be caused by flow-induced motion of the control rod against the guide card and the continuous guide. Although the design life of control rods is 15 years, Duke Power Company discovered extensive premature clad wear and started regular inspections of all 53 rod clusters in the core at each annual refueling outage. The inspection results are used to replace rods before the cladding is worn through, as well as to attempt to mitigate wear by re-positioning rods. To mitigate wear by re-positioning rods, it is necessary to identify the location of each wear scar and track its depth by using the inspection data. It is believed that through rod wear modeling and rod placement scheduling actual lifetime of rod clusters can be significantly extended and the inspection frequency can be reduced to every other outing. However, current methods of analysis of the inspection data are labor intensive, does not yield reliable and accurate recognition of wear features, and are sensitive to the amount of noise in the measurements.

In order to automatically process and interpret the inspection data EPRI Knowledge Based Technology Applications Center (KBTAC), jointly operated by Kaman Sciences Corporation and Syracuse University, is developing with Duke Power Company a neural net based software package. The functions of this software are to (1) identify and extract wear sections, (2) determine wear information including wear position, peak wear depth, cross-sectional area loss, and (3) organize this information in a database. In the next section our approach to this problem is explained.

Problem Requirements and Solution Approach

The inspection of the rods is performed by measuring the thickness of the clad using an ultrasonic sensor. The transducer rotates around the rod, while the rod is being moved up or down, hence measuring the thickness on a helical trace. During each turn thickness is measured at about 130 points. There are over 140,000 measurement points for each rod. In Figure 1 a large section of data with scar is shown. Figure 2 zooms in a couple of revolutions around a scar, where Figure 3 does the same for a region without scar. Although the data in those figures are extracted from a file that has a low "noise" content, one can observe the variations in the plots outside the scar regions. There are four sources of undesired variations in the measurements: (1) imperfections in the surface of the rod, (2) inaccuracies inherent in the ultrasonic distance measurement technology, (3) noise contributed by the measurement electronics, and (4) noise picked up from the environment by electromagnetic coupling. Out of these, the first one is not significant for the situation at hand. The other three make the recognition of scar a challenging task. In the rest of this paper, measurements from one revolution that includes a scar is referred to as the *scar data* and measurements from one revolution that includes no scar are referred to as the *noise data*.

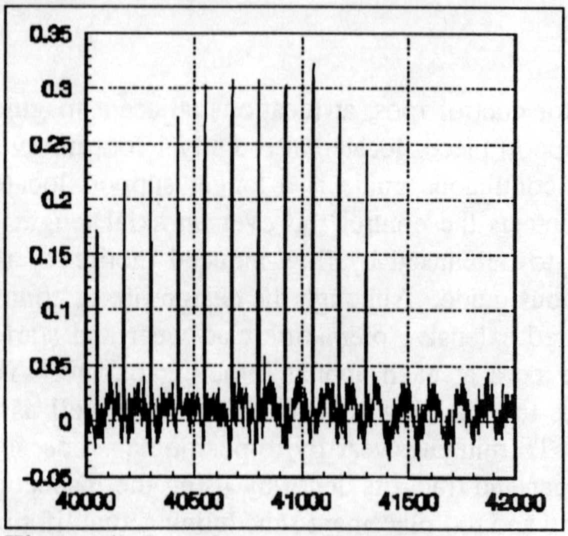

Figure 1 Scar from a guide card

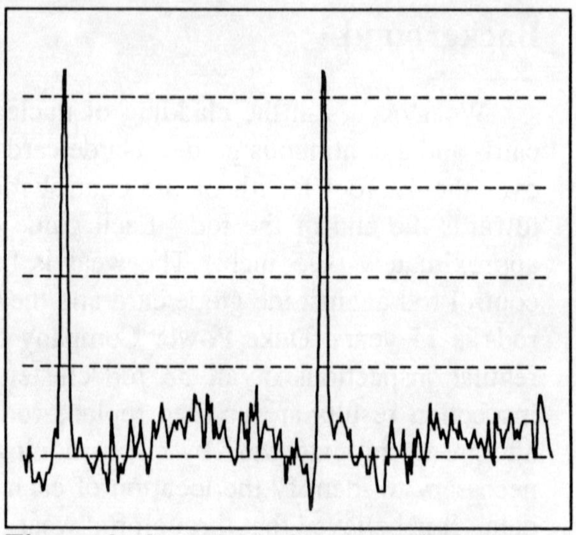

Figure 2 Scar data of two revolutions

Using the inspection data directly as input for a neural network was not considered to be a viable alternative since looking at data from one revolution would result in a network with over 100 input nodes. Instead, features that reflect the presence of symmetry in noise data and the lack of symmetry in scar data were extracted from raw data of ultrasonic measurements of the clad thickness. All four factors that contribute to the measurement noise are *zero mean*, that is, they cause equivalent deviations in positive and negative directions from the thickness measurements. Therefore, when there is no clad wear, the data are distributed pretty evenly above and below the *zero wear mark*, which we will name bias. When there is wear, there are more data points above the bias, and the amplitudes above the bias are larger than those below the bias. The features that indicate this symmetry situation include bias (the moving average of the noise signal), positive peak (maximum signal minus the bias), negative peak (minimum signal minus the bias), averages of signal values over the bias and below the bias, and the numbers of samples with signal values over the bias and below the bias. Eight such features were used as the input to a feed-forward neural net, that classified circumferential segments of data into two groups: noise and scar. As it can be seen in figure 1, support pieces cause scars that extend over several revolutions of the helical trace. This redundancy is exploited to eliminate possible misclassification of data from a single revolution, using a combination of rule-based and algorithmic routines. Segments of raw data that correspond to a wear scar are further analyzed using an algorithmic program to determine a variety of quantitative properties of the scar, and scar data is classified into groups that reflect the shape of the scar and the extent of the wear. At this stage of the research, initial feature extraction algorithm and the neural network for wear recognition have been implemented. The neural network groups all types of wear (single peak, double peak, etc.,) into one class for the initial recognition purpose. As a future project, different types of wear will be recognized using another network which processes those portions of the raw data file which are known to contain wear information. In the next section, the design of the neural network is described.

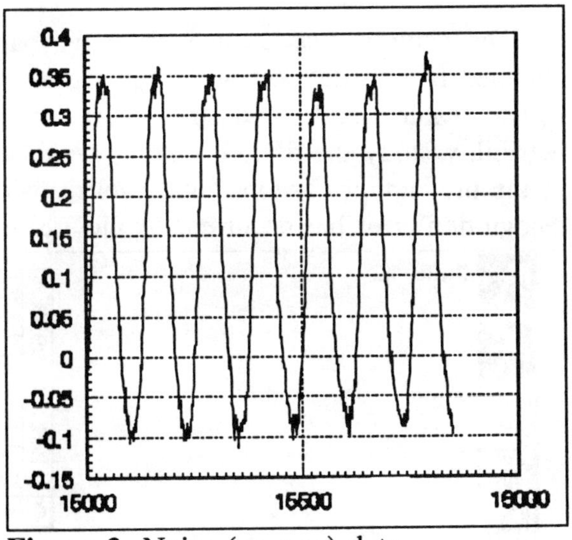

Figure 3 Noise (no scar) data

Wear Recognition

The character of measurement noise and the shapes of wear scars vary drastically from one inspection file to another. The recognition system is expected to adapt to such variations in the class features. Therefore, a neural network approach was chosen. The user is referred to Lippmann's paper [1] for a good introduction to neural networks, and to Kosko's book [2] for other signal processing applications. Using a neural network development tool, called Neuralworks [3], a feed-forward neural net with an input layer, two hidden layers and an output layer was constructed. The input layer has eight nodes, one for each feature, and an additional node which supplies the neural net with a constant term, called *threshold* or *bias* in the neural network literature (not to be confused with "bias" in feature extraction). The output layer has two nodes, one for the presence of scar and one for the presence of noise-only. Although it is possible to use one output node for a two-class recognition problem, the use of one node per class gives a better recognition accuracy. Instead of thresholding a single output value, we choose the node with the largest signal as the recognized class. As a result of parametric studies, the network with 6 and 4 neurons in the two hidden layers was found to give a good convergence speed and high recognition rate.

The network was trained using the backpropagation algorithm [1, 4], and with features obtained from a data set with high signal to noise ratio. Each training record included 8 real valued features and the desired neural net outputs corresponding to those features. For this feasibility study the desired outputs were determined by visually inspecting the raw data file. In the production version, high level decision making will be utilized to determine whether a recognition error has occurred. A picture of the computer screen is shown in Figure 4.

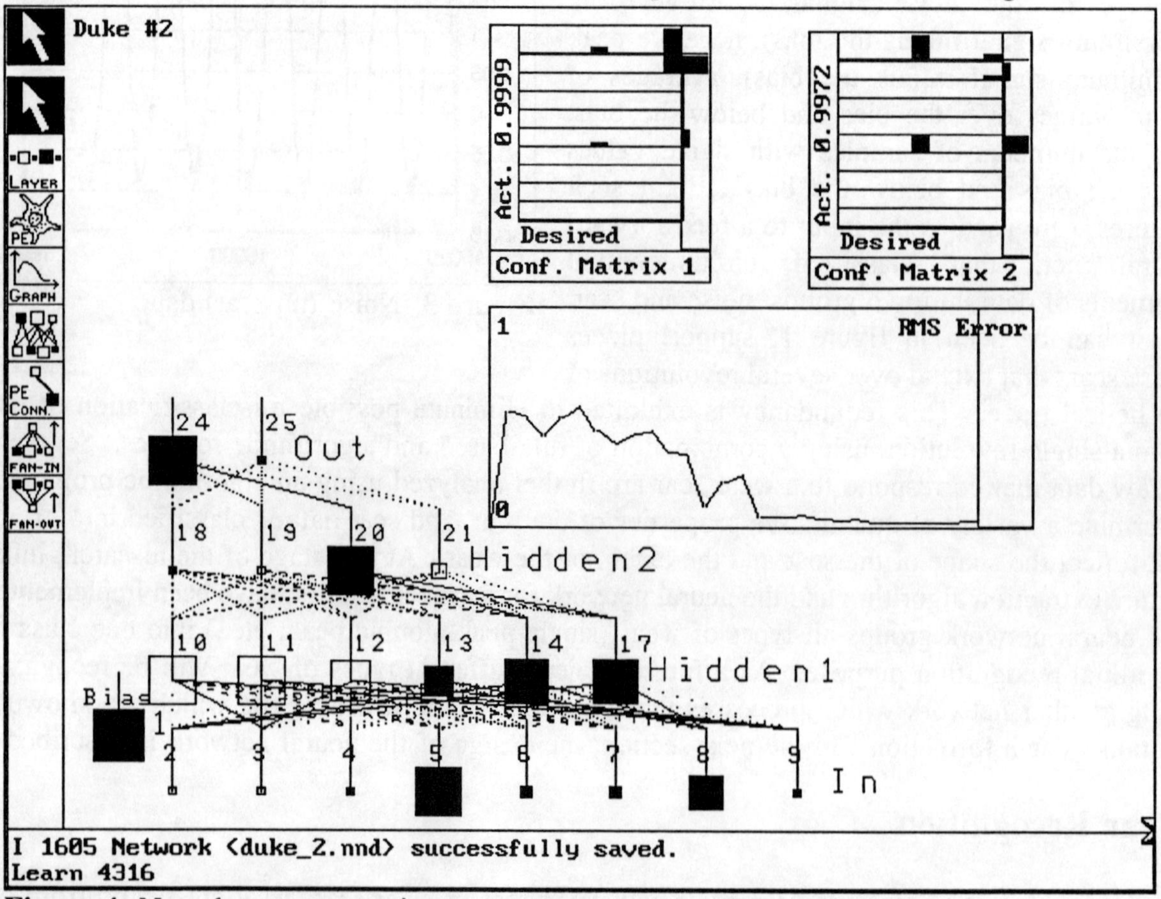

Figure 4 Neural net to recognize wear scars

The error plot in that screen shows how the rms output error reduced down to a 5% level in about 2500 iterations. The confusion matrices (#1 for the scar recognition, #2 for the noise recognition) can be interpreted as follows. The top of the horizontal axis is the histogram of the desired output, and the right hand side of the vertical axis is the histogram of the actual neural net output. The correspondence of the desired and the actual outputs are shown inside the matrix. Since there are only lower left corner (desired and actual values are both zero) and upper right corner (desired and actual values are both one) there is a good correspondence between the desired and actual output values. After the convergence, no recognition errors were made in the training set of features. Then, the same net was tested with no adaptation with a set of features with low signal to noise ratio. Although in some cases the net outputs approached the 50% recognition threshold, there were no misclassifications. When adaptation was allowed on the test data, the error margins improved. Another test that was performed was to initially train the net with the noisy data, and to test it with the clean data. Although the recognition with the noisy data was done without any errors, there were recognition errors with the clean data. As a result of this test, it has been decided to train a net with a set of features from a high signal to noise ratio file, and use this as a starting point for all signal processing (called standard net). Even if adaptation is allowed while processing inspection data files, the connection weights should be set to those of the standard net at the beginning of each new data file.

Conclusion

The results summarized here were obtained from a brief proof-of-concept study conducted by KBTAC and Duke Power. The results have been found very encouraging by Duke Power and EPRI, who are the co-sponsors of the research. During the next phase, an integrated software package that can handle binary raw data files, extracts the features from each revolution, runs the neural net, computes the wear properties and places the results in a database will be developed.

Long term tasks include intelligently updating the database, use of the earlier inspection results in the database to improve scar identification, construction of a dynamic model for rod clad wear and the use of that model for repositioning and wear forecasting.

References

[1] R.P. Lippmann, "An introduction to computing with neural nets," *IEEE ASSP Magazine*, April 1987, pp. 4-22.

[2] B. Kosko, *Neural Networks for Signal Processing,* Prentice Hall, Englewood Cliffs, N.J., 1992.

[3] *NeuralWorks Professional II/Plus Reference Guide*, Neuralware Inc., Pittsburgh, PA, 1991.

[4] D.E. Rumelhart and J.L. McClelland, eds, *Parallel Distributed Processing:Explorations in the Microstructure of Cognition*, Vol 1, MIT Press, 1986.

SAMSON
Severe Accident Management System Online Network

Rick Doremus
ARD Corporation
9151 Rumsey Road
Columbia, Maryland 21045
(410) 997-5600

1.0 Abstract

SAMSON, Severe Accident Management System Online Network, is a computational tool used in the event of a nuclear power plant accident. SAMSON examines over 150 status points monitored by nuclear power plant process computers after a severe accident has begun and makes predictions about when core damage, support plate failure, reactor vessel failure, and containment failure will occur. These predictions are based on the current state of the plant assuming that all safety equipment not already operating will fail. The status points analyzed include radiation levels, flow rates, pressure levels, temperatures and water levels. SAMSON uses neural networks trained with the back propagation learning algorithm to make predictions. Previous training on data from accident analysis code has allowed SAMSON to associate different states in the plant with different times to critical failures. The accidents currently recognized by SAMSON include steam generator tube ruptures (SGTR) and loss of coolant accidents (LOCA), with breaks ranging from less than 0.001 square feet in size to breaks greater than 3.0 square feet in size. SAMSON contains several neural networks for each accident type and size, and chooses the correct network after accident classification by an expert system. SAMSON also provides information concerning recovery strategies and the status of plant sensors.

2.0 History

The concept for SAMSON was first tested four years ago with the design of neural networks used to predict the number of ruptured tubes in a SGTR accident[1]. Upon successful completion, a demonstration application was created for a Macintosh computer. This application used networks trained to recognize various LOCAs. The neural networks predicted the time until support plate failure, the time remaining for recovery, and the future values for several key parameters. A working prototype was developed on a Sun SPARCstation 2GX and presents the user with an easy to use graphical user interface running under the Motif window environment. This prototype was delivered in February, 1992 to Commonwealth Edison Company with final delivery scheduled for December, 1992.

3.0 Training the Neural Networks

The neural networks were trained on data generated by the Modular Accident Analysis Program (MAAP) from Fauske and Associates, Inc. Numerous MAAP scenarios were generated based on the Individual Plant Evaluation completed for Zion Unit 1[2]. The scenarios were initially broken into 11 categories: LOCAs of 8 different sizes and SGTRs of 3 different sizes. MAAP reported specific sensor information approximately once per minute and when key events, such as the onset of core damage, occurred. Forty-seven different parameters were investigated for training the neural networks. Twenty-one parameters were used for initial training, but the number of parameters used in subsequent training depended upon the prediction of the network and the type of the accident.

Once all the necessary data were obtained, the data set was formatted for network training. Each minute of data was used as an input training pattern. The target output used for each pattern was the time to failure for the scenario minus the time since the start of the accident. Consider an accident scenario in which all the safety equipment is actuated at the start of the accident and core damage begins 100 minutes after the reactor scram. For a pattern taken 20 minutes into the accident, the target failure time would be 80 minutes. Now, consider a MAAP scenario in which no safety equipment is actuated at the beginning of the accident and only one injection pump actuates one hour into the accident. The time until failure for this accident is not continuous. For the first hour this scenario's target times are identical to those in the scenario in which no equipment is actuated. At one hour, the target times change to match a scenario in which one injection pump is actuated at 60 minutes. For example, suppose the failure time for a scenario with no safety equipment is 80 minutes and the failure time for the scenario with one injection pump starting at one hour is 100 minutes. A pattern used for training taken 20 minutes into the accident where one injection pump actuates at one hour would have an target failure time of 60 minutes (80 minus 20). A pattern taken 70 minutes into the accident would use a target failure time of 30 minutes (100-70).

As seen in the previous example, creating the data sets with the proper targets was complicated by the fact that SAMSON bases its predictions only on the current data received, and not by actions that might occur. As another example, Zion's high pressure safety injection (HPSI) pumps can not inject water until the pressure in the primary is less than 1535 psia[3]. Prior to this pressure being reached, SAMSON makes predictions based on no HPSI injection flow. To properly create targets for the portion of the scenario where the HPSI pump is not actuated, a separate scenario must be created where the HPSI pumps do not actuate throughout the accident. For any of the 35 unique MAAP scenarios, 2 or 3 other scenarios have often been used to create the proper targets.

Once the targets had been assigned for each data set, a global data set was created for training networks. Sixteen different networks were trained for the LOCAs, two for each of 8 distinct break size where was one network to predict the time until the onset of core damage and one to predict the time until support plate failure. When training a network for one LOCA size, data were also used from LOCAs which were one size larger and one size smaller to ensure that the network could predict accurately over a range of sizes. For example, the networks trained for a 1.0 sq.ft. break LOCA were also trained on 3.0 sq. ft. break data and 0.5 sq. ft. break data. Emphasis was placed on the LOCA size of interest by doubling the amount of training data from that LOCA. By training the networks on a range of data, the networks not only performed well on the LOCA size of interest, but also performed adequately on a wider range of LOCA sizes. This was done to ensure the predictions made by SAMSON during an accident are accurate when the exact size of the break is unknown.

The neural network used was a three layer, fully connected back propagation network written by Dr. Jim Howard of ARD[4]. The network used a learning rate of 0.01 and a momentum value of 0.5. Random values between -0.5 and 0.5 were used to initialize the connection weights. The inputs were normalized between 0 and 1 while the output was the predicted time until failure. A sigmoidal squashing function was applied to the hidden layer, which consisted of between 10 and 15 nodes. The network was trained for approximately 1500 cycles through a global training set that consisted of approximately 15,000 patterns. Training patterns were presented randomly to the network. The network weights that produced the lowest error, based on a validation set where back propagation correction was not allowed, were used in SAMSON.

4.0 Using SAMSON

SAMSON continually monitors the plant, waiting for a start-up signal. This signal can be a closure of a main steam isolation valve, a feedwater pump trip, a turbine trip, a safety injection actuation signal, or a reactor scram. Once the signal is received, SAMSON auto-initiates, opening the five windows shown in Figure 1. The first task SAMSON completes is classification of the accident. A rule-based system is used to classify the accident into a state recognized by SAMSON. SAMSON next calls the appropriate neural networks and begins to make predictions about the time until the onset of core damage and the time until support plate failure. These predictions are displayed in Figure 1 in the upper left window called 'ZION1 System Status.'

The 'Zion System Status' window displays the time since the start of the accident, the accident type as classified by SAMSON or as overridden by the user, and the time until core damage and support plate failure. Also displayed is the time until reactor vessel failure, which is fixed at one minute after support plate failure. The various predicted times are shown in a digital and an analog form. The displayed time until the onset of core damage is 7 hours, 57 minutes. To the right of the analog meter for each prediction is a rate meter. Bars below the hash mark indicate a negative rate while those above indicate a positive rate. (Note: The negative rates are displayed in red, the positive rates are green.) The size of the bar indicates the magnitude of rate of change. Once SAMSON has determined a failure has occurred, the portion of the window displaying the prediction is grayed out.

From this main window the user can also open the 'Accident Override'window, the 'Containment Failure' window, the 'Recovery Strategies' window, and the 'Core Map' window. Opening the accident override window allows the user to override the accident class determined by SAMSON. When the accident classification is overridden, SAMSON will call the appropriate networks based on the user input. The containment failure window displays the predicted time until hydrogen burn or containment overpressurization occurs. These times are determined using the predicted time until core damage, the time since the start of the accident, and information from the Zion IPE. The user can also query a database of recovery strategies. These strategies, based on the IPE completed at Zion, indicate what equipment must be functional and the latest time to recover the equipment to prevent further damage. The core map window contains a color coded map of all 65 core exit thermocouples. The color of the sensor indicates its relative temperature and whether the sensor has failed.

Directly below the system status window is the 'Predicted Time to Failure (TTF) History Graphs.' In this scrollable window, the histories of predictions made by SAMSON can be seen. In this example, you can see a large increase in the TTF for core damage and support plate failure at approximately 30 minutes. Using analysis tools currently under development, it can be shown that this increase in TTF is due to the actuation of a HPSI pump at 1/2 hour.

Below the TTF history window is an 'Events Log' that records when key events occur and other information. Information about accident classification is displayed in this window. To the right of the main window are the 'Sensor History Graphs.' This scrollable window can display the historical values of all sensors monitored by the plant's process computers. The last window, 'Sensor Summaries,' contains information about the individual sensors. SAMSON uses redundancy checking and information from the plant computer to determine the status of all sensors. The circles are shaded green or red depending on whether the data received from the sensor is good or bad. The number of circles corresponds to the number of sensors for a given parameter. Also displayed is the current value of all 'good' sensors.

5.0 Future Developments

Improvements to SAMSON involve efforts to allow the user more latitude in the operation of SAMSON. This includes the ability to override decisions made by SAMSON, such as forcing SAMSON to continue making predictions even when failure has already been predicted. Work also continues on improving the accuracy of predictions made by SAMSON and increasing the number of accident types recognized by SAMSON.

References

1. Phillips, Michael. "The feasibility of Using Artificial Intelligence Technology for Nuclear Power Plant Accident Management." (Written for Commonwealth Edison, Company) Columbia, Maryland: ARD Corporation, 1990.

2. Zion Station, Individual Plant Examination Submittal Report. (Written for Nuclear Regulatory Commission) Chicago: Commonwealth Edison, Company, 1992.

3. DiGiovanna, P. "Emergency Core Cooling Systems." Initial Licensed Operator Course, Chicago: Commonwealth Edison, Company, 1991.

4. Rumelhart, David E., Geoffrey E. Hinton, and Ronald J. Williams. "Learning representations by back-propagation errors," Nature, 323, 1986: pp. 533-536.

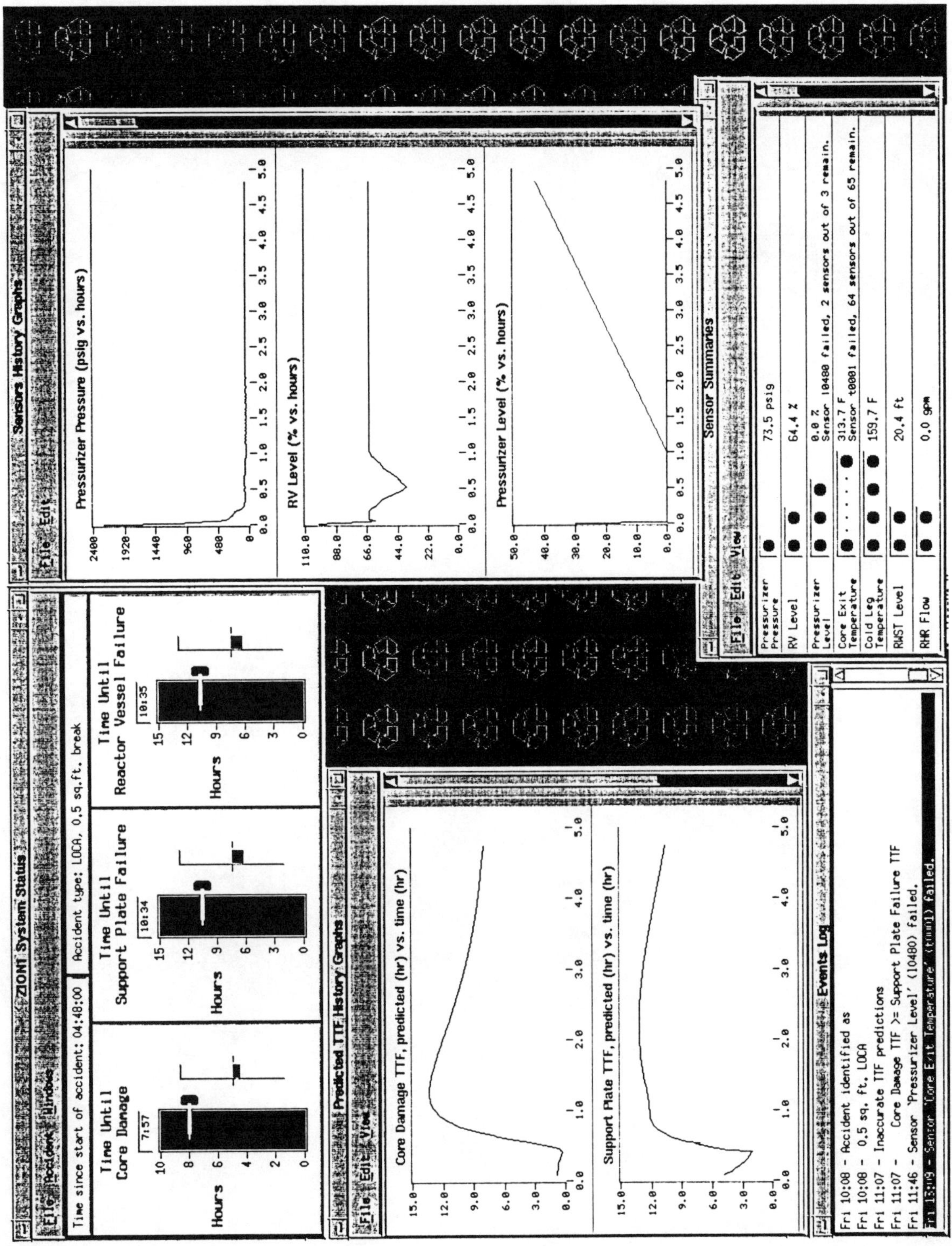

ZIONI System Status
Time since start of accident: 04:48:00
Accident type: LOCA, 0.5 sq.ft. break
Time Until Core Damage
Time Until Support Plate Failure
Time Until Reactor Vessel Failure
Hours
Predicted TTF History Graphs
Core Damage TTF, predicted (hr) vs. time (hr)
Support Plate TTF, predicted (hr) vs. time (hr)
Events Log
Fri 10:08 - Accident identified as
Fri 10:08 - 0.5 sq. ft. LOCA
Fri 11:07 - Inaccurate TTF predictions
Fri 11:07 - Core Damage TTF >= Support Plate Failure TTF
Fri 11:46 - Sensor 'Pressurizer Level' (10480) failed.
Sensors History Graphs
Pressurizer Pressure (psig vs. hours)
RV Level (% vs. hours)
Pressurizer Level (% vs. hours)
Sensor Summaries
Pressurizer Pressure 73.5 psig
RV Level 64.4 %
Pressurizer Level 0.0 % Sensor 10480 failed, 2 sensors out of 3 remain.
Core Exit Temperature 313.7 F Sensor t0001 failed, 64 sensors out of 65 remain.
Cold Leg Temperature 159.7 F
RWST Level 20.4 ft
RHR Flow 0.0 gpm

Figure 1.

D. Power System Operation

Comparison of Dynamic Load Models Extrapolation Using Neural Networks and Traditional Methods

He Ren-mu

North China Institute of Electric Power
100085, Beijing, P.R.China
Fax: +86 1 291 31 26

Alain J. Germond

Ecole Polytechnique Fédérale de Lausanne
CH-1015 Lausanne, Switzerland
Fax: +41 21 693 46 60

Abstract

Up to now the representation of load dynamic characteristics continues to be an area of great uncertainty and it becomes a limiting factor of power systems dynamic performance analysis. A mjor diffuculty, both for component-based and measurement-based methods, is the lack of data for dynamic load modeling. A way of solving this problem for measurement-based method is to interpolate and extrapolate the models identified from wide voltage variation data recorded during naturally-occurring disturbance or field experiments. This paper deals with groups of data measured in Chinese power systems using two models : a multi-layer feedforward neural network (ANN) using back propoagation learning, and difference equations (DE) using recursive extended least square identification. Analysis and comparison for the two approaches have been carried out. The results show that the DE model interpolation and extrapolation are nearly linear, they cannot describe the voltage-power nonlinear relationship of load dynamic characteristics. However, The ANN models can represent well this nonlinear relationship. They are promising dynamic load models.

1. Introduction

Transient stability analysis plays a significant role in power system planning and operating. Especially in weakly coupled system the results of this analysis are usually taken as the basis to define the system structure, the protections and emergency control measures. Thus the analysis accuracy will directly affect power system security and economy.

Up to now, the representation of load dynamic characteristics continues to be a limit to power system dynamic performance analysis. The most difficult things for either component-based methods or measurement-based methods is the lack of necessary data. However, for measurement-based load modeling it is possible to identify dynamic load models using large voltage variations recorded from power systems during naturally-occurring disturbances or field experiments, and to interpolate or extrapolate these models to other voltage variation ranges As we know, the voltage-power relationship of dynamic load model is nonlinear,. Could the models be interpolated and extrapolated correctly ? This is the most essential question for dynamic load modeling and this is the subject of this paper.

Recently artificial neural net models based on the present understanding of biological nervous systems became an interesting area for research because of their potential high computation rates, nonlinear mapping and great robustness to noise. Some applications for power system analysis have been published. So, it is interesting to study, with groups of field data, the possibility to use the ANN models for dynamic load modeling and to compare them with the traditional models.

2. Data and Models

2.1. Data description

This paper deals with groups of data measured in the Northeast Power System of China, in November 1987. On the occasion of circuit breaker tests, with artificial short-circuits made by the ShenYang High Voltage Switch Factory, measurements were carried out in a 220 KV nearby substation named HUSHTI. Voltages on the 220 KV bus, frequency, distribution transformers active and reactive powers were measured. Six artificial short-circuits were carried out during one and half hours at midnight when we can consider that there is no change of the bus load characteristics. So the model drawn from each measurement data group has to be able to represent the same behavior as the models drawn from the other data groups' voltage-power relationship. The first three tests were three-phase short-circuits with voltage variations range at the substation of about 8%, and the others were two-phase to ground short-circuits with voltage variation ranges of only 2.5%. Six groups of data were obtained. The former three, named H_0, H_1, H_2 have large voltage variations and the later three named H_3, H_4, H_5 have small voltage variations. Every data group contains 512-800 measurement points and the duration between two points is 6 ms. The total measurement time is about 3-5s.

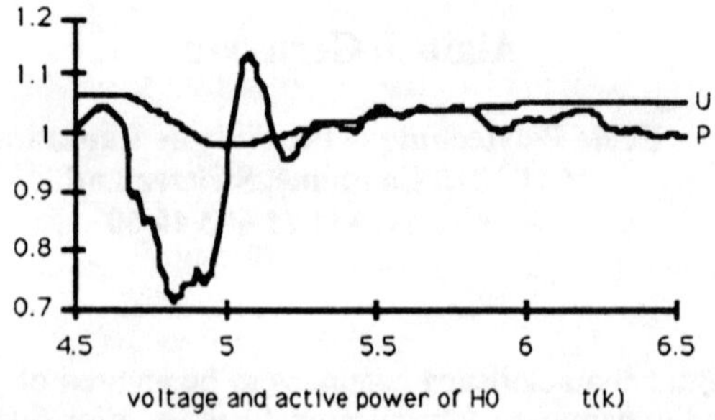

Figure 1

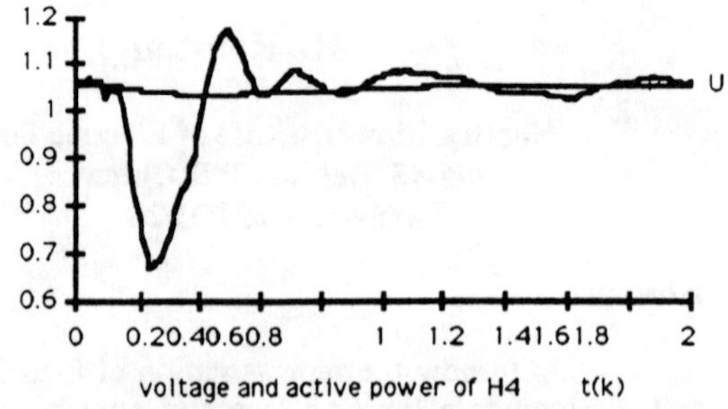

Figure 2

Figure 1 shows voltage and active power as function of measurement points for the H_0 data group which is the representation of the three wide voltage variation data groups. Figure 2 shows the same function for the H_4 data group which is the representation of the three weak voltage variation data groups. In this study, we use the H_0 data group to establish models then use H_3, H_4, H_5 to test the model interpolation ability. In opposite, we use the H_4 data group to establish models then use H_0, H_1, H_2 to verify the model extrapolation quality.

2.2. Two models and two identification methods

The measurement-based load modeling approach considers bus load as a black box connected to the bus. Inside the box load behavior is described by a stochastic system. The kind of models we have in the box is unimportant, it is sufficient that model depending outputs coincide with measured outputs. Two models respectively are put in the black box and two methods are used to determine their parameters.

A three layer perceptron is chosen as a model to represent dynamic load (ANN models) characteristics. The input layer, in our case, contains ten neurons. The hidden layer contains two neurons. The output layer, one neuron. In dynamic load model training data the input vector is:

$$x(k)=[U(k),U(k-1),U(k-2),U(k-3),U(k-4),U^2(k),U^2(k-1),U^2(k-2),U^2(k-3),U^2(k-4)]$$

The target output vector is $t_k=[P(k)]$. Here P is the active power, U is the voltage. k is the kth measurement point.

The back propagation training algorithm (BP) is used to minimize the square error in correcting the model weights and thresholds. At the end of the training the parameters become stable. Thus a dynamic load model is established.

In the load black box, we can also put difference equations (DE models) as load model. They are:

$$y(k) = H(k)\Theta(k) + v(k)$$

Here y(k) is the output vector, H(k) is the input vector, $\Theta(k)$ is the parameter vector and v(k) is the noise vector. In our case,

$$y(k)=[P(k)],$$

$$H(k)=[U(k),U(k-1),U(k-2),U(k-3),U(k-4),U^2(k),U^2(k-1),U^2(k-2),U^2(k-3),U^2(k-4)]$$

$$\Theta(k)=[a_0, a_1, a_2, a_3, a_4, a_5, a_6, a_7, a_8, a_9]^t$$

The Recursive Extended Least Square algorithm is used to identify model parameters.

3. Comparison of Interpolation and Extrapolation ability between the two models

In this paper, with two data groups H_0 and H_4 mentioned above, the ANN models and the DE models have been established. The model interpolation and extrapolation ability is estimated by the coincidence of two curves: the measurement curve and the model calculated curve.

3.1. DE models

3.1.1. Interpolation

For verifying the interpolation quality of DE models, the H_0 data group is used to establish a DE model . Therefore, the model interpolation estimation can be done with the later three data groups H_3 H_4 H_5 whose voltage variations range are less than that of H_0. The dotted lines represent measurement curves and the full lines represent

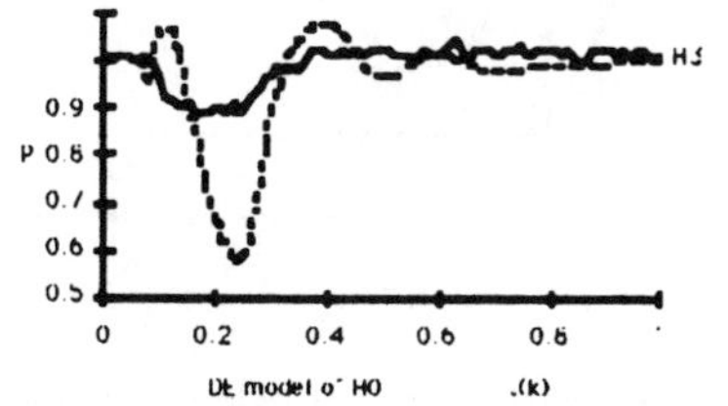

Figure 3

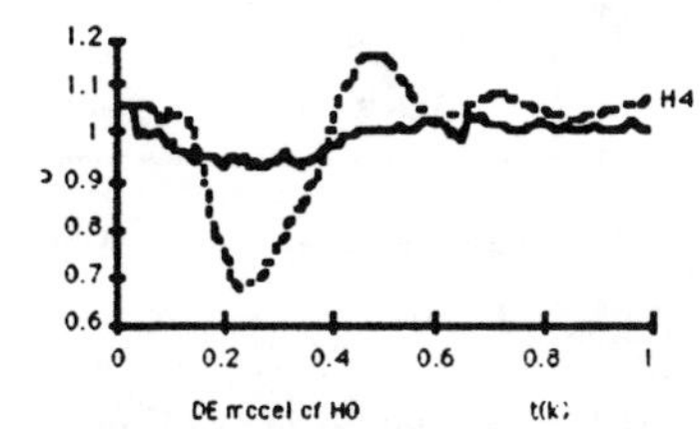

Figure 4

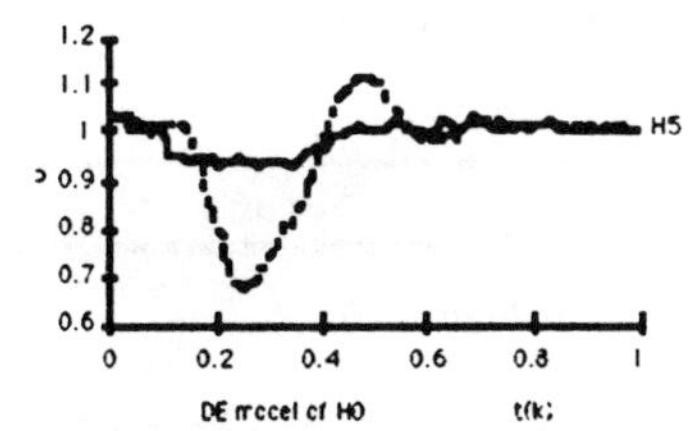

Figure 5

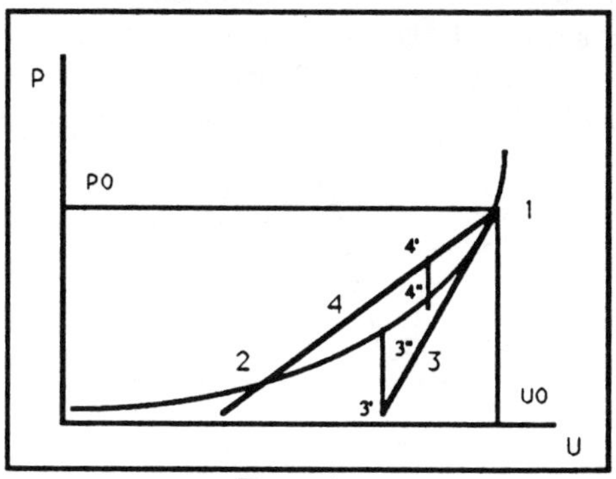

Figure 6

the curves calculated by the DE model obtained from the H_0 test. We note from these three figures that the active power variations calculated by the DE model are much smaller than the measured active power variation ranges in the all three cases. We suppose that the voltage-active power relationship of the dynamic load model is nonlinear somehow like the power function form (curve 1-2 of Figure 6) but the DE model is nearly linear. The H_0 data group has wide voltage variations, its DE model might be just like line 4. In the case of the DE model interpolation, the model calculated active power variations (4' defined by line 4) are certainly less than the measured active power variation range (4" defined by the curve 1-2). This explanation could also be made for the DE model extrapolation.

3.1.2 Extrapolation

The DE model identified from H_4 data group is used to test the model extrapolation ability. In using this model the extrapolation could be carried out with the data groups H_0, H_1, H_2 which have lager voltage variation ranges than H_4. The extrapolation results are shown in Figure 7, Figure 8 and Figure 9.

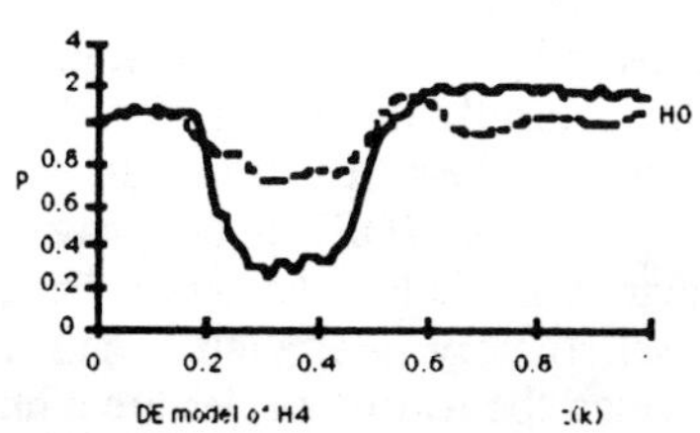

Figure 7

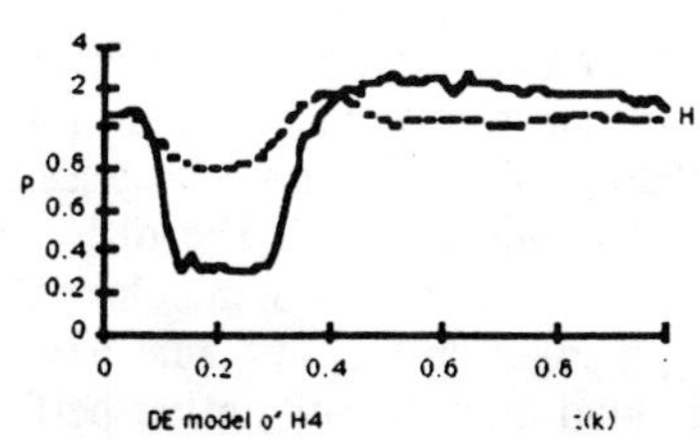

Figure 8

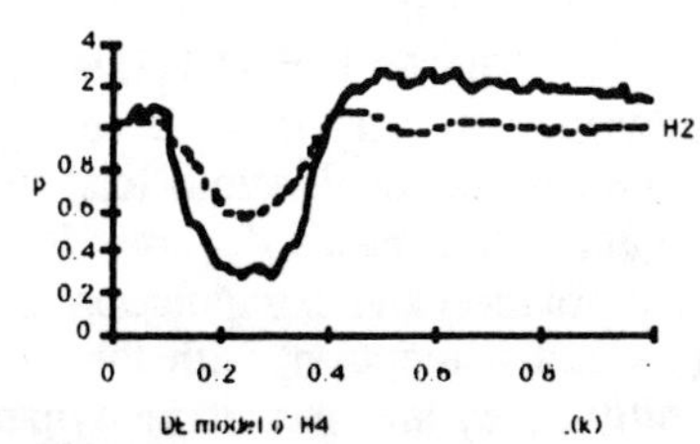

Figure 9

In this three cases, the active power variation calculated by DE model drawn from H_4 (full lines) are much larger than the measured active power variation ranges (dotted lines). We can find the explanation from the same nonlinear curve with interpolation (Figure 6). e.g. In the H_4 data group the voltage variation range is small. Its DE model might be like line 3. Following linear extrapolation of DE model, the active power variations calculated by DE model (3' defined by line 3) is certainly much larger than that of measurement (3" defined by curve 1-2). The DE model interpolation and extrapolation studies done above have proved our assumption, i. e. the voltage-active power relationship of dynamic load model is nonlinear somehow like a power function form. And the DE model interpolation and extrapolation are nearly linear. Therefore, the DE models cannot be used as dynamic load models.

3.2. ANN model

We note that the power function curve is similar to a part of sigmoid function . Thus, it is possible that ANN models could have better nonlinear features than DE models. The same way has been used for testing ANN models interpolation and extrapolation ability. The interpolation results are shown in figures 10 to 12 below:

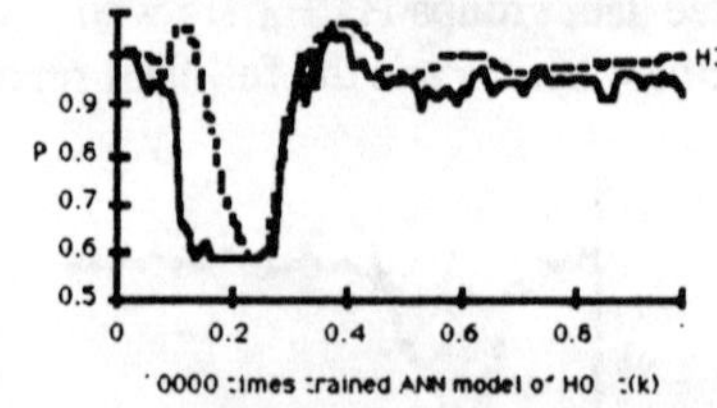

Figure 10

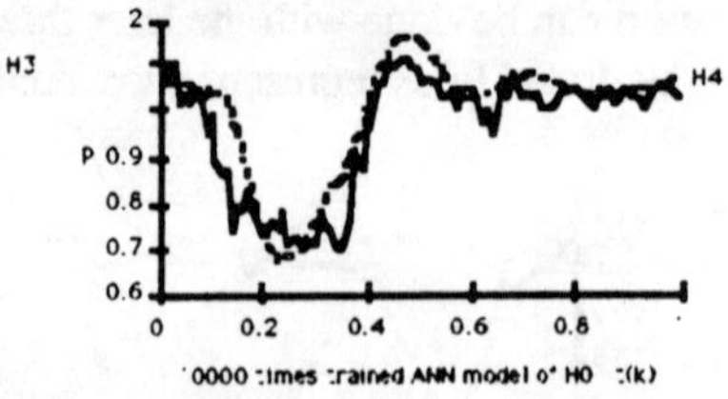

Figure 11

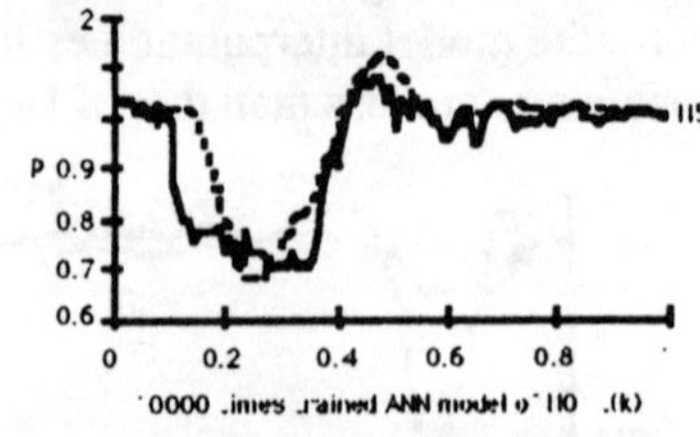

Figure 12

The results show that the ANN models' interpolation is not linear.any more The model calculated curves (full lines) coincide well with the three measured curves (dotted lines). The model training data H_0 has large voltage variation range and is situated in the curve part 2 (Figure 6). However the model can even interpolate load responses of H_3, H_4, H_5, which are placed in the curve part 1 (Figure 6). This shows a good ANNs' nonlinear interpolation.

The ANNs' extrapolation shows the same results, which are presented in the following figures 13 to 15. It is evident that all of the three model calculated curves coincide also well with the measured curves of

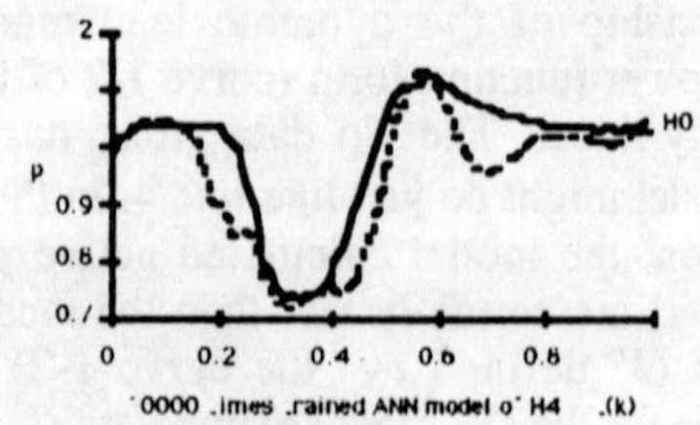

Figure 13

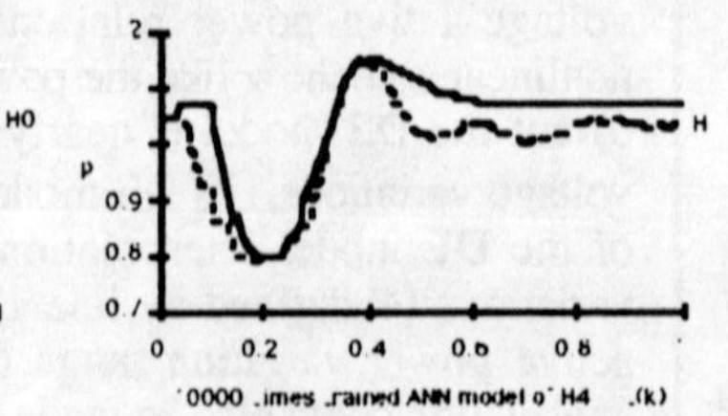

Figure 14

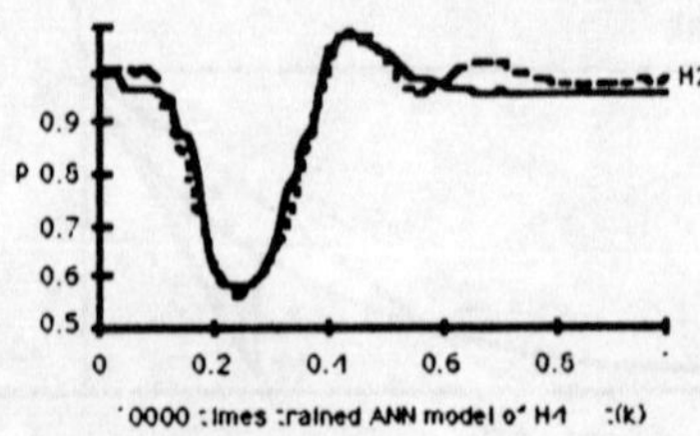

Figure 15

H_0, H_1, H_2. The ANN model training data H_4 has small voltage variations and is situated in curve part 1 (Figure 6). But the model can even extrapolate load responses of H_0, H_1, H_2, beyond the voltage variation data on which it is trained with satisfactory performances. This means the ANN models extrapolation is also nonlinear.

4. Conclusion

The ANN models interpolation and extrapolation possibility proved by the field data treatment in this paper allows using all the wide voltage variation data recorded during the naturally-occurring disturbance and field experiments for dynamic load modeling. Thus, it would be possible to use these real time data resources for a measurement-based dynamic load modeling method . The difference equations (DE) models have nearly linear interpolation and extrapolation features, and they cannot match nonlinear dynamic load behavior. The ANN models (layered perceptron) with the back-propagation learning rule have a satisfactory interpolation and extrapolation ability, they are promising dynamic load models. For further performance the learning rules are a major focus of neural net research. More researchs on the ANN models interpolation and extrapolation ability for dynamic load modeling have to be carried out with simulation data and field measurement data.

5. Reference

[1] IEEE Task Force Report "Load Representation For Dynamic Performance Analysis" Winter Meeting, New York, January, 1992
[2] R. P. Lippmann "An Introduction to Computing with Neural Nets" IEEE ASSP Magazine, Vol. 4.
[3] Yoh-Han Pao "Adaptive Pattern Recognition and Neural Networks" Addison-Wesley Publishing Company, 1989

On Neural Network Voltage Assessment

B. Avramovic, ECC, Inc

Introduction

Consider an arbitrary set of ordered pairs of objects with the first element of each pair coming from a set called input, and the second element coming from a set called output. Analytical results show that it is possible to construct an Artificial Neural Network (ANN) as an input/output network such that when the elements of the input set are presented to the ANN, ANN produces an output which is arbitrarily close to the corresponding element from the output set [1]. The significance of this result is that (practically) no constraints are placed on the functional relationship between the input and the output, and very importantly, this relationship does not even need to be specified. This suggests that ANN can accurately capture very complex and very nonlinear maps from the selected samples of this map, e.g. from the input/output samples of a physical process. This observation brings into focus two issues: one a proper sampling to insure that the relevant characteristics of the process are captured with as few samples as possible, and the other an ability of ANN to interpolate between the samples, which is often called the network's generalization ability, and which is determined by the network's architecture.

The ANN architecture consists of nodes, called neurons, which are interconnected with branches of different weights. The shape of the map between the input and the output is altered by adjusting the weights through a process called learning. The most widely used learning algorithm is called backpropagation [2,3,4]. The learning process, regardless of the algorithm used, tends to be time consuming.

Each of the neurons in an ANN receives one or more inputs through interconnections to outputs of other neurons or its own, and converts them into an output through either a static or a dynamic map - hence the name static or dynamic ANN. Also, depending on the arrangement of the weights, i.e. whether the signals going through the network loop back or go only in the direction from the input to the output, the networks can be feedback, or feedforward [2]. Feedforward (static) networks have a property that once trained they produce an output to a given input almost instantaneously.

The properties of ANN mentioned so far: the ability to capture a very complex, implicitly specified input/output map given through a set of samples, and a very quick response of the trained network to any given input are a motivation for considering ANN for voltage security assessment in power systems. Two issues facing this application, a long learning time and generation and sampling of system trajectories in order to obtain representative input/output sets, will be the primary focus of this paper. Examples of other applications of ANN to power systems can be found in [7].

Modified learning algorithm

Backpropagation [2] is the most widely used algorithm for tuning ANN weights. The weights w are adjusted through an iterative process using a simple formula

$$w^{k+1} = w^k + \alpha \nabla_w g$$

where $\nabla_w g$ designates the gradient in the backpropagation sense [2,3], and α is a user adjustable learning rate. Unfortunately, this algorithm is notoriously slow (an example in [5] quotes 1.28 million iterations on a divide by 3 problem). A modification due to Owens and Filkin [6], is motivated by an observation that above formula is much like the Euler's integration formula for

$$dw/dt = \nabla_w g \sim (w^{k+1} - w^k)/\alpha$$

Owens and Filkin further observe that the weights often go through periods of rapid changes, which motivated them to use a stiff differential equation solver. The differential equation based learning algorithm has the property that it converges very rapidly once it comes close to the solution - behavior quite opposite to that of the standard backpropagation method. We have based our training algorithm on the Owens' and Filkin's algorithm with a modification as follows:

$$dw/dt = \beta \nabla_w g.$$

Here β is an optional, user supplied gain, which permits even larger training speeds, as illustrated in Figures 1 and 2 for the standard XOR problem. From these figures it is clear that a gain of 500 results in a faster convergence of the weights than with the gain of 100. The limit on the gain is primarily posed by the robustness of the differential equation solver. We used the robust integration algorithm of the DASSL program [8].

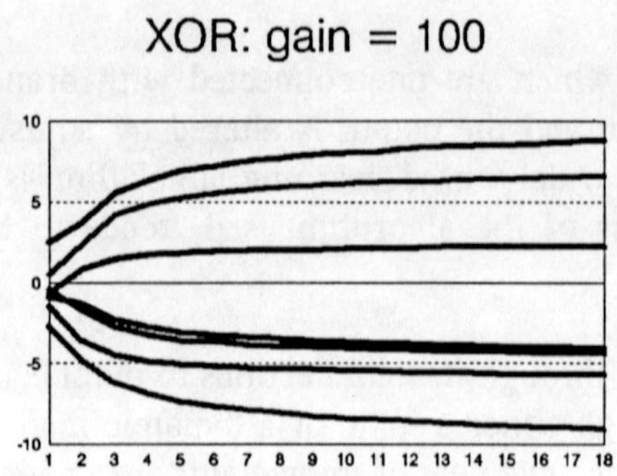

Fig.1: weights adjustment through time

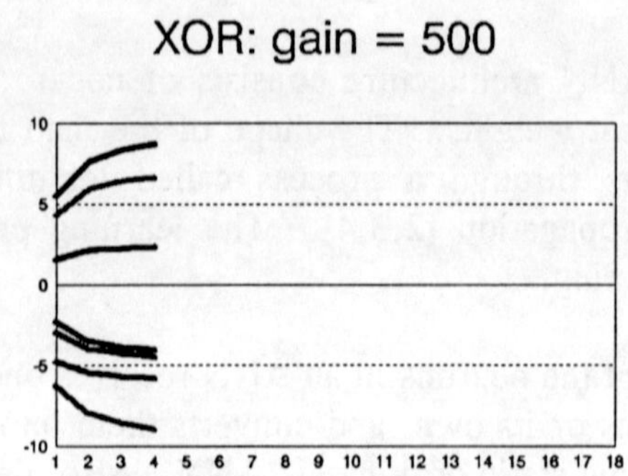

Fig. 2: weights adjustment through time

ANN approach to voltage security analysis

Postcontingency voltage assessment is an important operational problem in the context of secure operation of power systems [10]. It is also a challenging problem because of a very nonlinear relationship between independent variables that affect power system behavior, such as total system load, interchange pattern, generation levels, etc., and the resulting postcontingency bus voltages. The tools for mapping precontingency independent system variables to the postcontingency variables of interest (e.g. critical bus voltages) are well developed in terms of models and often in terms of solution algorithms, but they are still time consuming, limiting the extent of their use during on-line operation. The most notable example is the lack of on-line stability analysis tools.

By necessity then, a significant amount of security checking is performed off-line, and ahead of time; the resulting constraints on secure operation are typically expressed in terms of the precontingency system variables. If we save the results of these off-line simulation studies, it is then possible, in principle, to train an ANN on them to capture a relationship between independent system variables and the desired critical variables. Once trained the ANN can be used to interpolate between (and to a degree to extrapolate beyond) the samples used for training. Essentially this is a basis of the ANN approach considered in this paper. Its elements are depicted in figure 3 and applied to a small case study involving a 39 bus EPRI test system [9].

Figure 3: Components of the ANN based approach to security assessment

In our example "Simulate" block contains the slow voltage dynamics model of the system. Depending on the objective and operational concerns, this block may contain purely static models or detailed transient and long term stability models. In either case the function of the block is to simulate the response of the system to changes in independent system variables, and thereby "drive" the system through different parts of the space of possible operating points. Since this space is huge for any realistic power system model, it is of paramount importance to find a way to limit the investigation to only a smaller relevant region.

A way to limit the region of interest is as follows. Assume that the problem of interest is security assessment of a given system in the time interval of several hours into the future. A lot of variables characterizing such a system are known and can be specified as functions of time, either on the basis of planning (unit commitment, base unit loading), or on the basis of statistical behavior (total system load, load distribution to nodes). There are still associated uncertainties: operational - forecasting error for the load, possible interchange arrangement changes, changes in load factor, and structural - contingencies resulting in outages of lines, and/or generators. Our aim is to span and sample the region around the most likely operating trajectory defined by the uncertainties. The structural uncertainties are dealt with through contingency analysis in which a list of most severe outages is explicitly specified, and the effect of each contingency from the list evaluated in turn.

In our example we assume that the unit commitment is known; the hourly system load forecast is available, and that it is known with an error of +/- 5%. For this study we assume this load to rise from 5000 MW up to 7500MW, where it stays for a while before ramping down to 5000MW again.. The power factor (a ratio between the reactive and the active power at each bus) is assumed given with an error of +/- 5%. For each load change we assume that the generation is redispatched economically on the units under automatic generation control. For our test system outage of line 2-3 is the most severe single line outage. With the load as assumed and generators following it with economical distribution of their outputs, the voltage on bus 12 is the most critical, with voltage collapse being likely for loads in excess of 7000 MW if line 2-3 is out. The load as a function of time drives the simulator. At intervals of 1 second we apply a random (uniformly distributed) variation to total system load and to the distribution factors. Load tap changing transformers follow the system with their tapping dynamics (about 10 seconds between successive taps). We set up the economic dispatch program to execute every minute. Every two minutes the contingency list is processed to find out bus voltages right after the outage but before transformers have time to respond. We sample the output of the simulator at the rate of once every minute. In this way 60 samples are generated for each hour. It is possible to further reduce the number of training samples by passing the samples through a clustering algorithm. (Ours considers two samples to be in the same cluster if the Euclidian norm of their difference is below a threshold). The network was trained to relate the inputs to the outputs for the critical hour during which the load exceeded a safe 7000 MW level.

The rapid learning algorithm described previously is used for training. The ANN has the following configuration. There are only 17 inputs: the total system load plus eight generator active and reactive power outputs (the ninth is the slack generator). Note that the number of state variables of the load flow model for this system is 68! The outputs are the base case and the postcontingency bus voltage on bus 12. The number of hidden nodes is 4. With this configuration the total number of weights to be adjusted through training is 17*4 + 4*2 + 4 + 2 = 82. As can be seen from Figure 4, for the base case, and Figure 5, for the contingency case, the difference between the actual and the ANN predicted voltage is practically negligible for both the base case and the postcontingency voltage.

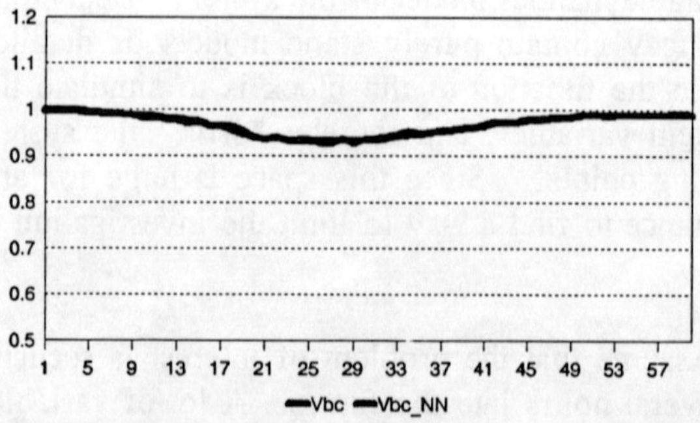

Fig. 4: Base case Bus 12 Voltage

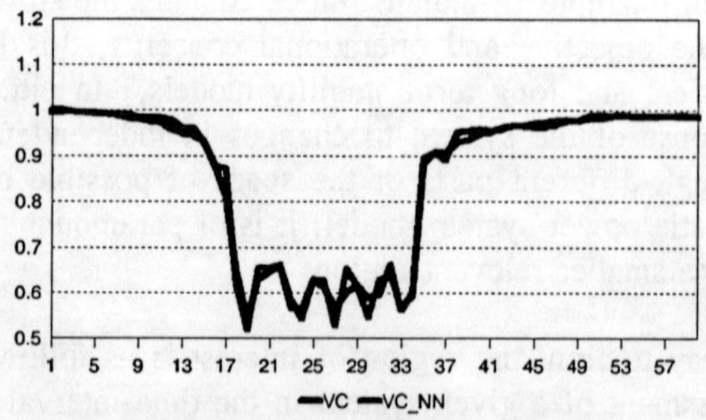

Fig. 5: Line 2-3 out; Bus 12 voltage

Conclusions

We have illustrated the capability of ANN to capture a very nonlinear relationship that exists between independent input variables and the resulting voltages in a highly stressed power system network. Once trained, ANN can be used in real time to compute almost instantly the output to a given input. Because of this feature, ANN seems to provide a convenient means for accurate tracking of postcontingency voltages (or any other variables of interest) as a function of the base network conditions, provided issues of scale can be resolved, considering that power systems are indeed large scale systems. An attempt to deal with the scale issue (in regards to the generation of samples) has been illustrated through a small case study. Ability of this ANN to generalize is an important topic of future research.

Acknowledgement

The work reported in this paper was supported in part by the National Science Foundation under Grant ISI - 9060250. The help of Mr. L.H. Fink in focussing on this problem is gratefully acknowledged.

References

1. R.Hecht-Nielsen: Theory of the backpropagation neural network, Proc. of the Intern. joint conf. on neural networks, Washington DC, June 1989, V.II, pp I-593-606.

2. B. Widrow, M.A. Lehr: 30 years of adaptive neural networks: Perceptron, Madaline, and Backpropagation, Proceedings of the IEEE, V. 78, No. 9, Sept. 1990, pp 1415-42

3. D. Rumelhart, D. Hinton, and C. Williams, Learning internal representations by error propagation, in D. Rumelhart and F. McCleland, eds: Parallel Distributed Processing, Vol. 1, Cambridge, MA, MIT Press, 1986

4. W. T. Miller, R. S. Sutton, P. J. Werbos, editors, Neural Networks for Control, MIT Press, 1990

5. Neural Network PC Tools, A practical guide, edited by R.C.Eberhardt and R.W. Dobbins, Academic Press, 1990

6. A. J. Owens and D.L. Filkin, Efficient training of the back propagation network by solving a system of stiff ordinary differential equations, Proc. of the Intern. joint conf. on neural networks, Washington DC, June 1989, Vol II, pp II-381- 6

7. M. A. El-Sharkawi, R. J. Marks II, Proceeding of the First International Forum on Applications of Neural Networks to Power Systems, Seattle, Washington, 1991

8.. K.E. Brenan, S.L.Campbell, L.R.Petzold, Numerical Solution of Initial-Value Problems in Differential-Algebraic Equations, North-Holland, 1989

9. B. Avramovic, L.H. Fink: Real-Time Reactive Security Monitoring; IEEE TRANS. v.PS-7, n.1, 1992, pp. 432-37

10. L.H.Fink, ed.: Proceedings Engineering Foundation Conference on Bulk Power System Voltage Phenomena - Stability and Security; Potosi, Missouri, September 1988; EPRI EL-6183.

NEURAL-NET SYNTHESIS OF TANGENT HYPERSURFACES FOR TRANSIENT SECURITY ASSESSMENT OF ELECTRIC POWER SYSTEMS

Dejan J. Sobajic and Yoh-Han Pao
Departments of Electrical Engineering
and Computer Science
Case Western Reserve University and
AI Ware Inc., Cleveland, Ohio 44106

Miodrag Djukanovic
Institut "Nikola Tesla"
Koste Glavanica 8A
Belgrade 11000
Yugoslavia

Abstract

In this paper a new direct method for transient security assessment of multimachine power systems is presented. A local approximation of the stability boundary is made by tangent hypersurfaces which are developed from Taylor series expansion of the transient energy function nearby a certain class of unstable equilibrium points (UEP). Artificial neural networks are used to determine the unknown coefficients of the hypersurfaces independently of operating conditions.

List of symbols and abbreviations

- n - the number of generator nodes;
- P_{mi} - mechanical power of the ith generator;
- $P_{ei}(0^+)$ - active power of the ith generator immediately after occurrence of a fault;
- E_i - electromotive force behind the transient reactance of the ith generator;
- G_{ij}, B_{ij} - components of the ijth elements of the reduced short circuit admittance matrix for the network;
- T_{ji} - inertia time constant;
- S_{ni} - rated apparent power;
- ω_s - synchronous speed of rotation;
- δ_i - rotor angle of the ith generator relative to the synchronously rotating reference frame;
- ω_i - angular velocity of the rotor of the ith generator relative to the synchronous velocity;
- CCT - critical clearing time;
- $M_i = (T_{ji}S_{ni})/\omega_s$ - inertia constant of the ith generator;

$M_{COI} = \Sigma M_i$

$\delta_0 = (M_T)^{-1}\Sigma M_i\delta_i$; $\theta_i = \delta_i - \delta_0$; $\omega_i = \omega_i - \omega_0$

$C_{ij} = E_iE_jB_{ij}$; $D_{ij} = E_iE_jG_{ij}$; $P_i = P_{mi} - E_i^2G_{ii}$

1. Introduction

Transient stability analysis of multimachine power systems is taking an increasingly important role in electric utility planning and operation. The classical numerical integration methods for power system transient stability analysis which need to solve multivariable non-linear differential equations are time consuming in terms of computation requirements [1]. Nevertheless, the detailed step-by-step time domain simulation approach is the most frequently accepted method by utility planners, because it is able to handle different models with different degrees of complexity of representation.

The method of tangent hyperplanes which are first order local approximations for the boundary of the region of stability (separatrix) is developed in [2]. Although this method gives better estimates of the CCT than the Lyapunov method, it requires the calculation of the eigenvalues and one eigenvector of (n-1) order square matrix. The approach is unreliable when the singular points are not type-1, and the method gives overestimates of the CCT. The method of tangent hyperplanes has been extended to the method of tangent hypersurfaces [3] introducing improved estimates of the CCT. Tangent hypersurfaces are second order approximations to the separatrix. Namely, the separatrix is normally convex, and consequently the intersection of the fault trajectory with tangent hyperplanes provides an optimistic estimate of the CCT.

From the computational point of view, the problem of evaluating the equations of tangent hyperplanes requires determining the positive eigenvalue and the corresponding eigenvector of the transposed Jacobian matrix with dimensions (n-1)x(n-1). Evaluation of the equation of tangent hypersurfaces requires solving the familiar Lyapunov matrix equation, or a set of simultaneous linear equations with $2n^2$-3n+1 unknowns. It is obvious that those computations are time consuming with the increase of the number of generators n.

All existing approaches are case-dependent and insufficient within the concept of power system vulnerability, where load levels and network topologies are treated as changeable. Speaking in the terms of adaptive pattern recognition, existing methods are not able to generalize accurately and to generate a set of operation rules applicable for new inputs.

This paper is concerned with straightforward construction of tangent hypersurfaces based on Taylor series expansion of the transient energy function (TEF) in the center of inertia reference frame. The necessity of finding eigenvalues, eigenvectors and the solving of the Lyapunov matrix equation is eliminated. Relevant TEF surfaces, at the associated type-1 points, take into account the fault location, through determination of singular points of interest. The numerical integration of the swing equations, until the intersection of the fault trajectory with the separatrixoccurs, is avoided through an iterative analytical procedure based on Taylor's series expansion. Finally, the generalization capabilities of artificial neural networks processing allow for these computations to be independent of operating conditions.[4,5]

2. Construction of Tangent Hypersurfaces

Throughout this paper a mathematical model of a power system commonly referred as the "classical model" is used. The equations of motion of the generators in the center of inertia reference frame are:

$$\dot{\theta}_i = \omega_i \qquad\qquad i = 1...n$$

$$\dot{\omega}_i = \frac{1}{M_i}\left(P_i - P_{ei} - \frac{M_i}{M_T}P_{COI}\right) = f_i(\theta,\omega) \tag{1}$$

Generally, a type-m singular point could be identified finding m eigenvalues (with positive real parts) of the Jacobian matrix evaluated at a singular point. All type-1 singular points necessarily lie on the stability boundary (or the separatrix), defined as,

$$\phi(x) = \sum_{i=1}^{2n} v_i x_i + \sum_{i=1}^{2n}\sum_{j=1}^{2n} w_{ij}x_i x_j + \sum_{i=1}^{2n}\sum_{j=1}^{2n}\sum_{k=1}^{2n} u_{ijk}x_i x_j x_k + ... = 0 \tag{2}$$

where:

$x = [\theta_1, \theta_2, \ldots \theta_n, \omega_1, \omega_2, \ldots \omega_n]^T$

The unknown constants v_i, w_{ij}, u_{ijk} can be determined by equating the coefficients of the like terms in the expression:

$$\frac{d}{dt}\phi(x) = \lambda_+ \phi(x) \tag{3}$$

where λ_+ is the positive eigenvalue of the Jacobian matrix $J_0{}^T = \partial f(\theta)/\partial\theta$ at the singular point. The transient energy function is defined as

$$V(\theta,\omega) = \frac{1}{2}M_{eq}\omega_{eq}^2 - \sum_{i=1}^{n} P_i(\theta_i - \theta_i^s) - \sum_{i=1}^{n-1}\sum_{j=i+1}^{n}\left[C_{ij}(\cos\theta_{ij} - \cos\theta_{ij}^s) - \int_{\theta_i^s+\theta_j^s}^{\theta_i+\theta_j} D_{ij}\cos\theta_{ij}\, d(\theta_i+\theta_j)\right] \tag{4}$$

where:

$$V_{KE_{Corrected}}^c = \frac{1}{2}M_{eq}\omega_{eq}^2 = \frac{1}{2}\frac{M_{cr}M_{sys}}{M_{cr}+M_{sys}}(\omega_{cr} - \omega_{sys})^2$$

is kinetic energy corrected for the energy that does not contribute to system separation.

We propose the iterative, hypersurface based, analytical method for determining the critical clearing time utilizing the Taylor series expansion of the transient energy function in the state space nearby an unstable singular point . This is achieved by treating the partial derivatives of the transient energy as functions of a single variable (fault clearing time-t). The computation of the state-space vectors θ and ω is speeded up using a limited Taylor series expansion of the fault-on trajectory. Three to four iterations of the Newton-Raphson procedure are sufficient in order to compute the CCT.

The hypersurface based approximation of the separatrix is constructed using a second order Taylor series expansion in the state-space as:

$$\phi(x) = d^2V(\theta,\omega) = \alpha\omega_{eq}^2 + \sum_{i=1}^{n}\beta_i(\theta_i - \theta_i^u)^2 + 2\sum_{i=1}^{n-1}\sum_{j=i+1}^{n}\gamma_{ij}(\theta_i - \theta_i^u)(\theta_j - \theta_j^u) \tag{5}$$

where the coefficients of tangent hypersurfaces α, β, γ are defined as follows:

$$\alpha = M_{eq}$$

$$\beta_i = \sum_{\substack{j=1 \\ j\neq i}}^{n}(C_{ij}^{pf}\cos\theta_{ij}^u - D_{ij}^{pf}\sin\theta_{ij}^u) + \frac{M_i}{M_T}\sum_{\substack{j=1 \\ j\neq i}}^{n} 2\, D_{ij}^{pf}\sin\theta_{ij}^u$$

$$\gamma_{ij} = \sum_{\substack{j=1 \\ j\neq i}}^{n}(-C_{ij}^{pf}\cos\theta_{ij}^u + D_{ij}^{pf}\sin\theta_{ij}^u) + \frac{M_i}{M_T}\sum_{\substack{k=1 \\ k\neq j}}^{n} 2D_{jk}^{pf}\sin\theta_{jk}^u \tag{6}$$

Consequently, the proposed analytical procedure follows the sequence outlined below:

1. For an initial estimation of the time of tangent hypersurface crossing, calculate rotor angles and speeds $\theta(t)$ and $\omega(t)$ using an approximate fault-on trajectory obtained via a Taylor series expansion.
2. Construct an approximation of the tangent hypersurface nearby an unstable equilibrium point using second order coefficients (equation (5)).
3. Calculate CCT as the time instant when the fault-on trajectory approximation crosses the tangent hypersurface using the Newton-Raphson iterative procedure.

3. Neural-Net Synthesis of Coefficients of Tangent Hypersurfaces

In this section we present a possibility to utilize the generalization capabilities of feedforward neural networks to determine (directly) the coefficients of tangent hypersurfaces under variable operating conditions. The input features which properly reflect:

- the impact of fault on individual generators,
- the electrical distance between individual generators and a faulted bus,
- changes in network topology,

are defined using the following expressions:

$$F_i = \left\{ [P_{mi} - P_{ei}(0^+)]/M_i - \left[\sum_{i=1}^{n} P_{mi} - \sum_{i=1}^{n} P_{ei}(0^+) \right] / M_{COI} \right\} (Y_{if}/Y_{ii}) \qquad i=1....n \tag{7}$$

where: Y_{if} - is the admittance distance from generator i to the faulted bus,
Y_{ii} - is the self-admittance of generator bus i.

In this study a three-phase short-circuit fault is simulated at the line 3-4 in the four-machine system (shown in Figure 1), and the fault clearing policy is to restore the pre-fault system topology.

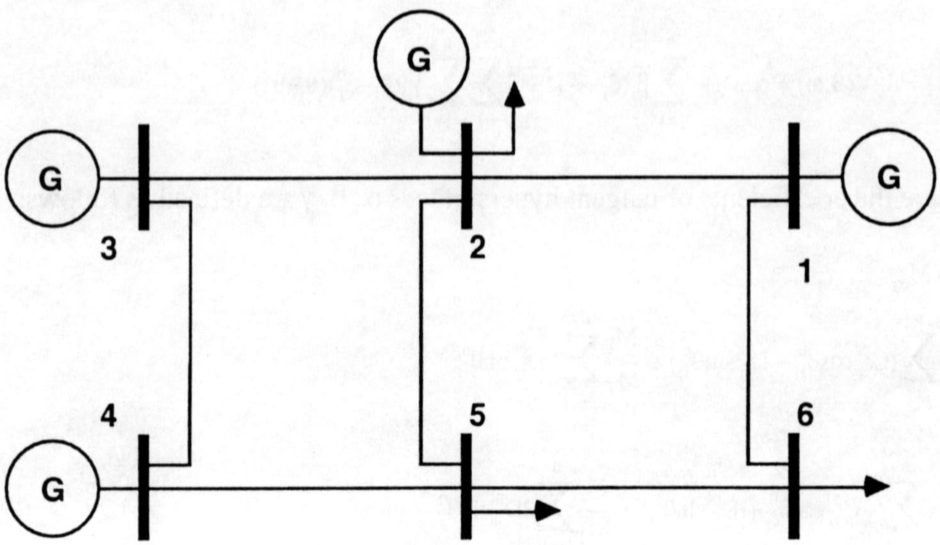

Figure 1. One-line diagram of the 4-machine

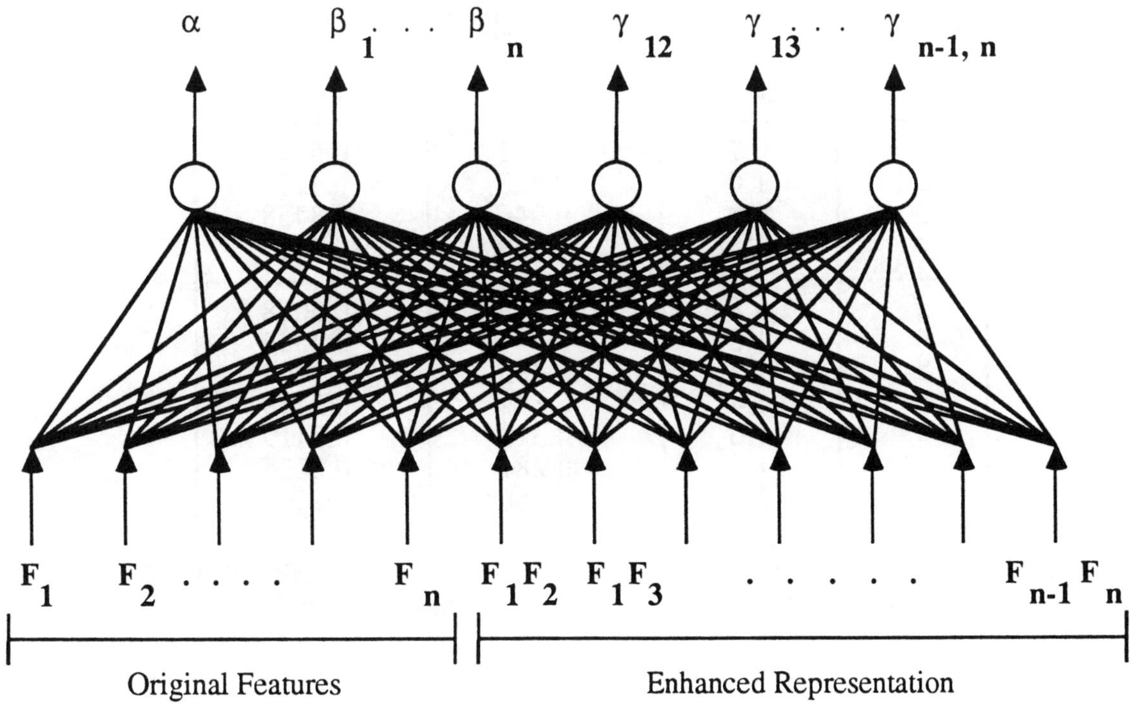

Figure 2. FLN used for synthesis of tangent power system hypersurface coefficients.

In order to generate the training set patterns two system topologies were used; topology 1 with all 7 lines in service, and topology 2 with line 3-4 removed. For the first topology 15 different loading conditions were selected with the loading level of the system in the range from 0.6 to 2.0 relative to the nominal operating point. For the second topology 10 loading conditions were selected.

The FLN-net (Figure 2) was allowed to train itself for 1750 iterations to learn the desired mapping. The parameters in the training process were: the learning rate η=0.7, the momentum term α=0.5. The least squares error E is reduced to 0.000926 with a maximum pattern error 0.00953 and maximum pattern error 0.000027. The critical clearing time has been estimated with coefficients of tangent hypersurfaces generated by the neural-net and the results have been compared with CCTs obtained with analytically evaluated coefficients. The results of the comparison are shown in Table 1.

It should be noted that examples 1-4 are from the second topology and with the operating conditions not presented to the neural-net during the training session, while the results 5-11 are from a new topology obtained by removing line 2-5. As seen from the above test cases the critical clearing times estimated by the neural-net agree with the analytically evaluated CCTs indicating the possibility that neural-nets can be successfully used in classifying the status of power system transient stability.

Example	Load Level [p.u.]	Analytically Based Coef. CCT[s]	ANN Based Coef. CCT[s]
1	0.7	0.541	0.5412
2	1.0	0.423	0.4234
3	1.3	0.346	0.3458
4	1.8	0.259	0.2586
5	0.8	0.505	0.5058
6	1.0	0.437	0.4383
7	1.2	0.385	0.3867
8	1.4	0.344	0.3458
9	1.6	0.310	0.3119
10	1.8	0.281	0.2836
11	2.0	0.256	0.2590

Table 1. The comparison of CCTs obtained with analytically calculated and neural-net estimated coefficients of tangent hypersurfaces.

4. Conclusion

The ANN based method for synthesis of tangent hypersurfaces have been proposed to provide quick assessment of power system vulnerability to large disturbances under variable operating conditions. Features suggested for ANN inputs relate closely to the power system's transient behavior enabling adaptability of network outputs and a direct fast assessment of tangent hypersurface coefficients. Numerical results exhibit close agreement between CCT parameters obtained through an analytical method and by using neural-net based estimates of coefficients of tangent hypersurfaces.

References

1. M. Ribbens - Pavella, and F.J. Evans, "Direct methods for studying dynamics of large scale electric power systems-a survey", Automatica, vol.21, no.1, 1985, pp. 1-21.

2. H. Yee, and B.D. Spalding, "Transient stability of multi-machine power systems by the method of hyperplanes", IEEE Transactions on Power Apparatus and Systems, vol. PAS-96, no.1, Jan/Feb 1977, pp. 276-284.

3. P.A. Cook, and A.M. Eskicioglu, "Transient stability analysis of power systems by the method of tangent hypersurfaces", Proc. IEEE, vol. 130, Part C, No.4, July 1983, pp.183-192.

4. D.J. Sobajic and Y.H. Pao, "Artificial neural-net based security assessment for electric power systems", IEEE PES WPM-1988, paper # 211-5, IEEE Transactions on Power Systems, vol. 4, No. 4, February 1989, pp. 220-228.

5. M. Djukanovic, D.J. Sobajic, Y.H. Pao, "Neural-net based unstable machine identification using individual energy functions", Electrical Power and Electrical Systems, vol.13, No. 5, Oct. 1991, pp. 255-262.

POWER SYSTEM STATIC SECURITY ASSESSMENT USING THE KOHONEN NEURAL NETWORK CLASSIFIER

Dagmar Niebur*, Student Member   Alain J. Germond, Member
Department of Electrical Engineering
Swiss Federal Institute of Technology, Lausanne
CH-1015 Lausanne
Switzerland

KEYWORDS: Power systems, static security assessment, artificial neural networks, Kohonen's self-organizing feature map

ABSTRACT

The operating point of a power system can be defined as a vector whose components are active and reactive power measurements. If the security criterion is prevention of line overloads, the boundaries of the secure domain of the state space are given by the maximal admissible currents of the transmission lines.

This paper presents the application of an artificial neural network, Kohonen's self-organizing feature map, for the classification of power system states. This classifier maps vectors of an N-dimensional space to a 2-dimensional neural net in a non-linear way preserving the topological order of the input vectors. Therefore, secure operating points, that is vectors inside the boundaries of the secure domain are mapped to a different region of the neural map than insecure operating points.

These mappings are studied using a non-linear power system model. Choice of security criteria and state space are discussed.

1. INTRODUCTION

An important task in power system operation is to decide whether the system is currently operating safely, critically or unsafely. If the system is safe, it is further of interest to know whether following an unforeseen outage of one or several lines, transformers or generators the system will still remain in a safe operating state. Important criteria for security assessment are violation of constraints for voltages, power flows, islanding etc. Contingency analysis for the case of one or several presumed outages of lines or generators (N-1 or N-M contingencies respectively) tests the constraint violation for load flows.

There exist powerful algorithms like fast decoupled load flow calculations which treat (N-M) contingencies completely [1]. However, the systematic computation of all failures required by a N-M security analysis (*i. e.* the simultaneous failure of M out of N elements) leads to an exponential increase of the computation time. For large networks, these multiple failures in the electrical network cannot be treated in real time if conventional computers are used to detect critical situations by simulations. Other methods like contingency ranking give fast results but do not detect all critical states [2].

Statistical pattern recognition techniques for security analysis have been proposed to take human experience into account in [3].

* currently at Jet Propulsion Laboratory, California Institute of Technology, Pasadena, CA 91109, USA

The processing power available at the time when these methods were developed was insufficient for their use in an industrial environment. More recently the application of associative memories for static security assessment of power system overloads has been studied in [4] and [5].

Today other Artificial Intelligence methods are studied for this purpose. Two main approaches to solve problems in power system analysis can be distinguished. Firstly, expert system techniques that explore human knowledge formulated explicitly as rules have been investigated for static security analysis in [6].

Secondly, artificial neural network techniques for security analysis have been proposed for the estimation of critical clearing time for the transient stability problem [7]. For static security assessment, a multi-layered network and the back-propagation algorithm were applied to a 4 machine - 8 bus - 14 line system in [8]. Classification of contingencies of a 4 machine - 17 bus - 35 line network has also been achieved with the Hopfield network [9]. A survey on the state of the art is given in [10].

However, approaches which simulate backpropagation networks are extremely time-consuming and necessitate a huge amount of training data. They do not take into account the properties of the system space but try to conceive the problem as a functional interpolation task. Due to the high dimensionality of the function being interpolated, accuracy is highly dependent on the number of training points. Further, each contingency has to be viewed as a different function. Therefore the combinatorial aspect of this task is still preserved. This argument is also valid for the Hopfield net approach. To our knowledge, none of the proposed models has actually been implemented in hardware.

The objective of this paper is to study Kohonen's self-organizing feature map for the classification of security states of power systems with respect to contingency analysis. The network determines similarities of system states in an unsupervised manner and successfully identifies line loading patterns for unknown as well as for trained system states. In contrast to the supervised learning of the backpropagation algorithm, unsupervised learning does not require the target class to be specified.

To our knowledge the study of self-organizing techniques for load forecasting was the first application of this type of methods to power system problems [11]. For security analysis and monitoring a 2-step technique was evaluated for a system of neural networks based on the functional link net architecture in [12]. An unsupervised learning step was used for feature clustering and a supervised learning step for feature identification.

In the next section a short introduction to Kohonen's self-organizing feature map is given. In section 3 the application of this classifier to power system security assessment is presented and simulation results are discussed.

2. KOHONEN SELF-ORGANIZING NEURAL NETWORK

2.1. Architecture of Kohonen Network

In accordance to experimental results it has been postulated that information in the brain is stored on a two-dimensional surface, and that related information occupies neighboring locations on that surface.

Mathematical models describing these *self-organizing neural networks* have been proposed initially by von der Malsburg [13] and developed further by Kohonen [14]. Throughout this paper the terms *Kohonen neural network*, *Kohonen classifier* and *self-organizing feature map* are used as synonyms for this type of network.

There is no target output and no "teacher" for evaluating an error function in these self-organizing networks. The learning of the synaptic weights is ***unsupervised***, which means that, upon presentation of new input vectors, the network determines these weights dynamically, such that input vectors which are closely related (in a sense to be precisely defined below) will excite neurons which are close together, or clustered.

The **Kohonen network** is an array of **M** processing elements (neurons). If these M neurons are arranged on a grid in a plane, the network is called two-dimensional, since this network maps high dimensional input vectors onto a two-dimensional surface. For a given network, the input vectors **x** have a fixed dimension **N**. The **N** components of the input vector **x** are connected to each neuron in the array. A synaptic weight w_{ji} is defined for a connection from the ith component of the input vector to the jth neuron. Therefore, an N-dimensional vector $\mathbf{w}_j$ of synaptic weights is associated with each neuron j. The neurons themselves are also connected in a predefined manner according to a given neighborhood relation.

Figure 1 shows a Kohonen network which maps 3-dimensional input vectors on a 2-dimensional map containing 16 neurons. A neuron in this square network has four direct neighbors, called neighbors of first order, except for the neurons along the borders of the grid. Two neurons i and j are neighbors of order k if i has a direct neighbor h who is a neighbor of j of order k-1. In figure 1 the complete neighborhood for neuron 5 is shown. Neighbors of order 1 are neuron 1, 4, 6 and 9, neighbors of order 2 are neuron 0, 2, 7, 10, 13 and 8, neighbors of order 3 are 7, 12 and 14 and finally neuron 15 is a neighbor of order 4.

The number of neurons does not depend on the dimension of the input vector nor on the size of the training set. However a small number of neurons can form only a small number of coarsely discriminating clusters, each one presenting a large set of training vectors. Depending on the application this coarse feature discrimination may not present a sufficiently detailed classification for a large training set.

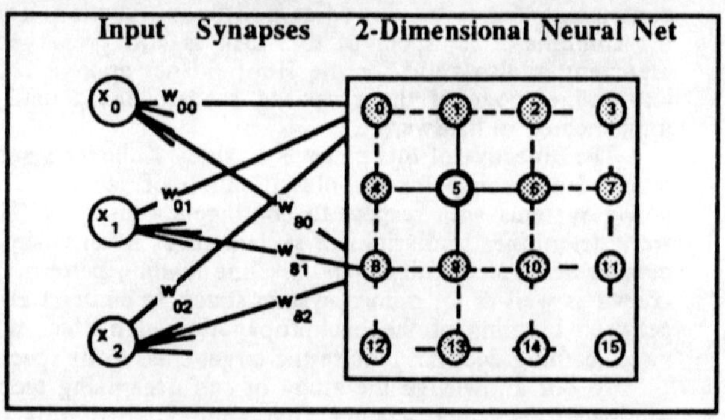

Figure 1: Kohonen network

2.2. The Self-Organization Algorithm

During the learning phase the input vectors are presented randomly. At each presentation of an input vector, *i. e.* at each step of the learning process, every neuron of the network calculates a scalar activation function which depends on the input vector and on its own weight vector **w**.

This function is chosen to represent a distance ||...|| between the input vector and the weight vector of the neuron under consideration. Possible choices are the Euclidian distance or the scalar product.

In biological systems, there is a distance-related feedback from every neuron to a given neuron. In the original model studied by Kohonen, the neighboring neurons have a positive feedback, whereas the more distant neurons have a negative feedback.

When Kohonen investigated self-organization, the distance-related feedback, implying a connection between every neuron of the map and truly parallel behavior of the system, could not be realized in hardware. Kohonen demonstrated experimentally that a simplified equivalent procedure can be used for numerical simulation. in [15]

Instead of calculating the neuron activation by distance-related feedbacks for every neuron, only the neuron with the weight vector closest to the input vector will be excited, and its weight vector as well as the weight vectors of its neighbors will be adapted. In biological systems the size of the neighborhood is determined by the strength of the positive and negative feedback, whereas in simulated systems the size of the affected neighborhood is specified in advance and it decreases as the learning goes on.

Therefore a neighborhood function $\alpha(t, i, c)$ is defined such that $\alpha(t, i, c)$ is monotonically decreasing in training time t and in the order of neighborhood of neuron i and neuron c.

Let now M be the number of neurons with weight vectors $\mathbf{w}_i \in R^N$, $0 \le i < M$ and $X = \{\mathbf{x} \in R^N \mid \mathbf{x} \text{ training vector}\}$ the training set. Let the time t be discrete and $\varepsilon > 0$ be small. The training algorithm can be formulated as follows:

Phase A: Unsupervised Learning

1: $t := 0$: initialize all w_{ij} randomly
2: Choose input vector $\mathbf{x} \in X$ randomly in the training set
3: Determine the neuron c such that its weight vector $\mathbf{w}_c$ is the closest to the input vector
$\|\mathbf{w}_c(t) - \mathbf{x}\| = \min \|\mathbf{w}_i(t) - \mathbf{x}\|$ for all i
4: Update the weight vector $\mathbf{w}_c$ of the selected neuron c and the weight vectors $\mathbf{w}_i$ of its neighbors
$\mathbf{w}_i(t+1) := \mathbf{w}_i(t) + \alpha(t, i, c)\,(\mathbf{x} - \mathbf{w}_i(t))$
for $i = c$ or $i \in$ neighborhood of c
$\mathbf{w}_i(t+1) := \mathbf{w}_i(t)$ elsewhere
5: Increment the time $t := t+1$
6: if $\alpha(t, i, c) > \varepsilon$ goto 2 else STOP

Figure 2 illustrates how the neighborhood function $\alpha(t, i, c)$ decreases in time. In the beginning the nearest weight vector chosen in step 3 will be completely adapted to the input vector x. Its nearest neighbors will be modified with a weaker factor $\alpha(t, i, c)$. For large t, $\alpha(t, i, c)$ becomes very small, the value of the input vector influences the modification of the weight vectors only slightly. The convergence criterion ε determines when $\alpha(t, i, c)$ is considered to be sufficiently small.

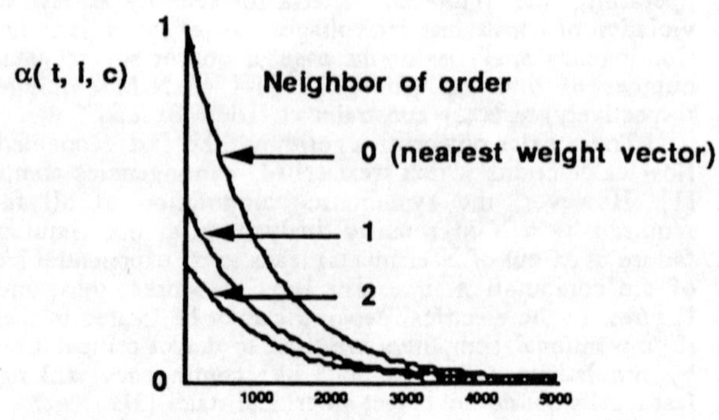

Figure 2: Example of neighborhood function $\alpha(t, i, c)$ for $i = 0, 1, 2$

Note also, that for small t basically all weight vectors will be adapted. The order of neighborhood only determines the strength of the adaptation. ($\alpha(t, i, c)$ for $i > 2$ is not shown in Figure 2.) For large t however the neighborhood decreases. In Figure 2 for example, neighbors of 2nd order will no longer be adapted after 3000 steps.

Phase B: Classification

1. Get an input vector **x**
2. Choose $\mathbf{w}_c$ such that
$\|\mathbf{w}_c(t_{max}) - \mathbf{x}\| = \min \|\mathbf{w}_i(t_{max}) - \mathbf{x}\|$ for all i, $0 \le i < M$

For each vector in the N-dimensional input space, whether

presented during the training phase or not, there is one responding neuron in the two-dimensional array of the Kohonen network, which has been excited most since its weight vector is the closest to the input vector (in the sense of a specified distance). One neuron may respond to several input vectors. Input vectors are mapped onto the weight vectors.

It has been demonstrated in [15] that this mapping preserves the topological order of the input vectors which is not necessarily known a priori. In fact the weight vectors themselves are constructed as weighted sums of a certain number of input vectors. If the input vectors form a vector space (in the mathematical sense) the weight vectors are elements of the same space representing a set of vectors of the same region of the vector space.

The choice of the input vector, the number of neurons and the neighborhood function will be discussed in section 3 with respect to the application to security analysis.

The adaptation process can be modeled by a Markov process. Stability and convergency properties of the learning algorithm of the self-organizing feature map have been studied in [16] and [17]. Analog implementations of self-organizing feature maps are examined in [18]. The implementation of a neuron performing the self-organization mapping with learning capabilities in CMOS technology is discussed in [19].

2.3. Technical Applications

Kohonen networks have successfully been applied in robotics to learn the inverse kinematic transformation of a robot with two degrees of freedom in [20].

A promising application of self-organizing network is the representation in two dimensions of input vectors of larger dimension. That technique has been used for the design of integrated circuits in [21].

3. APPLICATION OF THE KOHONEN CLASSIFIER TO SECURITY ASSESSMENT

3.1. Representation of the Operating Space

In this section the problem of static security assessment is formulated as a classification task. The operating state is described by a vector consisting of the active and reactive power of the power transmission lines. This vector is an element of an N-dimensional complex vector space called operating space where N is the number of lines. The operating point of a single contingency is an element of an N-1 dimensional subspace. The general concept of security regions and other definitions of operating points have been studied in [22] and [23].

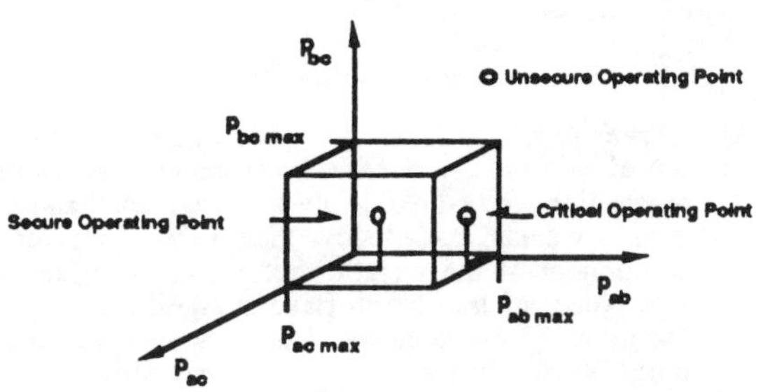

Figure 3: Operating space of a 3 bus - 3 line linear power system model

The boundaries of the secure state space are given by the limits of the buses and lines according to different security criteria. This paper treats the case of line overload prevention where the limits are given by the maximal power supported by these lines: $\|P_{ik} + jQ_{ik}\| \le S_{ik\,max}$, where $P_{ik} + jQ_{ik}$ is the complex power flow.

A safe operation point always lies inside the boundaries of the secure space. Critical points lie inside but near the boundaries and unsafe points are outside the boundaries. In case of the *linear* power system model the secure region of the operating space is convex. Figure 3 illustrates this concept.

3.2. Simulation of a 5 Bus - 7 Line Power System

The Kohonen network is used to classify line loading patterns resulting from single and double contingencies for a 5 bus - 7 line power system defined in [24] and represented in Figure 4.

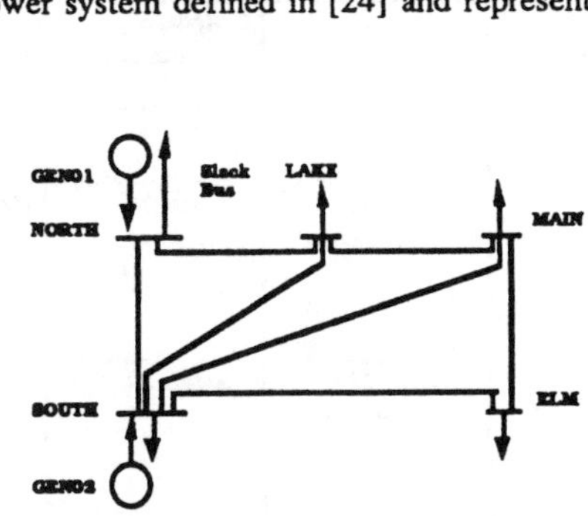

Figure 4: 5 bus - 7 line system

For 7 lines the system state is defined by 7 complex respectively 14 real components, the 7 active and reactive line power flows. These input vectors were obtained by off-line load-flow simulations. One load and generation scenario was defined for the base case. The base case, 7 single line contingencies, 21 double line contingencies, 2 single generator contingencies, 1 double generator outage and 2x7 single generator - single line outages were simulated forming a set of 46 input vectors. These 46 cases can also be viewed as 46 different line loading patterns corresponding to 46 different power system topologies.

Reallocation of power on the generators was not performed, with the following exceptions:

- in the case of outage of one generator, the missing power is reallocated to the other generator.
- in some cases of double branch contingencies, where one or several loads cannot be supplied, the generator power is rescheduled.

3.3. Simulation Data

A simulator for the Kohonen learning algorithm was developed in the language C on an Apollo DN 3500 work station. All the parameters can be defined and changed interactively: size of the input vector and of the Kohonen network, neighborhood function parameters and their decrease with the learning steps.

Vectors are drawn randomly from the set of 46 training vectors, and are presented as inputs to a two-dimensional Kohonen network with 49 neurons. The neurons are arranged on a 7x7 grid and numbered, row by row, from 0 to 48. The components of the weight vectors associated to each neuron have been randomly initialized between 0 and +1 MW and 0 and 1 MVAR respectively. Further, we define a *neighborhood order* function

$$k(t) := \left\lfloor k_0 \exp\left(\frac{-t}{\tau_k}\right)\right\rfloor$$

where $\lfloor x \rfloor$ denotes the largest integer less or equal to x. The *neighborhood set* $N_d(t, c)$ of neuron c is determined such that neuron $i \in N_d(t, c)$ if i is a neighbor of c of order d with $d \le k(t)$.

Finally we define the *neighborhood* function (see also figure 2):

$$\alpha(t, i, c) := \begin{cases} \dfrac{\alpha_0 \exp\left(\frac{-t}{\tau_\alpha}\right)}{1+d} & \text{for } i \in N_d(t, c) \\ 0 & \text{elsewhere} \end{cases}$$

The constant values have been chosen as

$$k_0 = 49,\ \tau_K = 1000,\ \alpha_0 = 1 \text{ and } \tau_\alpha = 1000.$$

Note, that the weight vector of a neighbor of order d will only be adapted if $d \le k(t)$. The neighborhood order function k(t) indicates after how many steps the adaptation process stops for all neighbors of order d. Because of the parameter choice,

Table 1: Classification of 46 cases by 49 neurons

Neuron 0 N-S & L-M	**Neuron 1** N-S & S-L	**Neuron 2** N-S & M-E	**Neuron 3**	**Neuron 4**	**Neuron 5** S-E & M-E	**Neuron 6** S & N N-L & N-S
Neuron 7 N-S & S	**Neuron 8** N-S & S-M	**Neuron 9** N-S N-S & N	**Neuron 10**	**Neuron 11**	**Neuron 12**	**Neuron 13**
Neuron 14	**Neuron 15**	**Neuron 16**	**Neuron 17**	**Neuron 18** S-M & S-L	**Neuron 19**	**Neuron 20**
Neuron 21 N-S & S-E	**Neuron 22**	**Neuron 23**	**Neuron 24**	**Neuron 25** S-M S-M & N S-M & M-E	**Neuron 26** Base Case N M-E	**Neuron 27** S-L S-L & N S-L & L-M S-L & M-E
Neuron 28 S-E & S-M	**Neuron 29**	**Neuron 30**	**Neuron 31**	**Neuron 32**	**Neuron 33** S-M & L-M	**Neuron 34** S-L & S L-M L-M & N L-M & M-E
Neuron 35	**Neuron 36**	**Neuron 37** S-E & S	**Neuron 38**	**Neuron 39** S-M & S	**Neuron 40** L-M & S	**Neuron 41**
Neuron 42 S-E S-E & N	**Neuron 43** S-E & L-M S-E & S-L	**Neuron 44** S-E & N-L	**Neuron 45**	**Neuron 46** S M-E & S	**Neuron 47** N-L & S-M	**Neuron 48** N-L N-L & N N-L & S N-L & S-L N-L & L-M N-L & M-E

$\alpha(t, i, c)$ decreases with the same time constant τ_α ($= \tau_\kappa$) as k(t). Therefore $\alpha(t, i, c)$ will be very small for large t and large d. By stopping the adaptation for these weight vectors the number of numerical operations on these vectors are reduced and therefore speed is gained.

After 4000 steps this network is well organized, *i. e.* the weight vectors do not change significantly.

All parameters have been determined experimentally. Satisfying results have already been achieved with a 4x4 network and faster decreasing neighborhood functions. However it is evident that the number of clusters can not exceed the number of neurons. Further a slow decreasing neighborhood function will enhance the generalization capabilities of the network since for a large number of steps all neurons will be adapted to all input vectors. The number of neurons of the example network has not been chosen to equal the square of the number of lines. The interaction between neural net size, neighborhood relationship and power system size will be explored in future work.

3.4. Interpretation of the Results

3.4.1. Classification of Training Vectors

Table 1 shows the classification results for 46 training vectors. For the table the following notations have been used: All buses are labeled by their first letter, all lines by the pair of buses which they connect. Therefore N denotes generator outage at bus North, S denotes generator outage at bus South, N-S denotes a single line outage of line North-South, N-S & L-M denotes a double line outage at line North-South and line Lake-Main and N-S & S is the double contingency of line North-South and generator outage at bus South. The "Base Case" denotes the case without any line or generator outage. S & N is the case where no generator is working and thus no load is supplied. Neurons containing empty cases do not classify any of the training vectors. That means their weight vector has never been chosen as the closest vector to any of the training vectors.

Six different clusters can be distinguished. Four out of six are surrounded by "empty" neurons like neuron 3 which do not classify any training vector. These clusters can be clearly distinguished from the neighboring clusters. The decision that the regions of cluster E and F form 2 different clusters and not 1 or even 3 has been made based on our interpretation of the classified cases. These 2 clusters do merge gradually into another.

Table 2 illustrates these 6 different clusters present in table 1. The numbers correspond to neurons 0 to 48.

Table 2: Cluster map for classification of 46 trained vectors

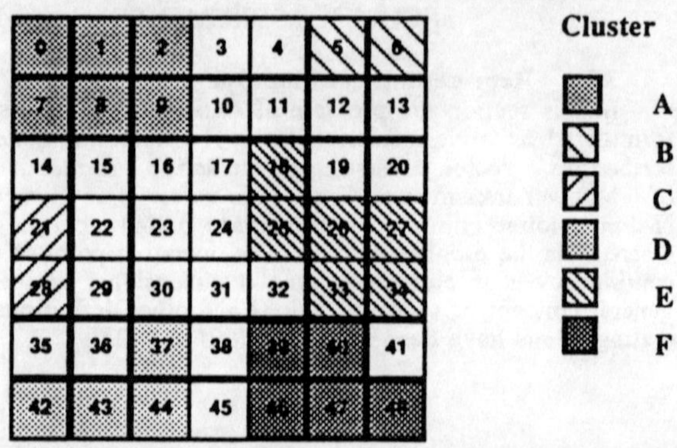

A) Cluster A groups 7 out of 9 cases where line North-South is out of service. These cases correspond to a loading pattern where line North-Lake is always overloaded and line Main-Elm is weakly loaded (less then 30%). Therefore the first component of the weight vector corresponding to the active power of line North-Lake is significantly greater than the other 13 components. The components 3 and 4 of the input vector corresponding to the active and reactive power of line North-South are both zero since North-South does not carry any power. Therefore components 3 and 4 of the weight vector are relatively small. Finally component 13 and 14 corresponding to a weak load in Main-Elm are also small.

The remaining 2 cases of North-South outages have been classified by neuron 6 of cluster B and neuron 21 of cluster C respectively.

B) Cluster B represents the set of double outages where the load cannot be supplied totally. These are the special cases where bus North or Elm are isolated from the rest of the network and the case of the double generator outage. However, these cases do not represent an overload situation and their loading pattern is quite different from those of cluster A.

C) Cluster C corresponds to a loading pattern where as in cluster A line North-Lake is heavily loaded respectively overloaded. But in contrast to cluster A line Main-Elm is loaded around 66%. Further, line South-Elm is out of service in both cases.

D) Cluster D classifies cases where South-Elm is out of service and Main-Elm is loaded around 66%. In contrast to the cases of cluster C, North-Lake is normally loaded (up to 50%). North-South is either heavily loaded or overloaded (72% to 135%). South-Main is loaded around 50% for neurons 37 and 42, heavily loaded to overloaded for neurons 43 and 44. The decision that 37, 42, 43 and 44 form 1 cluster and not 3 depend on the importance of the loading pattern of line South-Main.

E) Cluster E represents the non critical cases where no overload is present and line Main-Elm is very weakly loaded (less than 10%).

One exception should be noted here: Contingency case South-Lake & generator South, classified by neuron 34 represents an overload of line North-South. For line South-Lake neurons 21, 27 and 43 handle the normal cases and neurons 34 and 48 the overloads. The fact that different neurons classify these cases shows that the behavior of line South-Lake is not an important feature of the network. This corresponds to its location in the network. Generally its outage does not provoke major changes.

For outages of line Lake-Main, cluster E also handles the non-critical cases whereas all critical cases provoking an overload of line North-South are classified by neurons 40 and 48 of cluster F.

F) In general cluster F represents loading patterns of weak load of Main-Elm and overload of North-South. Thus 7 out of 9 outages of line North-Lake corresponding to these patterns are classified by cluster F. (The other 2 outage cases of North-Lake & North-South and North-Lake & South-Elm do not fall in this category!)

To interpret the organization of the weight vectors, we use the technique proposed in [21].

Some of the neurons in the network are associated with several input vectors, and some to none of the input vectors. However, the weight vectors for those neurons appear to represent feasible states of the power system, which can be understood as generalization from the learned vectors. The weight vectors themselves represent the weighted sum of a certain number of input vectors which form a cluster. Therefore, with our choice of variables for the components of the input vectors, each component of a weight vector represents either an active or a reactive line power flow. For example, the first component of every weight vector corresponds to the active power in the branch North-Lake. Table 3 shows the comparison between the 14 components of the weight vector w_9 of neuron 9 and the 14 components of the input vector corresponding to the 7 complex line power flows of the case outage North-South. This case is classified by neuron 9.

Table 3: 14 weight vector components $w_{9\,2j+i}$ $(1 \le j \le 7,$ $i = 0, 1)$ of Neuron 9 and and 14 components of input vector corresponding to 7 complex line power flows for outage North-South

$w_{9\,2j}$	$w_{9\,2j+1}$	LINE	p [MW]	q [MVAR]	loading [%]
144.758	2.107	NORTH-LAKE	144.73	-3.66	131.79
3.262	-1.128	NORTH-SOUTH	0.00	0.00	0.00
42.248	16.632	SOUTH-ELM	39.36	17.47	39.20
-1.170	23.280	SOUTH-MAIN	-3.97	23.80	21.96
-19.033	34.136	SOUTH-LAKE	-15.38	29.91	30.62
61.807	-27.470	LAKE-MAIN	66.87	-32.62	71.07
19.397	-8.063	MAIN-ELM	21.89	-11.20	23.41

The Euclidian distance between weight vector and input vector (= 12.7) is the minimal distance between this input vector and any weight vector of the network.

If we assume an active power limit in each of these lines, the comparison of this limit to the corresponding weight vector component of neuron i indicates whether neuron i classifies system states that are likely to violate this limit or not. Figure 5 illustrates this concept for the active power of line North-Lake. The values of the first component of all 49 weight vectors are presented on a map, at the location of the corresponding neuron defined in Table 1. In Figure 5 the 49 neurons occupy the edges of the squares. Only neurons on the borders are labeled.

The features of these clusters are directly represented by the values of the weight vectors. For example large values of the active power of line North-Lake correspond to large values of the corresponding first component of the weight. vector. Figure 5 illustrates how these features can be detected.

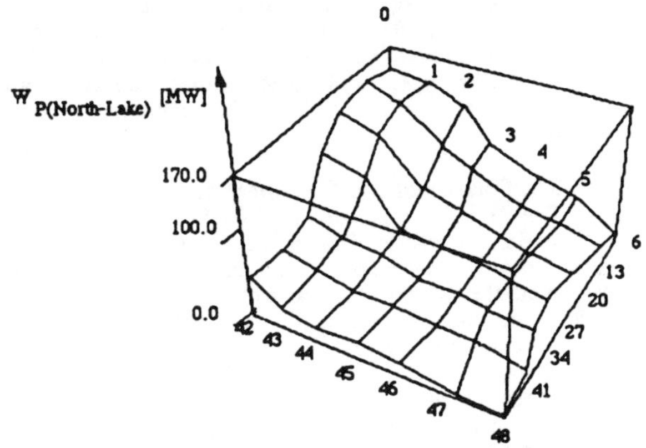

Figure 5: First component of the 49 weight vectors corresponding to the active power flow of line North-Lake

The limit for line North-Lake is at 100 MW. Therefore neuron 0, 1, 2, 3, 7, 8, 9, 10, 14, 15, 16, 21, 22 and 23 represent overload values whereas the values at neuron 47 and 48 corresponding to North-Lake outages are small.

3.4.2. Classification of Unknown Vectors.

In order to generate vectors which have not been trained but which are a "reasonable" distance from the 46 trained vectors, load and generation of the base case were uniformly changed to 90%, 95%, 105% and 110%. The 4x46 vectors obtained were presented to the Kohonen network trained on the 100% case. The claim of a reasonable distance is justified by the practical consideration that load and generation for a network usually vary around a scheduled case.

The following table shows the cluster map for the classification of vectors which have not been trained.

Table 4: Cluster map for classification of 184 untrained vectors

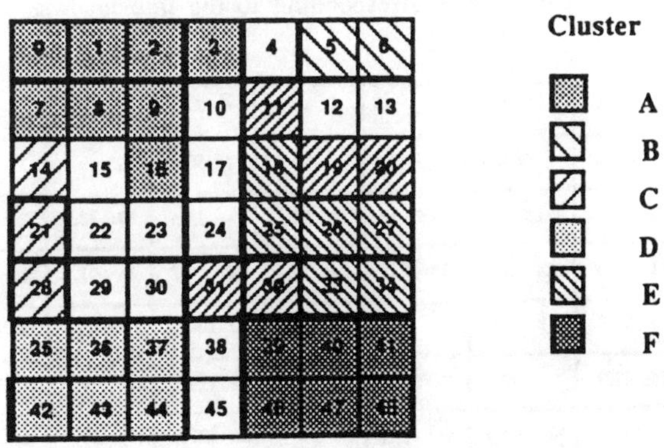

Instead of presenting in detail all 184 cases the main results will be presented.

1) All 6 clusters are still present and even more pronounced. For example, neuron 28 now classifies all 4 South-Main & South-Elm outage cases and neuron 21 classifies 3 out of 4 North-South & South-Elm outages.
2) The borders of the clusters are softened. For example neuron 14 now classifies the case of North-South & South-Elm at load level 110%. This corresponds to an intermediate load pattern where the load of line North-Lake gradually increases from minimal 115% at neuron 21, 163% at neuron 14 to maximal 214% at neuron 7. On the other hand, the load on line Main-Elm varies from maximal 71% at neuron 21 to 77% at neuron 14 down to 37% at neuron 7.

   Neuron 41 classifies some of the North-Lake outages thus linking clusters E and F. Neurons 11, 19 and 20 extend cluster E for the cases of 90% load.
3) The cluster B classifies the special cases (isolated buses and total generation outage). It is still separated by "empty" neurons from its neighboring clusters. Clusters A, C and D are now linked. They as well as cluster B are still separated from clusters E and F since neurons situated on a diagonal like neuron 5 and neuron 11 are not neighbors of first order.

3.5. Quality of the Classification

The main purpose of the classification process is the determination of overload situations. The Kohonen net not only determines whether a state is secure, critical or unsecure but also constructs the weight vector as a prototype of for a class of input vectors. As noted in section 3.4.1 under D), E) and F) neuron 43 as well as neuron 34 and all neurons of cluster F classify cases presenting an overload of line North-South. The differences between these cases have been discussed in 3.4.2. If the main interest lies only in the detection of overloads and not in the detection of specific features, a less detailed cluster map can be established. In table 5 the relative line loading for the 49 weight vector components corresponding to the line loading of line North-South is given

By fixing the limit value to exactly 100% loading, the borders of the insecure cluster are given by the "border neurons" 43, 37, 32, 33 and 34. However it is clear that these neurons will classify states heavily loaded (<=100%) as well as overloaded cases. For example neuron 34 classifies the South-Lake outage and generator South resulting in a 110 % overload of line North-South as well as the outage Lake-Main resulting in a 93% loaded line North-South. Neuron 32, 33 and 34 have been included in the secure cluster in table 2 and 4 because they classify 13 heavily loaded cases and only 10 overload cases. The heavily loaded cases still present critical system states! For heavily loaded cases which are classified as unsecure the term coarse classification will be used.

Misclassification on the other hand occurs for example for neuron 28, where the double contingency of line South-Elm and South-Main presents a 115% load of Lake-Main although the corresponding components of the weight vector present a load pattern of only 69%.

Table 5: Line loading in % for the 49 weight vector components corresponding to the line loading of line North-South.

0: 0.0	1: 0.2	2: 1.4	3: 10.1	4: 21.2	5: 22.1	6: 11.6
7: 0.1	8: 0.4	9: 3.5	10: 29.3	11: 50.1	12: 43.1	13: 36.5
14: 2.2	15: 2.5	16: 9.5	17: 54.1	18: 75.5	19: 73.9	20: 70.5
21: 23.0	22: 27.2	23: 45.6	24: 79.2	25: 85.7	26: 91.1	27: 91.7
28: 59.8	29: 61.5	30: 71.5	31: 97.5	32: 100.7	33: 100.3	34: 100.9
35: 83.9	36: 99.5	37: 106.3	38: 122.0	39: 122.7	40: 127.7	41: 128.0
42: 86.9	43: 103.3	44: 117.0	45: 125.4	46: 127.5	47: 139.1	48: 145.7

This misclassification case is more difficult to detect than those by border neurons. A possible criterion for inaccurate classification is given by the Euclidian distance between input vector and weight vector, which is 79 MVA in this particular case. The distance between patterns and weight vectors classified by the neighbor neurons 21 and 35 are less than 25 MVA and less than 15 MVA respectively.

Note that only 1 training vector presented an overload situation for line Lake-Main whereas an overload for example for line North-South is given for 15 training vectors. Therefore the Kohonen network learnt more details for the North-South overload situation resulting in a larger cluster than for the Lake-Main case.

The following table summarizes **all** coarse classifications and **all** misclassifications for the 46 trained and 184 untrained cases with respect to overload situations for at least one out of all 7 lines:

Table 6: Incorrect classified cases

Neuron	Heavily loaded	Overloaded	Coarse Classification	Misclassification
28 (secure)	1	4	0	4
31 (secure)	0	1	0	1
32(unsecure)	1	0	1	0
33(unsecure)	5	3	3	0
34(unsecure)	8	7	8	0
37(unsecure)	1	3	1	0
43(unsecure)	3	4	3	0
Total			16	5

Therefore out of 230 cases, 21 are incorrectly classified, but only the 5 misclassifications present a danger for the power system. The total error rate is about 9% whereas the misclassification rate is only about 2%.

4. DISCUSSION AND FUTURE WORK

The information obtained by the classification of the 46 cases can be treated under 2 different aspects. Firstly the 46 cases are treated as 46 different base cases corresponding to 46 different power system topologies. The Kohonen classifier then serves as a monitoring tool which will indicate imminent critical or unsecure loading situations. It has been shown that the network correctly classifies 98% of the untrained similar states. Information on the Euclidian distance between weight vector and input vector characterizes the quality of the classification and helps to detect these misclassifications. It can be expected that loading patterns which are obtained by interpolation between the treated cases are classified correctly as well. Therefore only a comparatively small number of loading patterns has to be trained. Change of security criteria and application to dynamic stability assessment present directions for future research.

Secondly the 46 cases can be studied as 1 base case plus 45 contingencies. In this case, the Kohonen classifier serves as a contingency analysis tool which helps to identify consequences of existing loading situations. For the generation of large sets of training vectors, the combinatorial aspect of the enumerative treatment of contingencies presents the same obstacle as for all other classical methods. However, for the fast identification of critical contingencies of unknown loading patterns the generalization capability of the Kohonen classifier prevents time-consuming load flow calculations. The similarity of cases can be identified without information retrieval in large databases or costly contingency computation.

Based on the work of [15] and [16] the relation between number of clusters, number of neurons and size of the power system has to be investigated. Scaling issues for large size power systems have to be addressed in future work.

Like neural networks in general, this classification technique is supposed to be implemented ultimately in dedicated hardware in order to be used under real-time conditions.

5. CONCLUSIONS

This paper demonstrates the feasibility of classification of load patterns for power system static security assessment using the Kohonen self-organizing feature map. Our results are based on simulations of the Kohonen network performed on a work station and not on neural net hardware. Therefore performance comparison with existing algorithms is not the issue of this paper. While the off-line training is time consuming for large training sets, the classification steps execute fast even on the simulator.

Instead emphasis is put on the interpretation of the obtained results. Simulation results show that indeed the self-organizing feature map projects safe, critical and unsafe operation points to different clusters. Knowing which cluster relates to which security state and mapping an operational point to one of these immediately identifies whether the system is operating safely or not. Depending on the neural network size the mapping further splits up the safe, critical, and unsafe clusters into more specific subclusters corresponding to different load patterns and therefore to different regions in the security space.

The most important and original aspect of this neural net is its generalization property. Using a training set consisting of 46 different line loading patterns, the Kohonen map not only classifies successfully 184 unknown loading patterns which have not been presented during the training phase but also is capable of extending clusters for vectors which do not fall in the same category as any trained one. This powerful and versatile feature is especially useful for power system operation where it is unrealistic to expect that all possible system states can be simulated off-line.

6. ACKNOWLEDGEMENTS

The authors are grateful to H. Kirkham, Jet Propulsion Laboratory for helpful informations and discussions.

The research described in this paper was started at the Swiss Federal Institute of Technology, Lausanne (EPFL), sponsored by EPFL, and was completed at the Jet Propulsion Laboratory, California Institute of Technology, sponsored by EPFL. Publication support was provided by the Jet Propulsion Laboratory, sponsored by the U.S. Department of Energy through an agreement with the National Aeronautics and Space Administration.

Reference herein to any commercial product, process, or service by trade name, trademark, manufacturer, or otherwise, does not constitute or imply its endorsement by the United States Government or the Jet Propulsion Laboratory, California Institute of Technology.

7. REFERENCES

[1] Stott, B., "Review of Load-Flow Calculation Methods", Proceedings of the IEEE, vol. 62, no. 7, July 1974, pp. 916-929.

[2] Ejebe, G.C. and Wollenberg, B.F., "Automatic Contingency Selection", IEEE Trans. on Power, Apparatus and Systems, vol. PAS-98, no. 1, Jan/Feb 1979.

[3] Pang, C. K., Prabhakara, F. S., El-Abiad, A. H. and Koivo, A. J., "Security Evaluation in Power Systems Using Pattern Recognition", IEEE Trans. on Power, Apparatus and Systems, vol. PAS-93 no. 2, May/June 74, pp. 969-976.

[4] EPRI EL-2343, "Feasibility of Using Associative Memories for Static Security Assessment of Power System Overloads", investigator Y.-H. Pao, Project 1047-2, Final report, April 1982.

[5] Oh, S. Y., "A Pattern Recognition and Associative Memory Approach to Power System Security Assessment", IEEE Trans. on Systems, Man and Cybernetics, vol. SMC-16, no. 1, Jan/Feb. 1986, pp. 62-72.

[6] Sobajic, D. J. and Pao, Y.-H., "An Artificial Intelligence System for Power System Contingency Screening", Proceedings of the IEEE-PICA Conference, Montreal, May 87, pp. 107 ff.

[7] Sobajic, D. J. and Pao, Y.-H., "Artificial Neural-Net Based Dynamic Security Assessment of Electric Power Systems", IEEE PES 1988 Winter Meeting, New York, Jan 31-Feb. 5, 1988, paper 88WM 211-5.

[8] Aggoune, M. E. et al., "Artificial Neural Networks for Power System Static Security Assessment", Proceedings of IEEE-ISCAS, Portland, 1989, pp. 490-494.

[9] Chow, J.C. et al., "An Improved Hopfield Model for Power System Contingency Classification", Proceedings of IEEE-ISCAS, New Orleans, 1990, pp. 2925-2928.

[10] Pao, Y.-H., "Applications of Neural-Net Computing to Power Systems" Proceedings of the NSF Workshop on Artificial Neural-Net Methodology in Power System Engineering, Clemson University, Clemson, SC, April 8-10, 1990.

[11] Dillon, T.S. et al., "Short Term Load Forecasting Using Adaptive Pattern Recognition and Self-Organizing Techniques", Proceedings of the 5th PSCC, Cambridge, 1975, pp 1-16.

[12] Sobajic, D. J., Pao, Y.-H., Njo W.and Dolce, J. L., "Real-Time Security Monitoring of Electric Power Systems using Parallel Associative Memories", Proceedings of IEEE-ISCAS, New Orleans, 1990, pp. 2929-2932.

[13] von der Malsburg, C., "Self-Organization of Orientation Sensitive Cells in the Striate Cortex", Kybernetik, 14, pp 85-100, 1973.

[14] Kohonen, T., "Self-Organized Formation of Topologically Correct Feature Maps", Biol. Cybernetics no. 43, pp 59-69, 1982.

[15] Kohonen, T., "Self-Organization and Associative Memory", 3rd edition, Springer-Verlag, Berlin, 1989.

[16] Cottrell, M., Fort, J., "A Stochastic Model of Retinotopy: A Self-Organizing Process", Biol. Cybernetics no. 53, 1986, pp 405-411.

[17] Ritter, H. and Schulten, K. "Convergence Properties of Kohonen's Topology Conserving Maps: Fluctuations, Stability and Dimension Selection", Biol. Cybernetics no. 60, 1988, pp 59-71.

[18] Vittoz, E., Sorouchyari, E., Heim, P., Arreguit, X., and Krummenacher, F. "Analog VLSI Implementation of a Kohonen Map", Proceedings of the International Conference on Neural Networks, EPF-Lausanne, October 1989.

[19] Hochet, B., Peiris, V., Corbaz, G. and Declerq, M. "Implementation of a Neuron Dedicated to Kohonen Maps with Learning Capabilities", Proceedings of the IEEE CICC, Boston, 1990.

[20] Ritter H., and Schulten K., "Extending Kohonen's Self-Organizing Mapping Algorithm to Learn Ballistic Movements" in Eckmiller and von der Malsburg, Eds., "Neural Computers", Springer, Heidelberg, 1987, pp 393-406.

[21] Tryba V., Metzen S. and Goser K., "Designing of Basic Integrated Circuits by Self-Organizing Feature Maps", Proceedings of Neuro-Nîmes, 1988.

[22] Wu, F. F. and Kumagai, S., "Steady-State Security Regions of Power Systems", IEEE Trans. on Circuit and Systems, vol. CAS-29, no. 11, no.v 1982, pp. 703 ff.

[23] Fischl, R. *et al.*, "The Application of Decision Theory to Contingency Selection", IEEE Trans. on Circuits and Systems, vol. CAS-29, no. 11, 1982, pp. 712-723.

[24] Stagg, G. and El-Abiad, A., "Computer Methods in Power System Analysis", McGraw-Hill, New York, 1968, pp 284.

Voltage Stability Monitoring with Artificial Neural Networks

Hiroyuki Mori
Department of Electrical Engineering
Meiji University
Tama-ku, Kawasaki 214
Japan

Abstract - In this paper, an artificial neural net based approach to voltage stability monitoring is presented. Firstly, a hybrid neural network is proposed to estimate the index for voltage instability and track the power system transition. Secondly, this paper presents a method for considering changes of network topology. Thirdly, a decentralized neural network is proposed to handle a large scale power systems so that the curse of dimensionality is alleviated. The proposed method is demonstrated in a sample system.

I. INTRODUCTION

The voltage stability problem has been recently studied in several aspects such as analysis, control, monitoring, and planning. Because of the lack of reactive power sources, load characteristics, network nonlinearity, etc., it is not easy to make a strategy to suppress the problem. Indeed, the voltage cascading collapse has been unexpectedly experienced in some countries. The research topics may be classified into the following:

1) Qualitative Analysis
2) Development of indices
3) Evaluation of critical points
4) Load characteristics

Neural net computing is useful for complicated problems. Neural net based approaches have been developed in power system problems; static and dynamic security assessment [4], [5], [13], [14], [17], forecasting [12], [15] identification[6], [9], analysis[7], [10], [11], and control problems[8], [16]. This paper deals with item 2) to take preventive control using an artificial neural network based approach. Evaluation of the indices is important in understanding how close the power system to be studied is voltage instability. The indices serve as the operator's guideline for monitoring voltage instability. In this paper, the idea of VIPI [2] is employed to evaluate the margin. Since calculation of VIPI is a sort of combinatorial problems, it is time-consuming. In order to calculate the indices efficiently, artificial neural networks are used.

II. INDICES FOR VOLTAGE STABILITY

This section describes an index for voltage stability with a pair of multiple load flow solutions. The index is referred as to VIPI [2]. The index approaches zero since power system conditions become heavy-loaded. Since the problem is related to the nonlinearity of power systems, it is not sufficient to relay on only one index [3]. It is influenced by two aspects of state and parameter space. The two indices are based on the margin to critical conditions with a pair of multiple load flow solutions. Specially, they are expressed by the angle between the operational conditions and the nearest singular point. One is expressed in state space corresponding to the nodal voltage solution while the other is in parameter space corresponding to the specified value of the load flow calculation.

Now, consider $\mathbf{x}_1$ and $\mathbf{x}_2$, a pair of the multiple load flow solutions [1]. The pair can be expressed as

$$\mathbf{x}_1 := \mathbf{x}^* + \mathbf{b} \qquad (1)$$

$$\mathbf{x}_2 := \mathbf{x}^* - \mathbf{b} \qquad (2)$$

where, vector $\mathbf{x}^*$ is a singular vector in $\mathbf{x}$-space.

Fig. 1 shows the relationship between x_1 and x_2. Noting that the pair satisfies the same specified value y_S, we have

$$y_S = y^* + y_b \qquad (3)$$

where
y_S: specified value vector of the load flow calculation such as

$$y_S = y(x) \qquad (4)$$

($y(\cdot)$: nonlinear load flow equation)

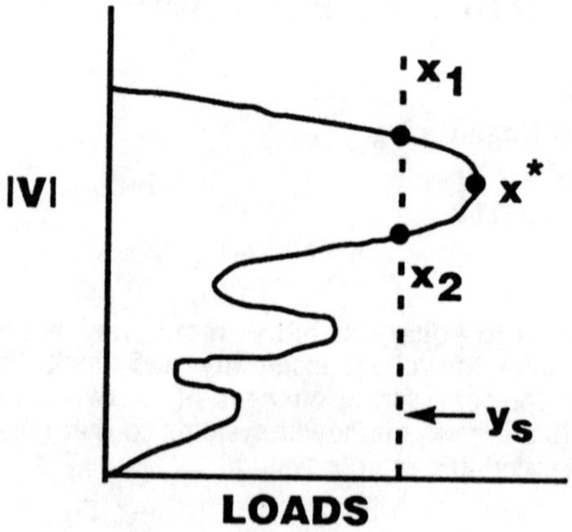

Fig. 1 Multiple Load Solutions

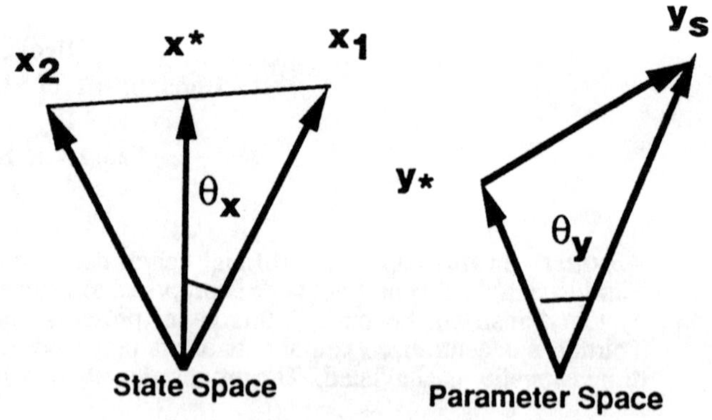

Fig. 2 Singular Vector in State and Parameter Space

$y^* := y(a)$, a singular vector in y-space.
$y_b := y(b)$

Fig. 2 shows the singular vectors in both state and parameter space. As power system conditions become heavy-loaded, x_1 and x_2 approach singular vector a in x-space. Therefore, indices for static voltage instability in state and parameter space can be written by Eqns. (5) and (6), respectively.

$$\theta_x := \cos^{-1} \frac{x_1^T x^*}{\| x_1 \| \cdot \| x^* \|} \quad [\text{deg.}] \tag{5}$$

or

$$\theta_y := \cos^{-1} \frac{y_s^T y^*}{\| y_s \| \cdot \| y^* \|} \quad [\text{deg.}] \tag{6}$$

The indices have an advantage that they are easily understood in a sense that they are expressed in degree. If the counterpart of the load flow solution obtained with the flat start is calculated by a method for finding multiple load flow solutions, the above indices are obtained.

III. HYBRID NEURAL NETWORKS FOR VOLTAGE INSTABILITY MONITORING

This section proposes a method for voltage instability monitoring with a hybrid neural network which consists of a multi-layer perceptron and the Kohonen neural net. The proposed method has a couple of the following functions: a multi-layer perceptron (MLP) is used to to estimate indices for voltage instability; the Kohonen net is employed to understand power system conditions with the feature extraction thought the self-organization. Fig. 3 shows the concept of the proposed method. Firstly, MLP is utilized to carry out pattern recognition of θ_x and θ_y το reduce the computational effort. The conventional method finds the multiple load flow solutions using the simplified exhaustive search method, where the conventional method means the simplified exhaustive search method. In an n-bus system, at most (n-1) pairs of multiple load flow solutions are calculated to find the closest counterpart. Among them, the pair with the smallest index was selected as the solution pair that expresses most critical power system conditions. The indices for voltage instability are estimated by the multi-layer perceptron. Since the neural network is based on the supervised learning, it is expected that the multi-layer perceptron provides more rigorous estimates than the Kohonen net. That is because the Kohonen net is involved in the category of the unsupervised learning. Concerning the multi-layer perceptron, input vector **u** is the power system conditions while output vector **z** is indices in space and parameter space, namely, θ_x and θ_y. Thus, these vectors may be written as

$$\mathbf{u} = (V_1, P_2, Q_2, P_3, Q_3, \ldots, P_n, Q_n, 1)^T \tag{7}$$

$$\mathbf{z} = (\theta_x, \theta_y)^T \tag{8}$$

Secondly, the Kohonen net is able to provide an insight into power system conditions from a standpoint of the feature extraction of input patterns. This is quite useful in power system monitoring because the different power system conditions from the normal ones are easily identified. The Kohonen net is effective for pattern

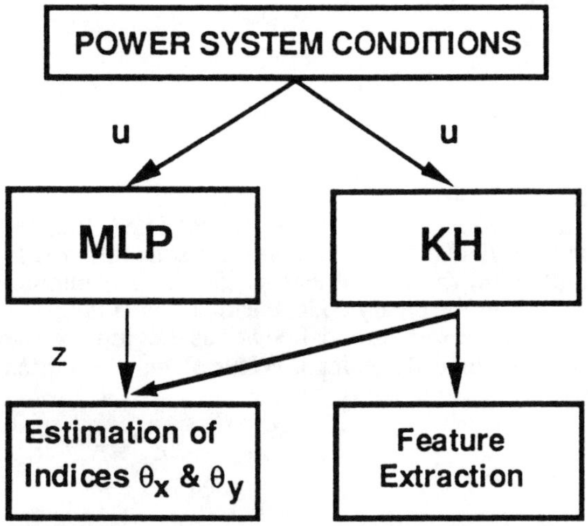

Fig. 3 Hybrid Neural Networks for Voltage Instability Monitoring

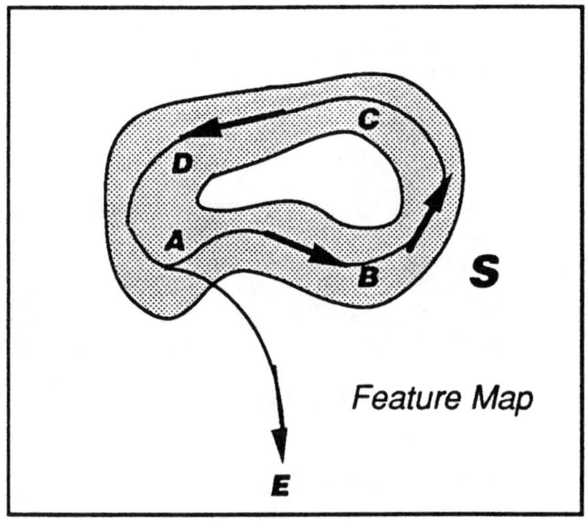

Fig. 4 Security Regions on the Kohonen Net Feature Mapping

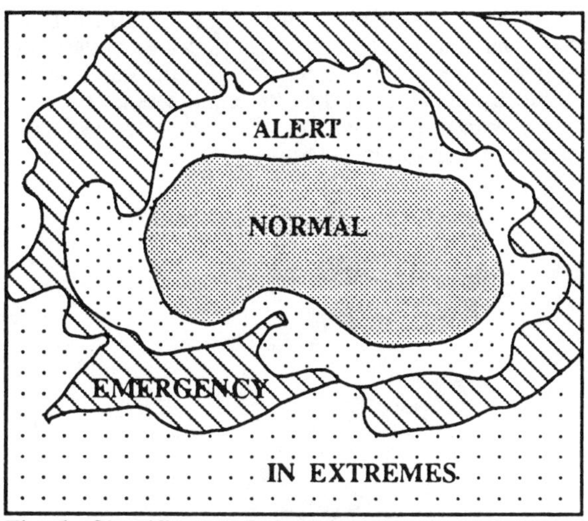

Fig. 5 Classification Security Regions on the Kohonen Net Feature Mapping

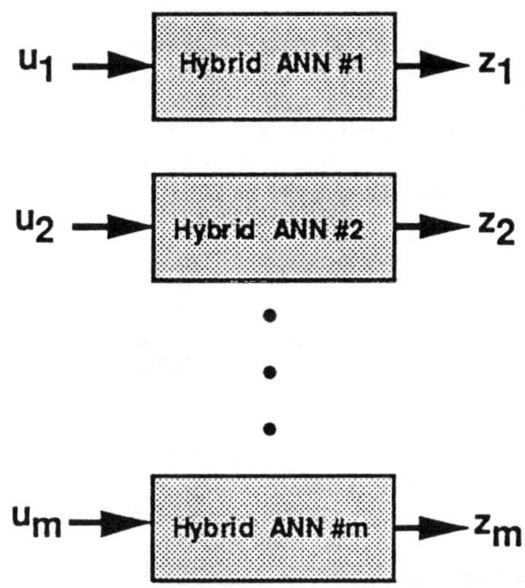

Fig. 6 Decentralized Neural Networks for Voltage Stability Monitoring

classification. Using the function, the Kohonen is used to estimate the indices. In addition, this paper focuses on the feature extraction of input vector. Unlike the other artificial neural networks, the Kohonen network has the function to extract features of input patterns. The network has a interesting function that high dimensional patterns are transformed into two or more dimensional output neurons. The mapping from input to output patterns with the self-organization can reflect the ordering of input patterns. In other words, input patterns with similar features are allocated to output neurons close to each other. Using the function, voltage instability monitoring is carried out by checking the trajectories on the feature map. For instance, consider the daily load in a power system. Since the load has daily periodicity with fluctuations, trajectories are formed on the feature mapping as shown in Fig. 4. In the figure, the shaded area, **S**, indicates secure power system conditions. The curve involving points A to D represent typical power systems conditions. If power system conditions move to different state, say, Point E, system operators is able to recognize what is happening in power systems through the mapping. Furthermore, power system operators can easily understand the extent to which power systems have a security margin with the feature mapping divided into security areas such as normal, alert, emergency, and in extremes (See, Fig. 5).

IV. DECENTRALIZED SCHEME FOR VOLTAGE STABILITY MONITORING

4.1 Consideration of Topology Change

There exists the nonlinear relationship between power system conditions and voltage stability indices. The relationship is identified with the artificial neural network.

$$\mathbf{z} = \mathbf{f}(\mathbf{u}) \tag{9}$$

where

z : estimates of indices
u: input variable vector
f(•): nonlinear mapping

The input vector is extended so that the changes of network configurations are reflected. It is necessary to consider power network configurations since they vary due to the operational scheduling. Most of the conventional methods with ANNs have not paid attention to the problem in spite of the importance. The problem is handled by expressing the change of topology as the binary code, 0 and 1 which implies on and off states of lines. This paper regards the branch power flows as well as nodal power as the input variables. The method gives us more informative than the conventional method. Thus, input vector **u'** may be written as

$$\mathbf{u}' := (\mathbf{u}_N, \mathbf{u}_B)^T \tag{10}$$

where

$\mathbf{u}_N$: nodal power quantities

$\mathbf{u}_B$: branch power quantities

$\bullet^T$: transposition of vector •

4.2 Decentralized Monitoring

A decentralized scheme for voltage stability monitoring is proposed with artificial neural networks. The centralized scheme mentioned is inclined to increase the number of input variables as power systems are larger. It is well-known that the voltage problem has significant local characteristics. Therefore, a decentralized scheme is presented to estimate the estimates of the indices and carry out the self-organization of input patterns. Fig. 6 shows a decentralized voltage monitoring scheme with m subnetworks. The indices at each subnetwork are evaluated by local quantities such as nodal and branch power variables. Namely,

$$\theta_{xk} := \cos^{-1} \frac{\mathbf{x}_{1k}{}^T \mathbf{x}^*{}_k}{\| \mathbf{x}_{1k} \| \cdot \| \mathbf{x}^*{}_k \|} \quad \text{[deg.]} \tag{11}$$

or

$$\theta_{yk} := \cos^{-1} \frac{\mathbf{y}_{sk}{}^T \mathbf{y}^*{}_k}{\| \mathbf{y}_{sk} \| \cdot \| \mathbf{y}^*{}_k \|} \quad \text{[deg.]} \tag{12}$$

where

θ_{xk}: index in state space for subnetwork k

θ_{yk}: index in parameter space for subnetwork k

$\mathbf{x}_{1k}$: element of voltage vector $\mathbf{x}_1$ corresponding to subnetwork k

$\mathbf{x}^*{}_k$: element of voltage vector $\mathbf{x}^*$ corresponding to subnetwork k

$\mathbf{y}_{1k}$: element of voltage vector $\mathbf{y}_1$ corresponding to subnetwork k

$\mathbf{y}^*{}_k$: element of voltage vector $\mathbf{y}^*$ corresponding to subnetwork k

Let us consider a neural network whose number of input, hidden, and output units are denoted as n_i, n_h and n_o, respectively. The number of the weights between units for the centralized scheme is equal to

$$n_{wc} = n_i \cdot n_h \cdot n_o \tag{13}$$

Also, let us assume that the network is uniformly divided into m subnetworks. As a result, the number of the weights for one subnetwork almost equals

$$n_{wd} \cong n_i \cdot n_h \cdot n_o / m^3 \tag{14}$$

From Eqns. (13) and (14), it is conjectured that the dimension of the weights for the centralized scheme is m^3

larger than that for the decentralized scheme. Hence, it is expected that the decentralized scheme is less time-consuming than the centralized one.

IV. SIMULATION

In the previous section, decentralized neural networks ware given to voltage stability monitoring. Thus, real-size power systems are decomposed into subnetworks and a hybrid neural network is constructed for each subnetwork. Namely, it does not lack the generality that a small-size hybrid neural network is examined. The proposed method was applied to a five bus system. The following conditions are used:

(a) The input vector representing power system conditions is given by

$$\mathbf{u}'(k) = (e_1(k),\ P_2(k), Q_2(k), ..., P_5(k), Q_5(k), 1)^T \tag{15}$$

Since the input vector has ten elements, there are ten input neurons. In this paper, 6000 power system conditions were created by the uniform random number generator in which 5000 power system conditions were the learning data and 1000 system conditions were the test data. The upper and lower bounds of the specified value $\mathbf{y_s}$ were set as follows:

$$\mathbf{u}'^{M} = 1.20\ \mathbf{y_s} \tag{16}$$

$$\mathbf{u}'^{m} = 0.80\ \mathbf{y_s} \tag{17}$$

where, $^{M(m)}$ denotes the upper (lower) bound.

Also, the slack voltage e_1 (k) has the following range:

$$e_1(k)^M = 1.02\, e_1 \tag{18}$$

$$e_1(k)^m = 0.98\, e_1 \tag{19}$$

For 6000 power system conditions, voltage instability indices θ_x, θ_y were calculated by the simplified exhaustive search method.

(b) The criterion of the convergence is 1000 iterations for MLP while it is 40 iterations for the Kohonen net.

(c) Output neurons were arrayed on the two-dimensional grid consists of 30 × 30 output neurons. The grid is expressed as rectangular lattice.

The estimation errors of the indices for MLP are 1.9 and 3.2[%] for θ_x and θ_y, respectively. On the other hand, those for the Kohonen net are 3.1 and 6.5 [%] for θ_x and θ_y, respectively. It can be seen that compared with the Kohonen net, MLP provides better estimates. Also, errors in the estimate of θ_x is smaller than those

Fig.7 Voltage Instability Index θ_x on Self-Organization Feature Mapping

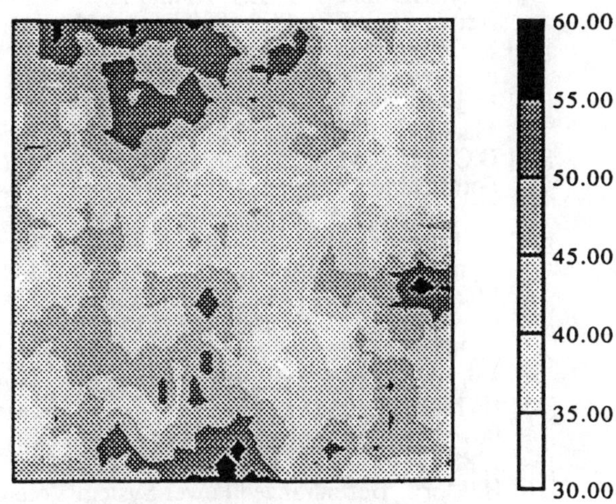

Fig. 8 Voltage Instability Index θ_y on Self-Organization Feature Mapping

in estimate of θ_y. It seems that changes in parameter give more significant changes in voltage instability index θ_y.

Figs. 7 and 8 show voltage stability indecis θ_x and θ_y on the self-organization feature mapping obtained by the Kohonen net with the rectangular lattice, respectively. In those figures, the darker areas represents secure power system conditions. Noting that the assignments of the output neurons in Fig. 7 is the same as that in Fig. 8, it can be found that securer areas in state space (**x**) do not imply those in parameter space (**y**). That is because of the nonlinearity of the load flow equation. This mapping can advise power system operators if power system conditions approach the critical conditions. Thus, it is expected that the representation is displayed on the CRT as an operator's guideline for voltage instability monitoring.

V. CONCLUSIONS

This paper has proposed an artificial neural net based method for voltage instability monitoring. Hybrid artificial neural networks were utilized to deal with the problem. One is the multi-layer perceptron to estimate voltage instability indices while the other is the Kohonen self-organization feature mapping to trace the trajectories of power system conditions on the self-organization feature mapping. Also, line flows are involved in the input vector to consider the change of network configurations efficiently. Furthermore, a decentralized scheme for voltage stability monitoring has been presented to deal with real-size systems.

REFERENCES

[1] Y. Tamura, H. Mori and S. Iwamoto, "Relationship between Voltage Instability and Multiple Load Flow Solutions in Electric Power Systems ", IEEE Trans. on Power App. and Syst., Vol. PAS-102, No.5, pp.1115-1125, May 1983.

[2] H. Mori and Y. Tamura, "On Voltage Security On-Line Index in Electric Power Systems ", Power System Engineering Committee of IEE of Japan, Paper No. PE85-41, June 1985 (in Japanese).

[3] H. Mori and S. Tsuzuki, "Estimation of Critical Points on Static Voltage Stability in Electric Power Systems", IFAC Proc. of 1990 Symp. on Power Systems and Power Plant Control, pp. 550-554, Seoul, Korea Aug. 1990.

[4] D.J. Sobajic and Y.-H. Pao, "Artificial Neural-Net Based Dynamic Security Assessment for Electric Power Systems", IEEE Trans. on Power Systems, Vol. 4, No. 1, pp. 220-224, Feb. 1989.

[5] M. Aggoune, M.A. El-Sharkawi, D.C. Park, M.J. Damborg, and R.J. Marks II, "Preliminary Results on Using Artificial Neural Networks for Security Assessment", IEEE Proc. of 1989 PICA, pp. 252-258, Seattle, WA, May 1989.

[6] H. Mori, H. Uematsu, S. Tsuzuki, T. Sakurai, Y. Kojima, K. Suzuki, "Identification of Harmonic Loads in Power Systems Using an Artificial Neural Networks", Proc. of Second Symposium on Expert Systems Application to Power Systems", pp. 371-378, Seattle, WA, July 1989.

[7] H. Mori and S. Tsuzuki, "Power System Topological Observability Analysis Using a Neural Network Model", Proc. of Second Symposium on Expert Systems Application to Power Systems", pp. 385-391, Seattle, WA, July 1989.

[8] N. Iwan Santoso and O.T. Tan, "Neural-Net Based Real-Time Control of Capacitors Installed on Distribution Systems", IEEE Trans. on Power Delivery, Vol. 5, No. 1, pp. 266-272, Jan. 1990.

[9] R.K. Hartana and G.G. Richards, "Harmonic Source Monitoring and Identification Using Neural Networks", IEEE PES 1990 Winter Meeting, Paper No. 90 WM 238-6 PWRS, Atlanta, GA, Feb. 1990.

[10] S. Ebron, D. Lubkeman and M. White, "A Neural Network Approach to the Detection of Incipient Faults on Power Distribution Feeders", IEEE Trans. on Power Delivery, Vol. 5, No. 2, pp. 905-914, April 1990.

[11] H. Mori and S. Tsuzuki, "Determination of Power System Topological Observability Using the Boltzmann Machine", IEEE Proc. of 1990 ISCAS, pp. 2938-2941, New Orleans, LA, May 1990.

[12] D.C. Park, M.A. El-Sharkawi, R.J. Marks, M.E. Aggoune, L.E. Atlas and M.J. Damborg, "Electric Load Forecasting Using an Artificial Neural Network", IEEE PES 1990 Summer Meeting, Paper No. 90 SM 377-2 PWRS, Minneapolis, MN, July 1990.

[13] H. Mori, Y. Tamaru and S. Tsuzuki, "An Artificial Neural-Net Based Technique for Power System Dynamic Stability with the Kohonen Model", IEEE Proc. of 1991 PICA, pp. 293-301, Baltimore, Maryland, U.S.A., May 1991.

[14] H. Mori, and Y. Tamaru,"An Artificial Neural-Net Based Approach to Monitoring Power System Voltage Stability ", Proc. of Nulk Power System Voltage Phenomena II: Voltage Stability and Security, pp. 347-358, Deep Creek Lake, Maryland, U.S.A., Aug. 1991.

[15] H. Mori, K. Itou, H. Uematsu, and S. Tsuzuki, "An Artificial Neural Net Based Method for Predicting Power System Voltage Harmonics", IEEE Trans. on Power Delivery, Vol. 7, No.1, pp.402-409, Jan. 1982.

[16] H. Mori, "Decentralized Power System Voltage Control Using Artificial Neural Networks", Proc. of 1992 IEEE ISCAS, pp. 1701-1704, San Diego, CA, U.S.A., May 1992.

[17] H. Mori, "Monitoring and Control of Power System Voltage Stability Using an Artificial Neural Networks", Preprints of 1992 IFAC/IFIP/IMACS International Symposium on Artificial Intelligence in Real Time Control, pp. 433-438, Delft, the Netherlands, June 1992.

INTELLIGENT LOAD SHEDDING

Damir Novosel[1]
Roger L. King

P.O. Drawer EE
Mississippi State University
Mississippi State, MS 39762

ABSTRACT - Previous work [1] has shown the viability of using artificial neural networks (ANNs) for the early detection and control of overloaded lines. This paper focuses on the development of an artificial neural network based scheme for intelligent load shedding. The load shedding scheme consists of localized detectors that consider various parameters of the power system during an emergency. The research was developed using the IEEE 30 bus test system which incorporates a design of net import and export regions. The approach is implemented in the import area of the system. The purpose of the scheme is to detect overloaded lines and make intelligent decisions about where within the topology of the system the load should be dropped and how much load to shed. The proposed scheme is designed to avoid unintentional separation of the system by fast, proactive, and adaptive control.

IMPLEMENTATION OF THE LOAD SHEDDING SCHEME - The modified IEEE 30-bus overhead test system, shown in Figure 1, was selected to demonstrate the intelligent load shedding scheme. This test system can be visualized as consisting of a net import area of power and one of a net export of power. A bus with intelligent load shedding is marked with a circle. The primary task of load shedding is to relieve heavily overloaded lines in the area. Special consideration is given to alleviating overloads on interconnections with the export area as an important requirement to preserve system integrity.

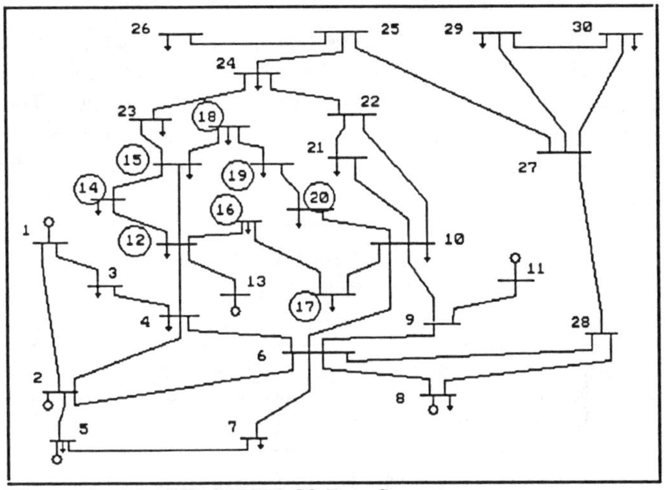

Figure 1. Modified IEEE 30-Bus System.

Every load shedding neural network is composed of an input layer, two hidden layers, and an output node. The input data that comprises the bus voltage and several active line flows is mapped to a single trip signal output that activates shedding of the chosen load at the bus. The input to the ANN acts as a description of the state of the power system and predicts the consequences of the action.

Coordination among the detectors is achieved by the training of all of the neural networks with carefully chosen training sets, extracted from the same emergency states. Each selected power system state provides simultaneous data for each neural network. The output from every ANN is a trip signal for shedding selected load. In the experiments, the load was divided into two priority levels. The ANN for a bus selects the lower priority level load to be shed at the bus.

The main priority was to alleviate line overloads just under the emergency rating without violating any other security constraints. An attempt was made to design the load shedding ANNs to shed the minimum necessary amount of the lowest priority load. Therefore, the most efficient shedding with the least amount of load is implemented. For example, a large amount of load on a distant bus will not be shed to reduce a small amount of

---

[1] Now with ABB Transmission Technology Institute, Raleigh, NC

overload on a local bus. Higher priority load in the vicinity of the overload will be shed in this case as well as in the case when it is the only way to relieve overload (e.g., heavy overload at radial line 25-26). Thus, a heuristic approach is applied to perform load shedding. A different load shedding scheme in another power system topology would require a different training of the neural networks.

The voltage "inequality" constraints and unavoidable reactive shedding that follows active load shedding is also considered in the performance of each detector. Thus, a voltage/reactive power relationship is embedded. Through sensing line flows that enter and leave the bus, bus load knowledge is embedded as well. The role of the proposed load shedding scheme is to alleviate overload fast and as optimally as possible.

The intelligent load shedding detectors are placed at buses 12, 14, 15, 16, 17, 18, 19, and 20, mainly to solve the problems in their vicinity or problems associated with the interconnection between buses 4 and 12. A comprehensive analysis of the whole system is performed with respect to these buses. This ensures the coordination of the localized intelligent load shedding with the rest of the import area control schemes. Thus, eight neural networks were trained by a selection of 65 training sets extracted from a much larger number of experimental sets. This implies that data are "expensive", requiring extensive off-line analysis. Problems associated with the limitations of ANNs are solved by implementing only binary outputs and by careful training. A good choice of the training sets is critical because the ANN cannot perform better than it is trained.

While the output of the detector at each bus is a single signal to trip load at the corresponding bus, the choice of the inputs is different for every bus. This choice is presented in Table 1. The goal was to enable every detector to perform well with the least number of inputs. The reason is to reduce measuring and communication requirements. Specified load to be shed was chosen to be 50% of the total bus load. The selection of the load priority depends on the specific system and more sophisticated implementation of load shedding priorities is usually required.

Detector at bus	Active flows for lines with direct connection to the bus	Bus voltage	Active flows for lines with no direct connection to the bus
12	4-12, 12-13, 12-14, 12-15, 12-16	12	
14	12-14, 14-15	14	4-12
15	12-15, 14-15, 15-18, 15-23	15	4-12
16	12-16, 16-17	16	4-12, 10-17
17	10-17, 16-17	17	6-10
18	15-18, 18-19	18	19-20
19	18-19, 19-20	19	6-10
20	10-20, 19-20	20	6-10

Table 1 Inputs to the intelligent load shedding ANNs.

The networks were trained with binary outputs of 1 and 0. However, in reality, the outputs during network performance will be analog values in a range [0,1]. To use the output information in an effective way, the following logic is used:

(a) Output above 0.7 is recognized as 1 (trip signal).
(b) Output from 0.3 to 0.7 is for operator notification.
(c) Output under 0.3 is recognized as 0 (no signal).

In case (b), if an overload still exists, notification to the operator can be helpful information to possibly shed this load.

CASE STUDIES - To test the performance of the intelligent load shedding detectors, 20 testing sets that are considerably different than the training sets, are introduced to the system. Thus, the adaptability of the scheme is tested. For all 160 outputs from the ANNs (8 detectors x 20 cases), not one is in the range [0.3, 0.7]. This means that all the actions for the test cases are recognized in a pattern space with a high degree of confidence. The recommended actions are implemented for the test cases and satisfactory performance of the entire coordinated scheme is achieved.

The comparison in active line flows before and after recommended load shedding is shown in Tables 2 and 3 for six test cases to illustrate the system's performance. If a line is overloaded above its emergency rating, an * is added. Only lines in the area with load shedding detectors are presented in Tables 2 and 3, although active flows on all lines in the IEEE 30-bus system are checked. The total amount of load that was shed for each case is also shown in the tables.

Line	Emergency rating (MW)	case 1 flow(MW)		case 2 flow(MW)		case 3 flow(MW)	
		before	after	before	after	before	after
4-12	40	41*	39	42*	40	42*	40
6-9	24	14	13	14	11	14	13
6-10	16	14	13	14	12	14	13
10-17	8	10*	6	9*	5	10*	7
10-20	16	16*	16	16*	16	13	13
10-21	40	18	18	18	18	20	20
12-14	16	OUT	OUT	6	6	8	8
12-15	32	22	21	16	14	19	19
12-16	16	3	2	4	4	3	1
14-15	8	-3	-3	3	2	2	2
15-18	16	6	5	7	4	2	2
15-23	16	4	4	4	4	4	3
16-17	16	-1	-2	OUT	OUT	-1	-2
18-19	16	0	0	1	2	-1	-1
19-20	16	-14	-14	-13	-13	-10	-11
Total shed load:		5.6 MW		7.1 MW		4.5 MW	

Table 2 Line Flow Comparison Before and After Load Shedding for Cases 1 to 3.

Analysis of case number 1: Overloads on lines 4-12, 10-17, and 10-20 are alleviated by the recommended load shedding at buses 17 and 20. While B17 output was "confident", outputs of B18 and B20 deviate from the closest binary value by more than 0.2 (0.287 and 0.782, respectively). Both networks hesitate in assigning a binary value because the load shed at any of these buses will alleviate overloads, but shedding load at both will be too much. This means that shedding the load at bus 18, instead of bus 20, would also alleviate overloads. The choice was correct because less load is shed with shedding at bus 20 and the overload on line 10-20 is alleviated just under its emergency rating.

Analysis of case number 2: Overloads on lines 4-12, 10-17, and 10-20 are alleviated by the recommended load shedding at buses 17 and 18. Compared to case 1, overloads on the same lines are alleviated with shedding at different buses. Instead of shedding load at bus 20, the load was shed at bus 18. This choice was necessary to alleviate all overloads because shedding at bus 20 instead of bus 18 would be insufficient to alleviate an overload on interconnection 4-12. Because of the sensitivity of the problem, output from B18 deviates from the closest binary value slightly more then 0.1 (0.858).

Analysis of case number 3: Overloads on lines 4-12 and 10-17 are alleviated by a recommended load shedding at bus 17. Analyzing and comparing the first three cases, shows that a good sensitivity is achieved with the total scheme because of the successful coordination achieved through training of the each detector.

ATTRIBUTES OF INTELLIGENT LOAD SHEDDING - Using artificial neural networks for intelligent load shedding appears to be promising. The attributes of the load shedding scheme are:

a) Adaptability - The analysis of test cases shows that ANN detectors are able to react properly to new situations. Unlike the existing load shedding schemes, the actions are adapted to the instantaneous power system state. Problems associated with load distribution are also overcome to a certain degree.

b) Proactive actions - The scheme reacts when the system approaches a serious problem, and not to a problem. Unlike underfrequency protection, there is no waiting for an "in extremis" crisis. Additionally, the disturbance is properly classified because of the usage of the artificial neural networks as a pattern classifier.

c) Accuracy - The results of comprehensive and accurate off-line analysis are considered. The intelligent load shedding detectors considered in this paper utilized bus voltage values, but did not consider bus voltage criteria in their decision process. However, problems associated with voltage and reactive power can be included in future training. Also, the case of unavoidable frequency drop can be considered by adding frequency as an input to the ANN.

d) Coordination - All detectors are coordinated among each other, and although local actions are performed, a problem that is spread throughout the whole system is considered. Coordination with other actions can be embedded into the ANN or can be obtained by sending or receiving additional signals. In this analysis, coordination of the load shedding scheme in the import area with the actions in the export area (generation shifting) is embedded through considering the flows through interconnecting lines. Finally, the aspect of time coordination, similar to common protective relays, can be an interesting topic for further research.

e) Speed - The speed of the trained ANN is one of the main advantages of the proposed intelligent load shedding detector. Rapid actions by the scheme should will enable the power system to return to a more secure state faster than otherwise possible.

CONCLUSIONS - This paper has described an approach to intelligent load shedding implemented on the IEEE 30-bus system. The primary goal of the system is to alleviate overloaded lines before an "in extremis" crisis occurs. Artificial neural networks are trained so that complex relationships among power system features are imbedded into them. Then, when a set of power system data are presented to the network (detector) an appropriate control action is recommended by the detector. Coordination of the detectors is achieved through training sets extracted from the same emergency states for each detector. This scheme adapts to an instantaneous system emergency state regardless of changes in the topology, load, or generation. Detectors in the scheme operate locally, but unlike the common local relays, they consider a much larger area. Problems associated with voltage/reactive flows and consequences of the actions are considered. Thus, this scheme is advantageous compared to many of the existing methods and technologies.

REFERENCES

[1] Damir Novosel and Roger L. King, "Identification of Power System Emergency Actions Using Neural Networks," Mohamed A. El-Sharkawi and Robert J. Marks II, eds., Proceedings of First International Forum on Applications of Neural Networks to Power Systems, University of Washington, July 23-26, 1991, IEEE 91TH0374-9, pp 205-209.

[2] Roger L. King and Damir Novosel, "An Architecture for the Intelligent Detection and Alleviation of Overloaded Transmission Lines," to be published in the Proceedings of the Third Biennial Industrial Electric Power Applications Symposium, New Orleans, LA, November 12-13, 1992.

CONSIDERATION IN INTELLIGENT ALARM PROCESSING

E.H. Chan	N.S. Markushevich	R. Adapa
UCI, Inc.	ECC, Inc.	EPRI
20833 Stevens Creek	1530 The Alameda	3412 Hillview Avenue
Suite 101	Suite 300	Palo ALto, Ca 94303
Cupertino, CA 95014	San Jose, CA 95126	

Abstract

This paper is a brief report of the results of the investigation to apply neural network technology to interpret bulk power system alarms. An experimental "Interpretive Alarm Processor (IAP)" was created to accomplish this task.

The objectives, scope, and methodology of the investigation, along with a description of the power system problems, operating conditions, and alarm sets used for this work are described.

Statistical analysis of the results of the tests, and the conclusions drawn from the tests are presented.

Introduction

This paper is concerned with the initial testing of this concept on bulk power system alarms. The objective of this work was to investigate the capability of a neural-network-based IAP to recognize the source power systems problems causing alarms, under conditions of delayed and partially-missing and erroneous information.

There are two classes of issues regarding this testing. The first, issues on a neural-network based IAP: overlapping alarm patterns; philosophy in accuracy; methodology in testing; limitation on predefining "problem cases". The second, technical issues: types of power system alarms; data presentation; "beginning" and "end" of the system problem "event"; detailed diagnostics versus broad-view problem identification.

Scope of Testing

An IAP based on ANN was created. A region the of Florida Power Corporation's bulk transmission network was chosen as a test system. The operating conditions of the power system were simulated by using a power flow computer model. Approximately 100 different "problem" conditions were included in the testing. These problem conditions were simulated individually under 24 different initial power system conditions and under 8 different delayed-alarm and missing-alarm situations, resulting in more than 18,000 different sets of alarms. These 18,000 sets of alarms were then submitted to the IAP to test the IAP's capability to recognize the individual conditions.
Statistical analysis of the recognition results was conducted.

Fault Categories in the Power System Selected for Testing.

For the purpose of testing the IAP, the following categories of power system problems were considered: i) Transmission line faults (normal fault isolation and that complicated by stuck breakers), ii) Transformer faults, iii) Generator trips, and iv) Bus faults.

Initial Conditions

For the purpose of this testing, an "initial operating condition" is defined to be the state of the power system prior to the occurrence of a set of alarms of interest. The approximately 100 system problems were simulated under four initial power system operating conditions.

- Basic initial condition of the power system: All breakers are closed and all generators and transmission lines are in service.
- Different generating conditions: Three different combinations of on-line/off-line generators were used. These generators were located in different parts of the power system, and their on/off statuses create different initial power flows in the power system network.
- Different connectivities: Ten 230 kV lines were disconnected, one line for each case in order to create different initial connectivity conditions (network configurations) for the same power system problems.
- Different combinations of on/off generators and line connectivities of the power system: One big generator and ten 230 kV lines were disconnected, one line at a time for each case.

Any one particular system problem would result in different alarms under these different initial operating conditions.

Alarm Information

Eight scenarios of completeness in alarm information were created for the tests:

- Full alarm information (All possible analog and status alarms are received.)
- Status alarm information only (Analog alarms are missing.)
- Failed telemetry from 6 stations, one at a time (All alarms are missing from a station.)

ANN Training

Only the alarm patterns with full information, reflecting only the problems under the "basic initial condition of the power system" were used in training the ANN-based IAP.

ANN Testing

All the same 100 problems were simulated under different, from the basic, initial conditions. The alarm patterns obtained from these simulation were presented by different completeness of information. Note: Even with full information presented, the list and values of the alarms in test cases are different from the training patterns.
This results in about 18,000 "previously not-seen" alarm patterns being submitted to the IAP versus approximately 100 "trained" patterns. Therefore, the relative size of the training set in comparison with the testing set is less than 1%.

Analysis of the Results of Testing

The IAP may respond to an alarm pattern with one or more problem classifications.

After submitting the 18,000 sets of alarm patterns to the IAP, the IAP responses were broken down into three main categories: i) Correct Response, ii) Incorrect Response, and iii) Inconclusive Response.

Obtaining either a correct or incorrect response means the IAP has made a determination of a power system problem. In 79% of the cases, a determination was made; and in 21% of the cases, the response was inconclusive.

The percentage of correct responses represents the accuracy of the determination. Correct responses were returned in 99.85% of the number of determined cases; and incorrect responses were obtained in 0.15% of the number of determined cases. An analysis of the inconclusive responses (i.e., more than 2 responses) showed that in 61% of the cases, the first response (highest ranked) was another system problem with a common protection zone with the submitted case. These results mean that by following the first response in the inconclusive cases, the operator would be close to the correct power system problems in about 61% of the cases. Because of this, in the cases with more than two responses, it would be reasonable to present to the operator a message such as: Uncertain, but probably Problem XYZ. It is most likely that these two problems are "members" of the "neighborhood" of the source problem.

The use of such a message will increase the number of correct actions in more than 10% of the total number of cases.

The percentage range of correct recognition for all initial conditions and for different completeness of information can be seen in the following figure.

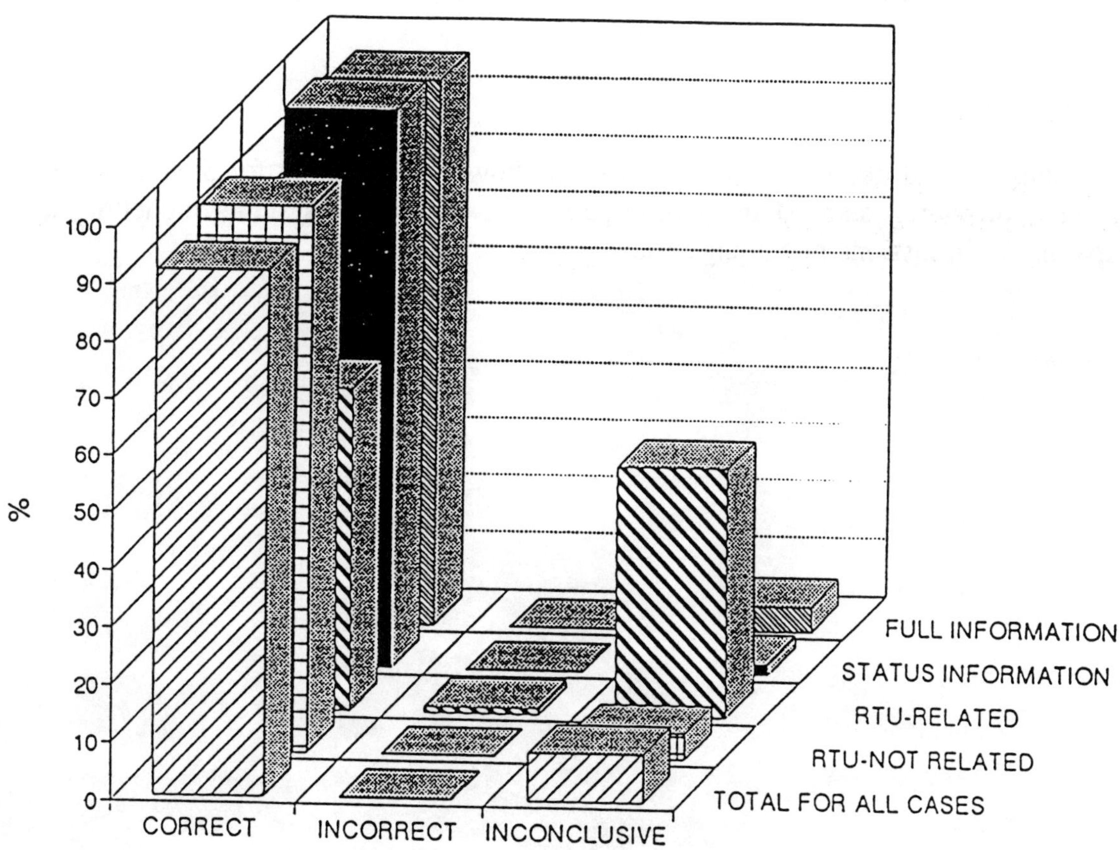

On this figure, the RTU-related cases mean power system problems which occur at a station where the SCADA RTU is out of service, and no information from this station is in the alarm pattern.

The RTU-not related cases mean power system problems which do not involve equipment failures at the stations with RTU being out of service at the time of the problem.
The results presented in this figure include the "probable" messages.

Conclusions

An Interpretive Alarm Processor (IAP) based on an Artificial Neural Network (ANN) for steady-state power systems model was developed, and limited testing was performed.

A relatively small set of typical power system problems was used for the IAP memory training. A set of testing cases, about 180 times larger than the training set, was submitted to the IAP for recognition. The throughput is 79% and the accuracy is 99.85%. If a "probable" message is used, the average percentage of correct recognition of the problems would be 91.5% of all submitted cases.

Hence, the neural network-based IAP was capable of providing a high degree of determination of steady-state power system problems with a fairly high level of accuracy, under realistic diversity of operating conditions and completeness of information.

Future development and investigation of neural network based IAP are needed and should include:

- Investigation of alarm pattern behavior under dynamic operating conditions
- Categorization of the problems into a smaller number of cases
- Investigation of the capability of recognition of a wider range of problems
- Full scale system testing with a dynamic power simulator.

Acknowledgement

Authors would like to acknowledge Florida Power Corporation for their contribution to this project, providing detailed input information for the power systems modelling and for participating in discussions pertaining to this subject.

E. Modeling and Prediction

PREDICTIVE SECURITY MONITORING WITH NEURAL NETWORKS

Dejan J. Sobajic, Yoh-Han Pao and Dennis T. Lee

Departments of Electrical Engineering and Computer Science
Case Western Reserve University
and AI WARE Inc., Cleveland, Ohio 44106

Abstract

In this paper we demonstrate the feasibility of neural network methodology for carrying out real-time analysis of power systems. This methodology will be capable of characterizing the near-term transient stability of the system, as well as perform mid-term and long-term dynamic security analyses. We distinguish between the two tasks in the following manner. In the transient stability analysis, we are principally concerned with the question of whether the system can return to a steady state. In mid-term and long-term security analyses, we are also concerned with the manner in which the final steady state is reached, whether systems performance constraints are violated on the way and whether further protective actions might be triggered unexpectedly with undesired results. Being able to predictively identify such likely transient violations of constraints would allow the operator to take appropriate anticipatory action.

1. INTRODUCTION

Power systems are systems of great complexity. In approaching any specific task with a given objective, the operator is confronted with a very large number of options in his choice of actions. The past two decades has brought about progress in the number and power of computer programs available for simulation and analysis of the behavior of power systems. In principle, therefore, an operator can predict analytically the consequences of any action that he might want to exercise. Or alternatively all possible scenarios and their outcomes can be calculated off-line ahead of time for search and retrieval when required. In practice, these strategies require <u>astronomically</u> long computational times, or astronomically long computational times <u>and</u> <u>astronomically</u> large computer memory capacities to store <u>all</u> the calculated results. Neither of these two approaches are practicable.

Humans do not operate in either of these two modes, but rely instead on a combination of experience, understanding, judgement and "common sense", aided by judicious use of a certain amount of algorithmic computation.

The aim of the practice of Artificial Intelligence technology is precisely to emulate that type of effective human behavior, but hopefully also to implement it in repeatable, non-tiring, accountable form with nearly all the virtues retained and the faults removed.

The contemplated and ongoing Expert Systems efforts cover a wide range of tasks, including alarm processing, real-time guidance of operators, transmission network design, system restoration, load forecasting, equipment status monitoring, and so on. Typically, the Expert Systems rely on a judicious combined use of heuristics and algorithms for fast and reliable scanning of all appropriate data and options, use of heuristic rules to limit choices, or to guide search, execution of algorithms to predict or to infer results, and use of heuristic rules to interpret results and to plan next action.

In this work we show that use of neural-net technology can help us to extend our operational range to shorter computational time. This is critical for the task of dynamic security analysis [1].

2. EFFECTIVENESS OF NEURAL-NET TECHNOLOGY FOR GENERATOR DYNAMICS MODELING

The objective is to demonstrate that a neural-net can act as a fast and accurate simulator of a synchronous generator dynamics. A neural-net learns by observing actual real-time generator behavior and can then predict future consequent generator transient behavior under a wide set of different circumstances.

We combine the use of unsupervised and supervised learning neural-net system paradigms to configure a neural-net system for this task [2]. We use an enhanced supervised learning procedure, developed earlier to increase learning speed and accuracy. All the networks are modeled using the Functional Link net architecture. This approach is based on our experience and on results obtained in previous related work [3,4].

We have modeled a nonlinear multimachine power system where each machine is represented with transients processes in stator, rotor and damper windings. A hysteresis loop was used to represent the magnetic properties of the iron-core. All machines and loads are connected over a transmission network defined using static (lumped) parameters [5].

Each machine variable, for example the rotor angle, is viewed as a complex nonlinear function of the history of various machine variables. We modeled that mathematically and generated time sequences which were shown to a neural-net in a real-time manner. In other words, the neural-net was taught how to anticipate what the next value of δ should be.

Having learned that, that neural-net became, in fact, a remarkably accurate representation of the complex nonlinear dynamic relationship itself. In our case, we had access to the analytical relationship because we had formulated it in the first place. But, the point is that we have shown that a neural-net can acquire, through observation, a remarkably accurate representation of complex functions. We duplicate this feature for all relevant generator variables in the system and for all generators. In this approach, we would not need to impose arbitrary assumptions and limitations in our modeling, but merely allow the neural-nets to acquire the appropriate representations.

Specifically, in our illustration we let h(x) represent the history of the variable x and we let the output of the neural-net be $\hat{\delta}$ and let δ be the actual values. Then

$$\hat{\delta} = f_\delta\,(H(x)) \tag{1}$$

where

$$H(x) = \{h(\delta), h(\omega), h(I_d), h(I_q), h(i_f), h(V_d), h(V_q), h(P_m), h(e_f), h(u_e), h(p_c)\} \tag{2}$$

and

$$h(\delta) = [\delta(k), \delta(k-1), \ldots, \delta(k-m)] \tag{3}$$

and so on. In these expressions:

δ	=	rotor angle
ω	=	rotor speed
I_d, I_q	=	stator currents in d- and q- axis circuits in the network reference frame
i_f	=	field winding current
V_d, V_q	=	stator voltages in d- and q- axis circuits, in the network reference frame
P_m	=	mechanical power
e_f	=	field winding voltage
u_e	=	exciter field voltage
p_c	=	speed changer position
k	=	discrete time index
m	=	number of data history points
f_δ	=	neural-net map.

In Figures 1 and 2, we show the neural-net based predictions of the rotor angle, terminal voltage and armature current variations following a change in the mechanical power (P_m). In the same diagram, we display the response of the mathematical model and compare that to the neural-net predictions. The quality of the neural-net modeler and predictor is highly satisfying. Computing time ratio is of the order 2000:1.

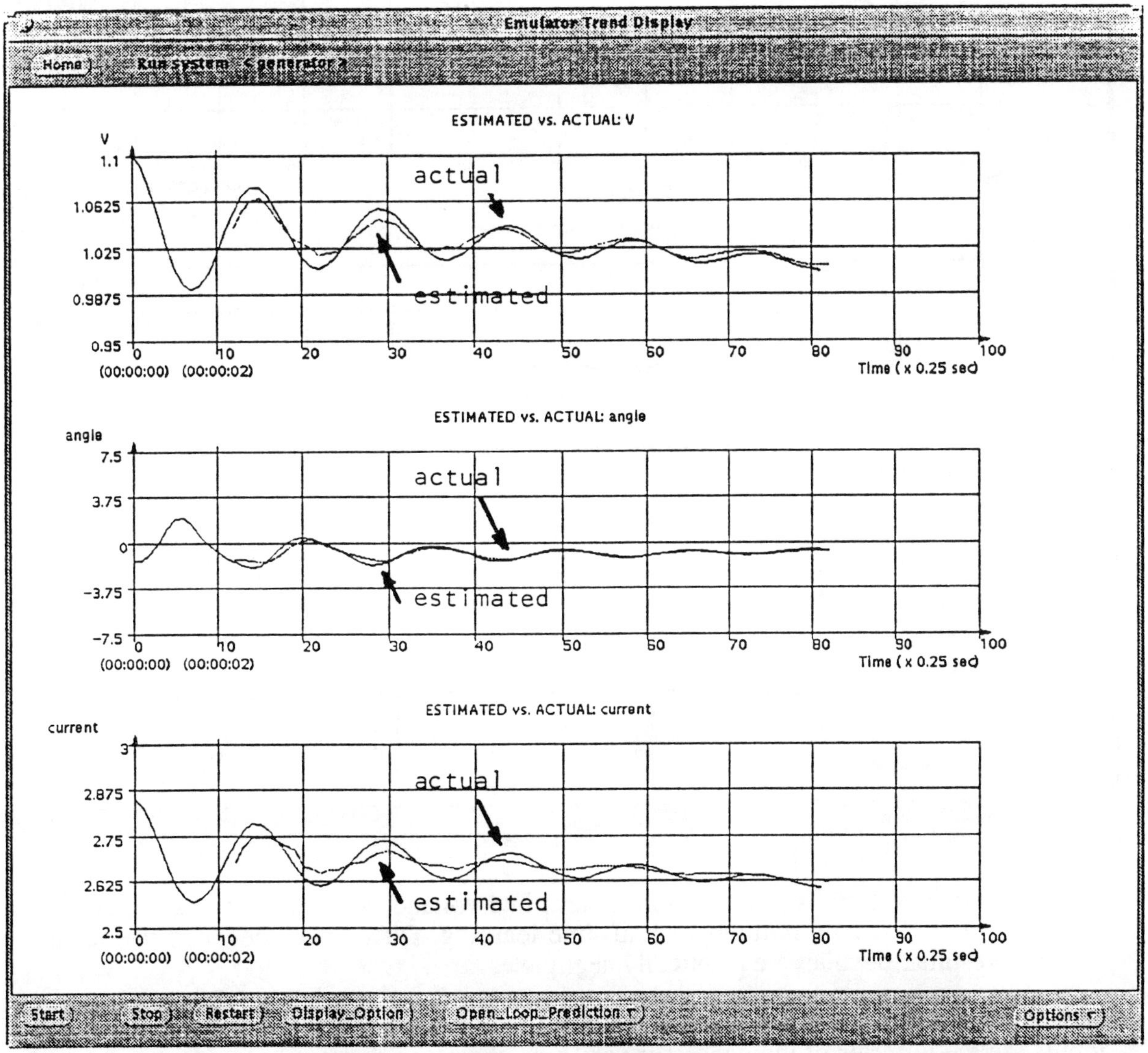

Figure 1. Time evolutions of the generator terminal voltage (V), rotor angle (δ) and armature current (I). The true values are labeled as *actual* and neural-net predictions as *estimated.* This diagram exhibits neural net predictions at the beginning of the learning process. The neural net system is set in the on-line (incremental) learning mode where the on-line measurements are processed in real-time [6].

3. COMBINED USE OF NEURAL-NET GENERATOR MODELS AND STRUCTURE-PRESERVING TRANSMISSION NETWORK MODELS

The essence of our approach depends on the fact that the reaction of the network is included in the inputs to the neural-net model of the dynamics of the generator. Thus, the

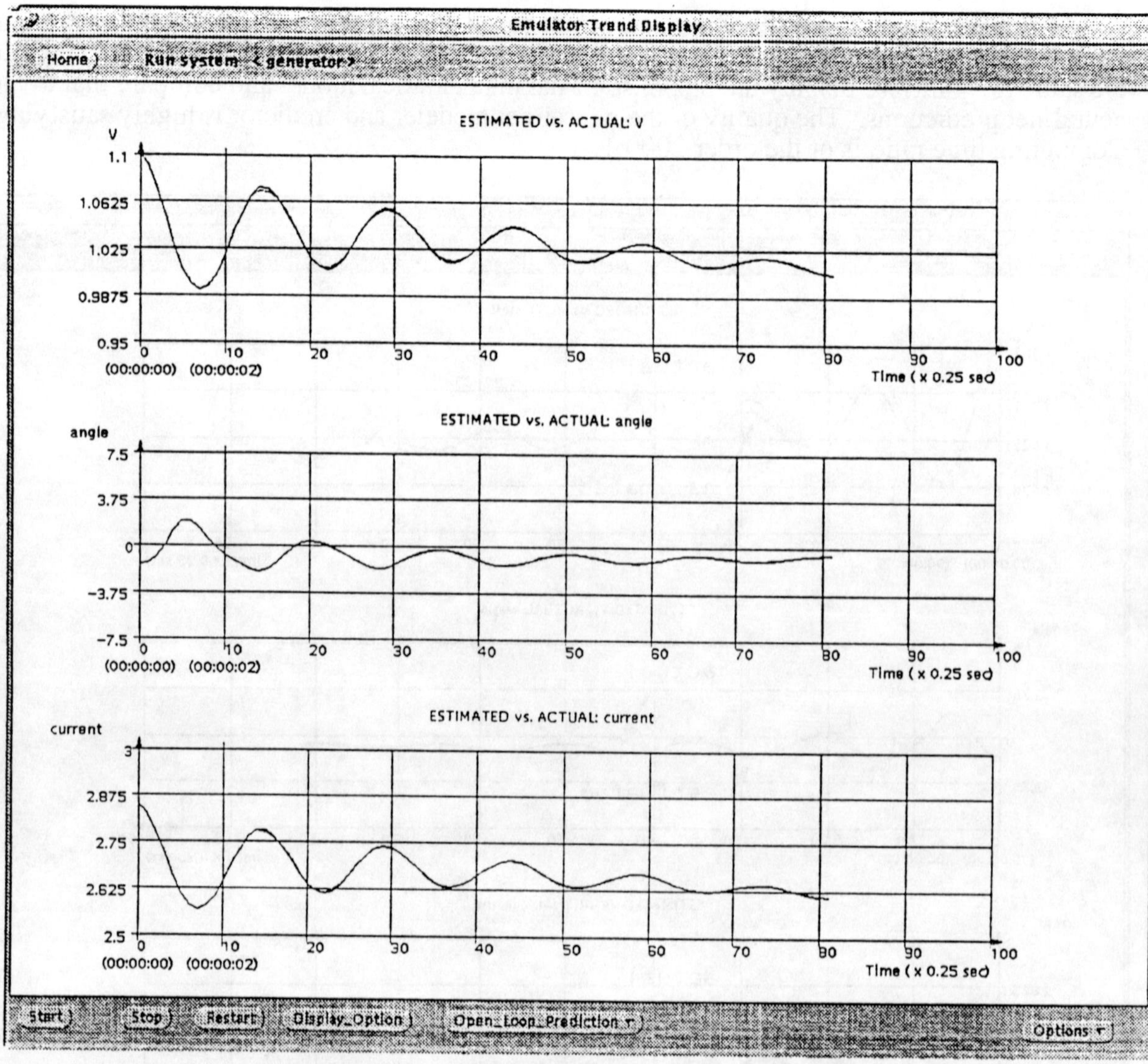

Figure 2. Time evolutions of the generator terminal voltage, rotor angle and armature current after (about) 6 minutes of real-time learning. Differences between the actual and estimated values are practically negligible.

inputs are the histories of the generator variables, the network reactions and controls, and the outputs of the neural-net are the predicted future values of the generator variables. The latter are used in network calculations which yield updated network reactions which in turn are used for the next round of predictions.

We can express these relationships for machine i as

$$y_i = f_i\,(H(x_i)) \qquad i = 1,\ldots, M \tag{4}$$

where

$$y_i = \{\delta_i\,(k+n),\ \omega_i(k+n),\ I_{di}(k+n),\ I_{qi}(k+n),\ i_{fi}(k+n)\} \tag{5}$$

and

$$H(x_i) = \{h(\delta_i),\ h(\omega_i),\ h(I_{di}),\ h(I_{qi}),\ h(i_{fi}),\ h(V_{di}),\ h(V_{qi}),\ h(P_{mi}),\ h(e_{fi}),\ h(u_{ei}),\ h(p_{ci})\} \tag{6}$$

M	=	number of machines in the system
n	=	predictor step

Each neural-net machine model is trained, stored and maintained in distributed manner with use of local data-bases. In that manner, the models are easy to maintain and may be modified and updated when necessary in real-time. Thus, another layer of parallelism can be introduced and that can be explored to effect.

The multimachine transient analysis comprises all individual neural-net machine models connected over the transmission network. We propose the use of the structure-preserving I-V network model as

$$\underline{I} = \underline{Y}_{bus} \underline{V} \tag{7}$$

or

$$\begin{bmatrix} I_d \\ I_q \end{bmatrix} = \begin{bmatrix} G & -B \\ B & G \end{bmatrix} \begin{bmatrix} V_d \\ V_q \end{bmatrix} \tag{8}$$

It should be noticed that relationship (8) is linear in terms of I and V and can be evaluated in parallel manner. This approach offers truly outstanding prospects for speed and parallelism.

4. USE OF PREDICTED DYNAMICS BEHAVIOR FOR DYNAMIC STATE ESTIMATION AND SHORT-TERM TRANSIENT ANALYSIS

The objective is to determine the effectiveness of utilizing the predictive capabilities of combined neural-net-analytic (CNA) multimachine power system models in validating real-time data sequences and estimating the state of the system, as well as in predicting short-term trend of system transients.

In real-time operation, the CNA power system model serves two different roles. In the first one, the CNA power system model is utilized to evaluate the transient system conditions. From the current system state and its history, we predict the imminent system's future. This is of importance to the system operator who may decide that stability control actions are necessary to alleviate undesired oscillations. However if predicted transients are bounded and acceptable, then the dynamic security assessment procedure is automatically enabled.

In a different role, it acts as the adaptive filter of raw dynamic measurements. Raw data usually contain a certain level of noise. Rather than predicting statistical properties of noise for signal/noise processing, we propose pattern-matching mechanism for signal interpretation and classification. This mechanism is fine tuned and is made sensitive to a noise threshold learned during training.

The unsupervised learning Functional Link net provides the basis for the pattern-matching process. The essence here is that during the training phase, honeycomb-like cluster families are established and represented by cluster prototypes. Each prototype carries a unique combination of pattern features. In the operation mode, the new pattern is matched against the cluster prototypes using a similarity rule defined by the Euclidean distance. We have developed this concept and tested it for transient system analysis in our previous work on critical clearing time estimation with good results.

5. IDENTIFICATION OF "CONTINGENCIES" EVOKED BY TEMPORARY CONSTRAINT VIOLATIONS

Coarse-scale CNA power system model can be effectively used as a multistep-ahead transient predictor in evaluating the effect of likelihood of over-limit excursions and the "contingent" actions likely to be evoked by these temporary constraint violations.

Mid-term and long-term dynamic behavior of power systems is affected by characteristics of prime mover-turbine systems, frequency and voltage regulating devices and possible state-dependent secondary events, such as the operation of protective relaying. Assuming that the aforementioned system components are appropriately defined, the CNA power system dynamic model can be effectively augmented and used for fast prediction of the mid-term to long-term transients resulting from complex interactions of current system conditions with system's operational capabilities.

The key to the successful completion of this task is the accuracy and speed of highly parallel computational organization of the CNA power system model, coupled with coarse-scale multistep-ahead prediction ability. The structure-preserving network representation provides necessary flexibility to incorporate effects of state-dependent switching phenomena.

Screening is accomplished through use of a combination of AI and neural-net techniques. Basically, in some cases there are clean cut indications that overlimit constraints will not occur. These cases are best screened out with use of rules. In other cases, neural-nets can be trained to provide estimates of the extent of constraint violation.

An AI knowledge-base will also be used to identify the corrective actions which are likely to be triggered by the constraint violations. Preventive actions to counter such undesired perturbations can also be identified and recommended, if desired by the operator.

6. CONCLUSION

In this paper we have provided a brief account of the conceptual basis for design and development of a memory-based system capable of performing dynamic security prediction in real-time. In the incremental learning mode a neural-net system learns system dynamics "on the fly". This learning mode allows for effective minimization of memory, storage and computational requirements. Numerical example is provided to illustrate the speed and accuracy of on-line modeling of synchronous generator dynamics which constitutes the critical task in practical implementation of the CNA power system model. By adjusting the coarseness scale of the CNA the speed of these computations can be tuned for real-time or even faster than real-time performance.

References

1. D.J. Sobajic and Y.-H. Pao, "Artificial Neural-Net Based Security Assessment for Electric Power Systems", IEEE/PES Winter Meeting, January, 1988, Paper #88WM-211-05.
2. Y.-H. Pao and D.J. Sobajic, "Combined Use of Unsupervised and Supervised Learning for Dynamic Security Assessment", PICA, Baltimore, MD, May 7-10, 1991.
3. D. J. Sobajic and Y.-H. Pao, "Neural Net Control in Power Systems", NSF Workshop on Applications of Artificial Neural Network Methodology in Power System Engineering, Clemson, SC, April 8-10, 1990.
4. D. J. Sobajic, Y.-H. Pao and D. T. Lee, "Autonomous Adaptive Power System Control", Special Issue on Expert System Applications in Power Systems of International Journal of Electrical Power and Energy Systems, Vol.14, No. 2/3, April/June, 1992, pp. 166-175.
5. M. Djukanovic, D.J. Sobajic and Y.-H. Pao, "Neural-Net Unstable Machine Identification using Individual Energy Functions", International Journal of Electrical Power and Energy Systems, Vol.13, No 5, October, 1991, pp. 255-262.
6. *FLN-Control Toolkit*, AI WARE Inc., Cleveland, Ohio.

Empirical Modeling in Power Engineering Using the Recurrent Multilayer Perceptron Network

Alexander G. Parlos
Department of Nuclear Engineering
Alton D. Patton
Department of Electrical Engineering
Texas A & M University
College Station, Texas 77843

Abstract

Electric power systems are characterized by increasingly more complex and more sophisticated daily operations and resource planning activities, necessitating the use of non-conventional approaches to address issues and concerns that have yet to be effectively resolved by conventional methods. This paper presents the application of the recurrent multilayer perceptron network, a hybrid (feedforward and feedback) architecture, as a model structure for empirical modeling in electric power systems. In particular, empirical modeling of operational power plant dynamics and long-term electric load forecasting is addressed. Transient response prediction results reveal the feasibility of high fidelity empirical power plant models. Furthermore, forecasting of the total energy usage by the U.S. manufacturing sector indicates that during and following significant structural changes, the proposed method appears to offer improved forecasting accuracy. Comparative forecasting results with conventional approaches are presented, demonstrating the potential benefits of empirical modeling using the recurrent multilayer perceptron as a nonlinear state-space model structure.

I. Introduction

The efficient and optimum economic operation and planning of electric power systems has occupied an increasingly more important position in the electric power industry. Prior to 1973, electric utilities in the U.S. spent approximately 20 % of their total revenues on fuel for electricity production. That number increased to 40 % of the total revenues by 1980 [12]. The amounts involved just for annual fuel expenditures in a large utility company can be on the order of a few billion dollars. Thus, even small percentage savings resulting from improved operation and planning could have significant economic impact on both the supplier and the consumer of electric energy.

Load forecasting has become one of the most important aspects of electric utility planning. The economic consequences of improved load forecasting approaches have kept development of alternate, more accurate algorithms at the forefront of electric power research. From short-term load forecasts used in the daily unit allocations [7], to mid-term load forecasts used for fuel budgeting, to the long-term load forecasts used for resource planning and utility expansion, a number of different techniques have been proposed and are currently in use [2],[3],[4]. These methods can be classified into two major categories: (1) time series approaches, treating the load pattern as a (dynamic) time series signal and predicting the future load using some time series analysis tool, and (2) regression approaches, which recognizes the strong effect of weather on electric load use, for example, and attempts to find a (usually static) functional relationship between the two. The major shortcoming of these approaches is their limited accuracy partially resulting from use of linear model structures for the former (nonlinear time series using conventional methods have been proposed, but they have not yet proven practical), and the use of static nonlinear functional relationships for the latter.

One of the main issues in power plant operations and control is the accurate, simple, and cost-effective modeling of the various plant components and of the overall plant. Modeling has become a major issue because power plants, and process systems in general, are characterized by poorly understood and not well-quantified dynamics, such as boiling phenomena and multi-phase behavior. Thus, in order to investigate and implement a number of advanced power plant automation features, it is imperative that accurate models useful in closed-loop operation be developed. Nevertheless, most of the available models are primarily based on first-principles and they are not easily adaptable for on-line use. Therefore, empirical models have been investigated in the literature, with most attempts remaining within the limits of linear model structures [5]. More recently, however, as a result of the renewed interest in Artificial Neural Networks (ANNs), a number of authors have attempted their utilization for the nonlinear identification of power plant models [6],[8],[10].

This paper presents an empirical nonlinear state-space model structure based on the technology of ANNs, and its preliminary application to problems of significant importance to power engineering. A brief discussion of the utilized empirical model structure is presented, followed by preliminary results from: (1) empirical modeling of power plant transients, and (2) long-term electric load forecasting. The paper ends with a brief summary and conclusions.

II. The Recurrent Multilayer Perceptron Network

In this study a recently developed hybrid (feedforward and feedback) ANN architecture, namely a Recurrent Multilayer Perceptron ($RMLP$), is used as the empirical model structure [8]. The major contribution of this development has been the improved modeling accuracy in the presence of noise and during multi-step ahead prediction, as compared to a purely feedforward or a purely feedback ANN architecture. For an $RMLP$ composed of $\mathcal{L}$ layers, with each layer containing $N(\ell)$ nodes, for $\ell = 1, ..., \mathcal{L}$, the response of its i-th node located at its ℓ- th layer is governed by the following equation:

$$x_{[\ell,i]}(k) = f_{[\ell]}(\sum_{j=1}^{N(\ell)} w_{[\ell,j][\ell,i]} x_{[\ell,j]}(k-1) + \sum_{j=1}^{N(\ell-1)} w_{[\ell-1,j][\ell,i]} x_{[\ell-1,j]}(k) + b_{[\ell,i]}), \qquad (1)$$

where $x_{[\ell,i]}(k)$ is the output signal of the i-th node at the ℓ-th layer at time step k; $b_{[\ell,i]}$ is the bias to the i-th node at the ℓ-th layer, provided that the node has bias; and where $w_{[\ell-1,j][\ell,i]}$, and $w_{[\ell,j][\ell,i]}$ denote the feedforward and feedback network weights. Furthermore, $f_{[\ell]}(\cdot)$ is the layer discriminatory function. It should be noted that the only inputs to the network are the manipulated input signals and any measurable disturbances acting upon the dynamic system being identified. Supervised training of this network is performed using a dynamic gradient descent or a static learning algorithm [8],[9].

III. Applications to Power Engineering

The potential applications of ANNs to power system engineering are numerous, and the possibilities have been widely documented in the literature [1]. Specifically, in the following paragraphs the applicability of the *RMLP* architecture to two problems relevant to power engineering is described.

III.1. Empirical Modeling of Power Plant Dynamics

The effectiveness of the *RMLP* architecture in identifying complex nonlinear systems has been demonstrated, via the accurate real-time prediction of a U-Tube Steam Generator (UTSG), a steam boiler, transient response [9]. UTSGs are complex dynamic systems encountered in the operation of certain nuclear power plants. They are characteristic of complex nonlinear system amenable to analysis. They are approximated by high dimensional, nonlinear models with unstable and non-minimum phase open-loop responses. Thus, only closed-loop identification is feasible. Furthermore, no simple known analytic forms of these approximate models is available, because they include a number of experimentally determined correlations and look-up tables.

An *RMLP* with five input nodes, three output nodes and two hidden layers, containing 10 and 12 recurrent nodes, respectively, is used for identifying a single ANN model for the entire normal operating regime of a UTSG [9]. The input and output layer nodes of the *RMLP* have linear discriminatory functions, whereas the hidden layer nodes use hyperbolic tangent. Training is performed using a dynamic gradient descent algorithm, developed for enhanced supervised learning in *RMLP*s [8]. Following off-line learning with a data set containing a number of positive and negative step and ramp inputs, the resulting empirical model is tested using numerous transients entirely different than the

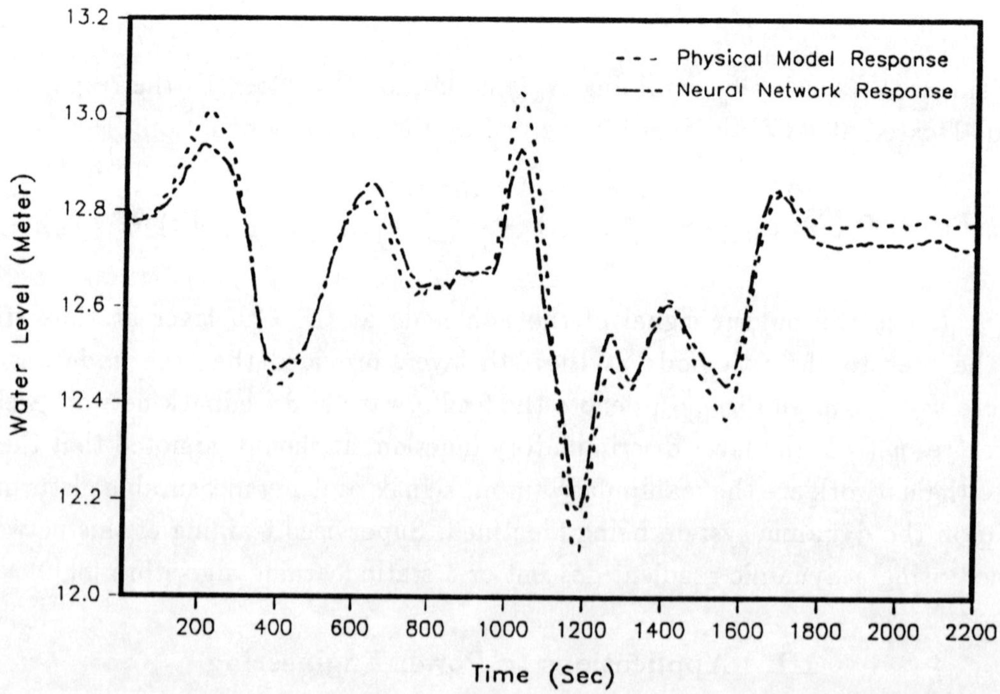

Fig. 1. Comparative UTSG Water Level Response Using a Physical and an *RMLP* Model.

ones included in the training set. The water level (an output) response of a plant validated model and of the identified *RMLP* model for a hot standby to full power ramp-up is shown in Fig. 1. The prediction accuracy is quite acceptable, in view of the fact that the *RMLP* weights have been frozen to their off-line values, i.e. no on-line learning has been allowed. On-line learning has been found to further improve the network predictive performance.

III.2. Long-Term Load Forecasting

As an example of long-term electric load forecasting, the total energy usage of the U.S. manufacturing sector has been forecasted using historical data. The key explanatory variables are the index of total industrial production, the price of electricity in 1972 cents per kWh, and, the average price of oil and gas in 1972 cents per $MBtu$. Two approaches were used in this forecasting study: (1) a double logarithmic regression model in which the kWh per unit industrial production is related to the industrial production, the price of electricity, and the price of oil and gas as follows: (kWh/industrial production index) = A_0 ·(industrial production index)$^{\beta_1}$ ·(electricity price)$^{\beta_2}$ ·(oil & gas price)$^{\beta_3}$ [11], and, (2) an empirical model based on an *RMLP* with the same three inputs and one output, as the first model.

Use of the *RMLP* architecture as a load forecasting tool involves pre-processing of the relevant historical data, training for estimation of the network weights while preventing

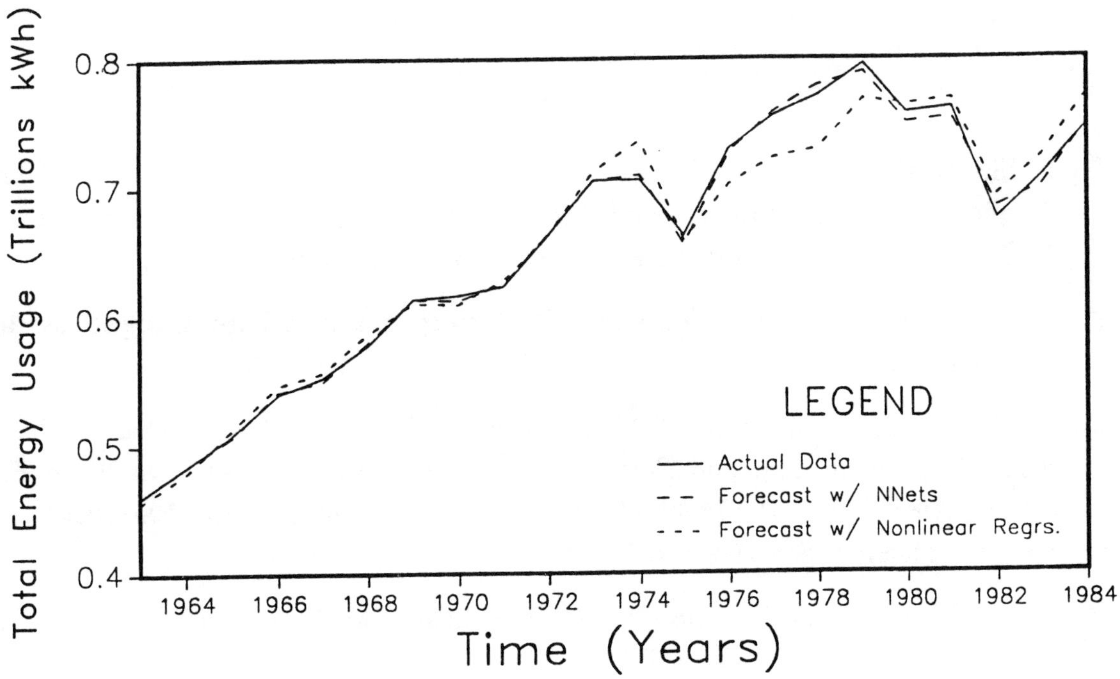

Fig. 2. U.S Manufacturing Sector Total Energy Consumption.

overparametrization, and forecast model validation to test the generalization properties of the network. In addition to the input and output layer nodes, the *RMLP* is composed of 1 hidden layer containing 12 recurrent nodes which are also connected with cross-talk. The training set consisted of the 1963 to 1978 annual data, whereas the test set included the 1979 to 1984 annual data. The *RMLP* network was trained using a static learning algorithm for 15,000 iterations [9]. Following the off-line learning, unconditional forecasting is performed using the aforementioned test set. The results of this comparative forecasting study are shown in Fig. 2.

IV. Summary and Conclusions

Empirical modeling using the *RMLP* network has been applied to two problems of interest in power engineering. In particular, transient modeling of power plant dynamics, and long-term electric load forecasting is addressed. Transient response prediction demonstrates the high fidelity of the resulting *RMLP* empirical model. Furthermore, it is apparent that the proposed method offers no substantial improvement in forecasting accuracy when trends do not change significantly. However, during and following significant changes in external conditions, the proposed empirical model structure appears to offer improved forecasting accuracy. These comparative results appear quite encouraging, requiring further investigation.

References

[1] El-Sharkawi, M. A., R. J. Marks, and S. Weerasooriya, "Neural Networks and Their Applications to Power Engineering," in *Control and Dynamic Systems,* Vol. 41, Part 1, C. Leondes, Edt., Academic Press, New York, NY, 1991.

[2] Goodrich, R. L., *Applied Statistical Forecasting,* Business Forecast Systems, Inc., Belmont, MA, 1990.

[3] Grady, W. M., L. A. Groce, T. M. Huebner, Q. C. Lu, M. M. Crawford, "Enhancement, Implementation, and Performance of an Adaptive Short-Term Load Forecasting Algorithm," 91 SM 448-1 PWRS, paper presented at the 1991 IEEE/PES Summer Meeting, San Diego, CA.

[4] Gross, G., and F. D. Galiana, "Short-Term Load Forecasting," *Proc. of the IEEE,* Vol. 75, No. 12, December 1987.

[5] Ljung, L. and S. Gunnarsson, "Adaptation and Tracking in System Identification - A Survey," *Automatica*, Vol. 26, No. 1, pp. 7-21, 1990.

[6] Narendra, K.S. and K. Parthasarathy, "Identification and Control of Dynamical Systems Using Neural Networks," *IEEE Transactions on Neural Networks,*, Vol.1, No.1, March 1990.

[7] Park, D.C., M.A. El-Sharkawi, R.J. Marks II, L.E. Atlas, and M.J. Damborg, "Electric Load Forecasting Using an Artificial Neural Network," *IEEE Transc. Power Systems,* Vol. 6, No. 2, 1991, pp. 442-449.

[8] Parlos, A., A. Atiya, K. Chong, W. Tsai, and B. Fernandez, "Recurrent Multilayer Perceptron for Nonlinear System Identification", *Proc. IJCNN-91*, Seattle, Washington, 1991.

[9] Parlos, A.G., A. Atiya, K. T. Chong, and W.K. Tsai, "Nonlinear Identification of Process Dynamics using Recurrent Neural Networks," *Nuclear Technology*, January 1992, pp. 79-96.

[10] Qin, S.-Z, H.-T. Su, and T. J. McAvoy, "Comparison of Four Neural Net Learning Methods for Dynamic System Identification," *IEEE Transactions on Neural Networks*, Vol. 3, No. 1, pp. 122-130, 1992.

[11] Stoll, H. G., *Least-Cost Electric Utility Planning,* John Wiley & Sons, New York, NY, 1989.

[12] Texas Public Utilities Commission, "Monthly Fuel Reports: Annual 1989".

Modeling and Identification with Neural Networks

Tariq Samad
Honeywell SSDC
3660 Technology Drive
Minneapolis, MN 55418

Abstract

This paper illustrates the diversity of modeling and identification applications in the electric power industry and discusses how neural networks can be used for addressing them. Two different approaches are described. First, a neural network can be trained on application data. The trained network then implements an accurate nonlinear "black-box" model of the phenomenon of interest. Second, a neural network can be trained in simulation as a model structure identifier and/or a parameter estimator. In this case, the network outputs aspects of a parametric model. Some topics of ongoing research in neural network modeling and identification are briefly noted.

Examples of Modeling Applications

Modeling problems are pervasive in the electric power industry. Examples include the following:

- How does the demand for electric power vary with the environmental temperature, the day of the week, and the time of day?

- How does plant heat rate depend on factors such as condenser backpressure and circulating water temperatures?

- How are slagging characteristics in coal-fired boilers affected by coal quality and load conditions?

- How does fuel flow impact main steam pressure dynamics in boilers?

Questions such as these can sometimes be answered by undertaking analyses based on physical and chemical laws. This is often not feasible for complex real-world applications, either because of our incomplete understanding of the phenomena or because the resulting physical models tend to be computationally demanding.

In many cases of practical interest, we must therefore resort to system identification (Ljung, 1987): models must be constructed primarily based on experimental data. Input/output data relevant to the application must be collected and analysed by automated model-fitting procedures. Data can be gathered either during regular plant or system operation or during specially conducted experiments.

Neural networks are fast becoming a preferred approach for system identification. This contribution briefly discusses two different neural-network-based modeling approaches and identifies some topics of ongoing research.

Black-Box Neural Network Models

The development of a neural network model from input/output data is an example of supervised learning. The data is referred to as a "training set" of examples, and the weights of the neural network are iteratively modified until an acceptable fit is achieved. Weight adjustment is usually performed with simple gradient-descent schemes.

As an example of a training set, consider the load prediction problem (Atlas et al., 1990). For this problem, training examples could be collected on an hourly basis over an extended time period. Each example might consist of the temperature reading and load at a given time of day, the time of day itself, the day of the week, and possibly temperatures and loads at previous hours (all of these would consitute the training input), together with the observed load at the next hour (the "desired output" of the network).

The use of neural networks for model-fitting is closely related to statistical methods (White, 1989; Geman, Bienenstock, and Doursat, 1992). In fact, neural networks are essentially a nonparametric (black-box), nonlinear statistical modeling approach with some significant advantages over many conventional statistical approaches: Neural networks are feasible for parallel hardware implementation, they are relatively easy to use in that extensive domain knowledge is not essential, and they are expressively powerful. We now know that the popular neural network models are "universal approximators"—an appropriate neural network can approximate any function to any desired degree of accuracy.

This combination of convenience and functionality is not obtained without cost. In particular, actually realizing a sufficiently accurate neural network model requires some care. It is often easy to obtain good performance on the training set, but not as easy to ensure that the network generalizes well on novel data. Good generalization is the true criterion of network performance, and the nonparametric nature of neural network models implies that it is not automatic.

Two approaches are commonly used for increasing the likelihood of good generalization. First, during training, accuracy on a "test set" is monitored. The test set is not directly used for updating network weights (e.g., gradients are computed only over the training set), but it is periodically used to exercise the network. Training is then terminated when accuracy on the test set starts to deteriorate, before "over-fitting" on the training set is encountered (Fig. 1).

Second, the smallest network capable of high accuracy is used. Network size is best measured in terms of the number of weights—the free parameters in the network. Smaller networks are more commensurate with the usually limited amount of data available for training. Large networks are "over-parametrized".

The final result of neural network training on application data is a black-box model. Such a model can be used in various ways:

• Prediction. A neural network can be trained to predict future values of relevant quantities. These predictions can then be used for planning and scheduling purposes. For example, with reliable predictions of the demand for electric power over hourly or daily timescales, distribution networks could be configured and units could be near-optimally committed. As another example, a neural network could be trained, using data collected from plant sensors, to predict ash deposition on boiler heat-transfer surfaces. An accurate ash deposition prediction model can be the basis for an on-demand soot-blowing advisory system and for determining optimized soot-blowing schedules.

• Process control. Conventional advanced control algorithms use process models to compute optimal control actions. The process models currently employed by these "model predictive control" algorithms are linear. With neural network models developed from dynamic input/output process data, nonlinear model predictive control can be realized. For example, if the actual nonlinear dynamical relationship between fuel flow and main steam pressure could be accurately modeled, tighter and more efficient control of megawatt output could be achieved. See (Samad, 1991) for an overview of other uses of neural network models in control systems.

• Sensitivity analysis. A trained network can be readily analysed to determine input/output sensitivities. Sensitivity analysis can help us identify how variables corresponding to the network output depend upon the various inputs. Unlike with linear models, the complex dependence of sensitivities on operating conditions can be captured.

• Plant or system wide optimization. Neural network models can be developed at various levels of abstraction. At higher levels, models of plants or systems of plants can be considered. For example, the effect of plant inputs on heat rate can be modeled (Guo and Uhrig, 1992). The resulting models can then be used to optimize an appropriate criterion (e.g., plant efficiency, cost savings).

Parametric Identification with Neural Networks

While black-box modeling is the most popular application of neural networks, others have also been developed and deployed. For example, neural networks can be used to identify physical, or physically meaningful, (and thus non-black-box) models, in particular for control purposes. With such models, it is common to distinguish between two identification aspects:

1. Model structure identification. For dynamic process models, for example, model structure refers typically to the form of (linear or nonlinear) differential or difference equations.

2. Parameter estimation. Given a structure, values for the associated parameters must be determined. For example, relevant parameters for a boiler model can include heat transfer coefficients of waterwalls and other interior surfaces.

In current practice, the model structure is usually identified either by a user or through some trial-and-error method. Parameter values are typically estimated by solving an optimization problem: values are sought for which the error between the model output and the actual system output is minimized. When, as is often the case, the parameters depend nonlinearly on the experimental data, the optimization problem is a nonlinear one and iterative algorithms must be used.

Neural networks trained through supervised learning can be used for both these aspects (Konar, Samad, and Foslien, 1992). As structure identifiers, they can be trained to select elements of a model structure from a predetermined set. These elements can then be composed into a legal structure. This requires that the space of likely model structures be known in advance. If this constraint is satisfied, neural network structure identifiers can thus automate what is currently a manual, expertise-intensive aspect of system identification.

Neural networks can also be trained to generate parameter values for a given structure or set of structures (Samad and Mathur, 1992), dramatically improving estimation times relative to iterative optimization algorithms. The iterative process, each iteration of which is computationally involved, is replaced by a feed-forward "one-shot" operation that is computationally simple.

The training of both neural network structure identifiers and parameter estimators is done off-line, using a generalized simulation system (Fig. 2). Thus no training is needed on actual data. Since data for network training is easily generated, generalization is facilitated. Generalization is problematic if data is sparse. With a virtually infinite supply of relatively cheap data, over-fitting and over-parametrization concerns are substantially alleviated.

Honeywell has developed a Hybrid Parametric Identifier that integrates neural networks for structure identification and parameter estimation with an iterative parameter identification algorithm (Konar, Samad, and Foslien, 1992). The target models in this case are linear differential equations, to be used in conjunction with a conventional model predictive control algorithm. Structure identification consists of determining the presence or absence of features such as delay or dead-time, an integral, one or two lags, inverse response, etc. The parameters are meaningful quantities such as the time delay, process gain, and process time constants (even for linear systems, these parameters are nonlinear functions of input/output data). Fig. 3 shows an identified model for a duct static pressure system. The model is a linear approximation to a mildly nonlinear process—hence the small errors in the steady-state values.

Topics of Ongoing Research

Applications of neural networks to modeling and identification problems are no longer purely a research topic. At the same time, research is continuing on further improvements to the technology. We note below three important topics of ongoing work:

• Dynamic neural networks. Most neural network applications rely on "feedforward" (algebraic) network structures. Thus the network itself has no nontrivial dynamics (instead, dynamical information is explicitly provided by including past values as network input). By incorporating dynamical elements into the network itself, it is anticipated that the use of neural networks for modeling dynamical systems will be greatly facilitated.

• Incremental learning. Current neural network learning algorithms are best used off-line. Data relevant to a modeling application is collected and the network fully trained before being put to use. An important research topic is the development of on-line learning techniques that allow rapid yet stable adaptation of the network.

• Neural network design. Whereas effective, automatic techniques for "learning" the weights of a given network architecture exist, the appropriate architecture for a modeling problem is still determined by manual trial and error. This is a time-consuming process, and it rarely produces network designs that are optimized for the application. A promising approach for solving this neural network design problem is genetic-algorithm-based optimization (Harp and Samad, 1992).

Conclusion

We have identified two approaches to using neural networks for modeling and identification applications. The two differ in the nature of the model produced. By training a neural network on data taken from an application of interest, a black-box model can be developed. In this case, the trained neural network is the model itself. In the second approach, a pre-trained neural network takes application data as input and identifies structural features and/or parameter values for a parametric model. Here the neural network outputs the model. However produced, models can be used for purposes such as prediction, control, and plant optimization.

Although research issues still remain to be addressed, this technology is now mature for practical application. Dramatic performance improvements and cost savings have been achieved in domains as diverse as financial forecasting and process control, and there appears to be no reason why the utility industry cannot reap similar rewards.

References

Atlas, L. et al. (1990). Performance comparisons betwnee backpropagation networks and classification trees on three real-world applcations. In *Advances in Neural Information Processing Systems 2*, D.S. Touretzky (Ed.), Morgan Kaufmann Publishers.

Foslien, W., A.F. Konar, and T. Samad (1992). Optimization with neural memory for process parameter estimation. *Applications of Artificial Neural Networks III*, S.K. Rogers (Ed.), SPIE Proc. 1709.

Geman, S., E. Bienenstock, and R. Doursat (1992). Neural networks and the bias/variance dilemma. *Neural Computation, 4*, pp. 1-58.

Guo, A. and R.E. Uhrig (1992). Use of artificial neural networks to analyze nuclear power plant performance. *Nuclear Technology, 99*, pp. 36-42.

Harp, S.A. and T. Samad (1992). Optimizing neural networks with genetic algorithms. *Proc. 54th American Power Conference.*

Konar, A.F., T. Samad, and W. Foslien (1992). Hybrid neural-network/algorithmic approaches to system identification. *Preprints of the IFAC Symp. on DYCORD+ '92*, College Park, Md.

Ljung, L. (1987). *System Identification: Theory for the User.* Prentice-Hall.

Samad, T. (1991). Neural networks for control: an overview. *Proc. ISA/91*, Anaheim, Calif.

Samad, T. and A. Mathur (1992). Parameter estimation for process control with neural networks. *International Journal of Approximate Reasoning.* (In press.)

White, H. (1989). Learning in artificial neural networks: a statistical perspective. *Neural Computation, 1*, pp. 425-464.

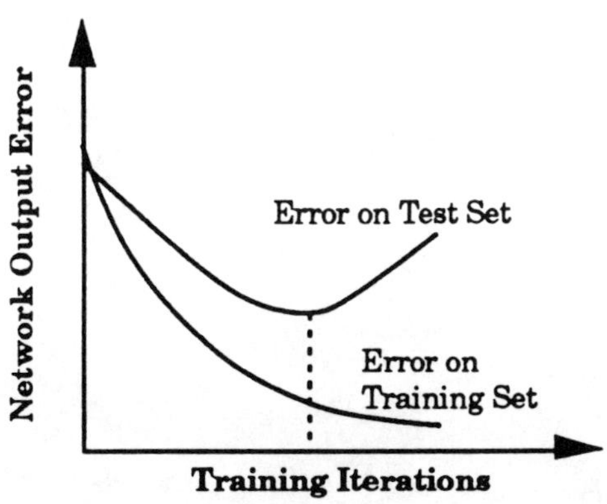

Fig. 1. Minimizing error on the training set can be counterproductive. A separate Test Set is often used to decide when to terminate training.

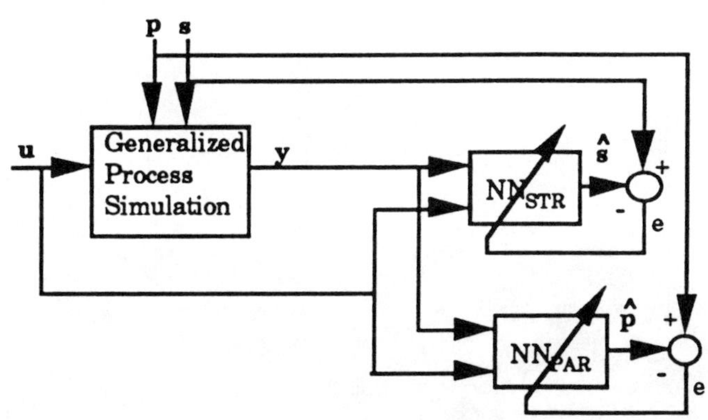

Fig. 2. Training neural network structure identifiers and parameter estimators for parametric modeling. Training is done in simulation, with training example generation integrated with neural network learning.

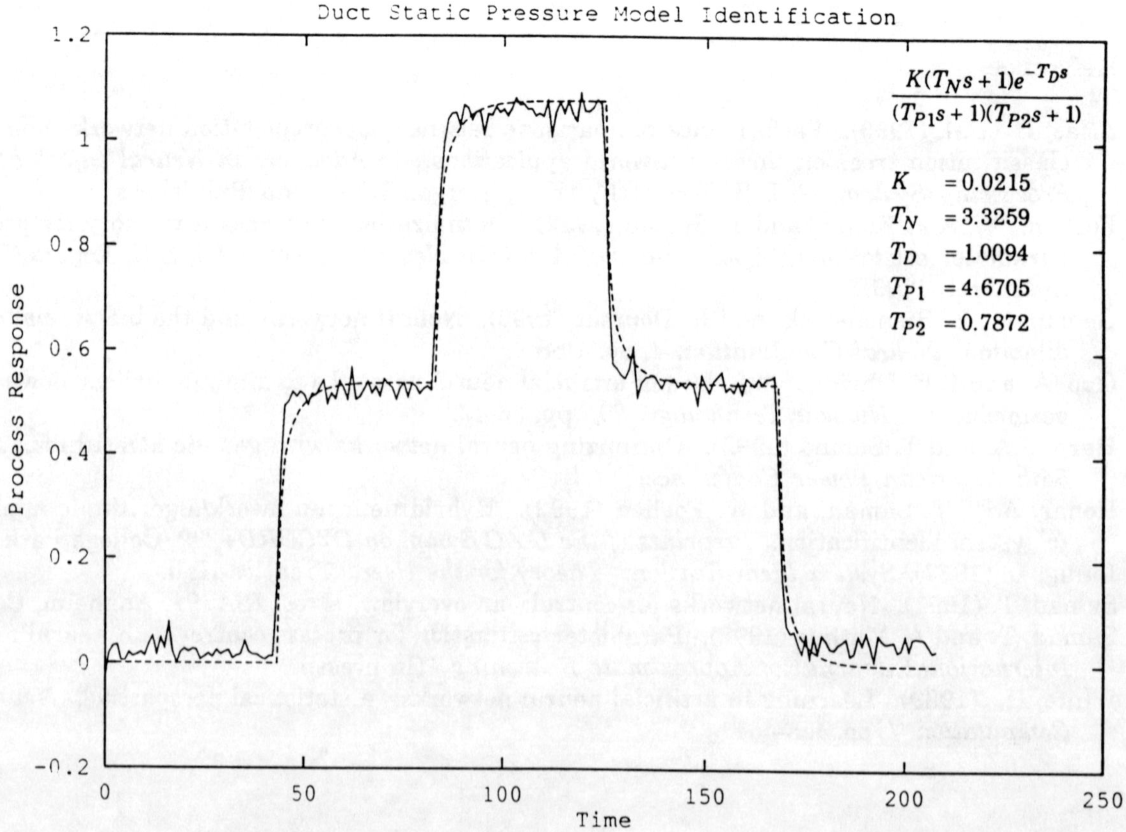

Fig. 3. Parametric model identification for a static pressure system. The identified model—structure and parameter values—is shown at top right. The dashed line is the model fit: a linear approximation to a slightly nonlinear process.

Autoregressive Neural Network Prediction: Learning Chaotic Time Series and Attractors.

Eric A. Wan *

Abstract

A neural network architecture, which models synapses as Finite Impulse Response (FIR) linear filters, is discussed for use in time series prediction. Various examples of chaotic time series are used to show the performance of the networks. Phase-space plots of the network dynamics are given to illustrate the reconstruction of underlying chaotic attractors. An example taken from the *Santa Fe Institute Time Series Prediction Competition* is also presented.

Introduction

The goal of time series prediction or *forecasting* can be stated succinctly as follows: given a finite sequence $y(1), y(2), \ldots y(N)$, find the continuation $y(N+1), y(N+2)$... The series may arise from the sampling of a continuous time system, and be either stochastic or deterministic in origin. The standard prediction approach involves constructing an underlying model which gives rise to the observed sequence. In the oldest and most studied method, a linear autoregression (AR) is fit to the data:

$$y(k) = \sum_{n=1}^{T} a(n)y(k-n) + e(k) = \hat{y}(k) + e(k). \tag{1}$$

This AR model forms $y(k)$ as a weighted sum of past values of the sequence. The single step prediction for $y(k)$ is given by $\hat{y}(k)$. [1].

Neural networks may be used to extend the linear model to form a *nonlinear* prediction scheme. The basic form $y(k) = \hat{y}(k) + e(k)$ is retained; however, the estimate $\hat{y}(k)$ is taken as the output $\mathcal{N}$ of a neural network driven by past values of the sequence. This is written as:

$$y(k) = \hat{y}(k) + e(k) = \mathcal{N}[y(k-1), y(k-2), \ldots y(k-T)] + e(k). \tag{2}$$

Note this model is equally applicable for both scalar and vector sequences. The use of this nonlinear autoregression is motivated by two key concepts. First off, *Takens Theorem* [2] implies that for a wide class of deterministic systems, there exist a nonlinear finite embedding of the form $y(k) = g(y(k-1), y(k-2), \ldots y(k-T))$, which predicts the series exactly. Secondly, it has been shown [3, 4] that a feedforward neural network $\mathcal{N}$ with an arbitrary number of neurons is capable of approximating any *uniformly* continuous function. These arguments, taken together, imply that a neural network should be capable of implementing an optimal predictor.

Network Architecture and Prediction Configuration

The neural network architecture used in this study is the FIR (Finite Impulse Response) Neural Network. An FIR Neural Network resembles a standard feedforward network but replaces each synapse with

*The author is with the department of Electrical Engineering, Stanford University, Stanford, CA 94305-4055. This work was sponsored by NASA under contract N0014-86-K-0178 and EPRI under contract RP8010-13.

[1]The more general AutoRegressive Moving Average (ARMA) model includes a weighting on past samples of the error and is written as $y(k) = \sum_{n=1}^{T_n} a(n)y(k-n) + \sum_{l=1}^{T_l} c(l)e(k-1)$. For long term prediction, only the AR term becomes relevant (see [1]).

a adaptive FIR linear filter. A complete description of the architecture along with the training algorithm can be found in [6].

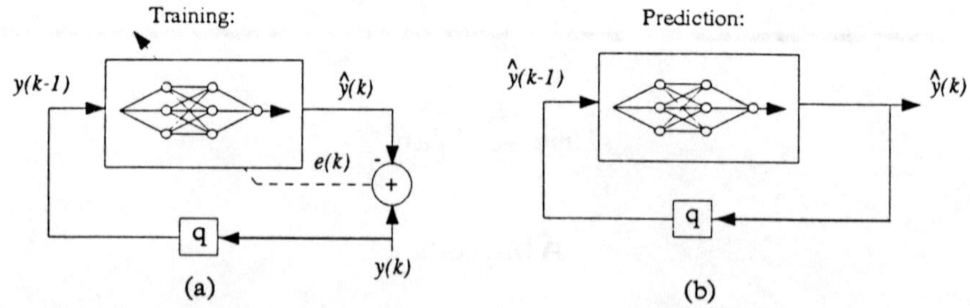

Figure 1: Network Prediction Configuration: (a) Training. (b) Iterated Prediction.

Fig. 1a illustrates the basic predictor training configuration. At each time step, the input to the FIR network is the known value $y(k-1)$, and the output $\hat{y}(k) = \mathcal{N}_q[y(k-1)]$ is the single step estimate of the true series value $y(k)$. Our model construct is thus: $y(k) = \mathcal{N}_q[y(k-1)] + e(k)$. Since the FIR network has only a finite memory of past samples, $\mathcal{N}_q[y(k-1)]$ is equivalent to a finite nonlinear regression on $y(k)$. During training, the squared error $e(k)^2 = (y(k) - \hat{y}(k))^2$ is minimized by using the *temporal backpropagation* algorithm to adapt the network [5]. Note we are performing *open-loop* adaptation; both the input and desired response are provided from the known training series.

Once the network is trained, long-term *iterated* prediction is achieved by taking the estimate $\hat{y}(k)$ and feeding it back as input to the network: $\hat{y}(k) = \mathcal{N}_q[\hat{y}(k-1)]$. This closed-loop system is illustrated in Fig. 1b. The system can be iterated forward in time to achieve predictions as far into the future as desired.

Examples of chaos prediction and attractor reconstruction

In the examples that follow, networks of various dimensions are trained on time series generated by chaotic processes. The long term iterated prediction of the networks are then analyzed.

Hénon Map: For our first example, consider the rather benign looking Hénon equations: $x_{n+1} = 1.0 - ax_n^2 + y_n$, $y_{n+1} = bx_n$, with $a = 1.4$ and $b = 1.3$[7]. The iterated time series x_n is shown if Fig. 2a. The phase-plot x versus y reveals a remarkable structure called a *strange attractor* (see Fig. 2b). Increased magnification of the attractor would reveal ever finer detail in a fractal like geometry.

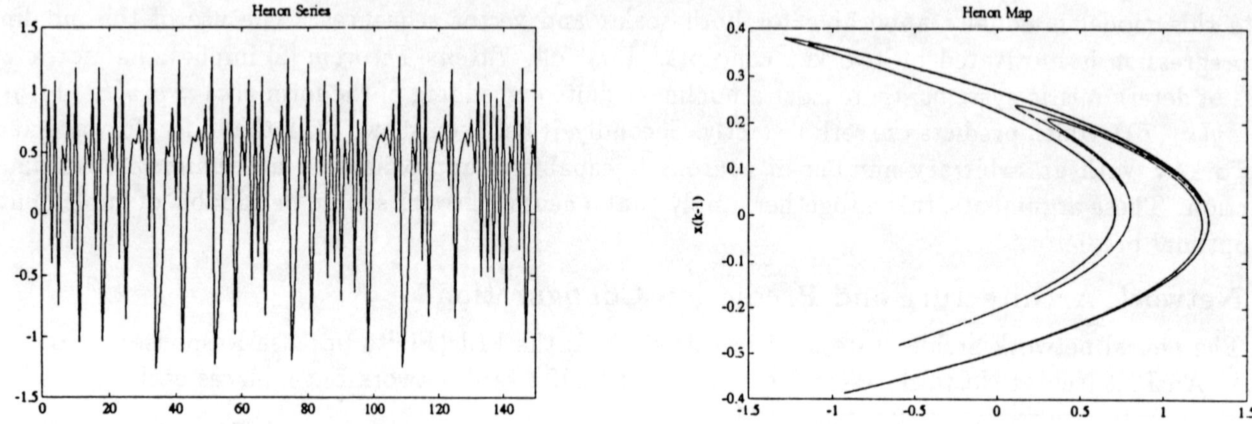

Figure 2: (a) Hénon Series. (b) Strange Attractor.

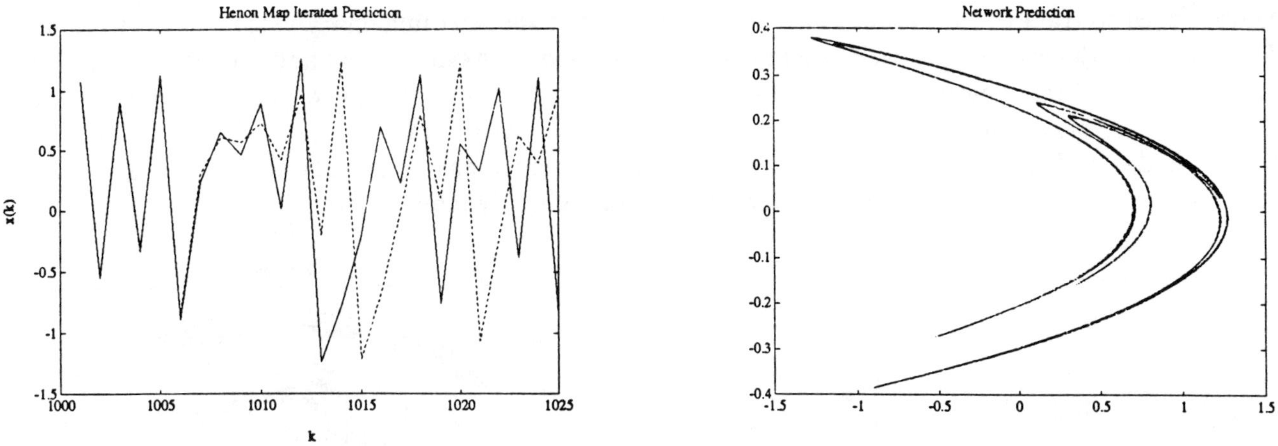

Figure 3: (a) Time Series Prediction (solid line is prediction, dashed line is actual series). (b) Network Attractor.

An FIR network was trained on the series using single step predictions. (Dimensions: 1x12x12x1 nodes with 2:1:1 order FIR filters in each layer)[2]. The iterated prediction versus the true series is shown in Fig. 3a. As can be seen, the prediction is exceptionally accurate starting out, but then diverges after around 10 time steps. This divergence is unavoidable and reveals one of the fundamental tenants of chaos theory. The network system, however, can still be iterated thousands of time steps into the future and then used to construct its corresponding attractor. As seen in Fig. 3b, the original Hénon attractor emerges indicating that the network has indeed captured the underlying dynamics.

Hénon Map Plus Noise: As a second example, we again use the Hénon map, but this time add Gaussian noise to the training series. This has the result of smearing the original attractor as shown in Fig. 4a. The network never sees the noise free attractor, yet when the converged network is iterated forward in time the resulting attractor forms sharper representation of the original attractor (see Fig. 4b).

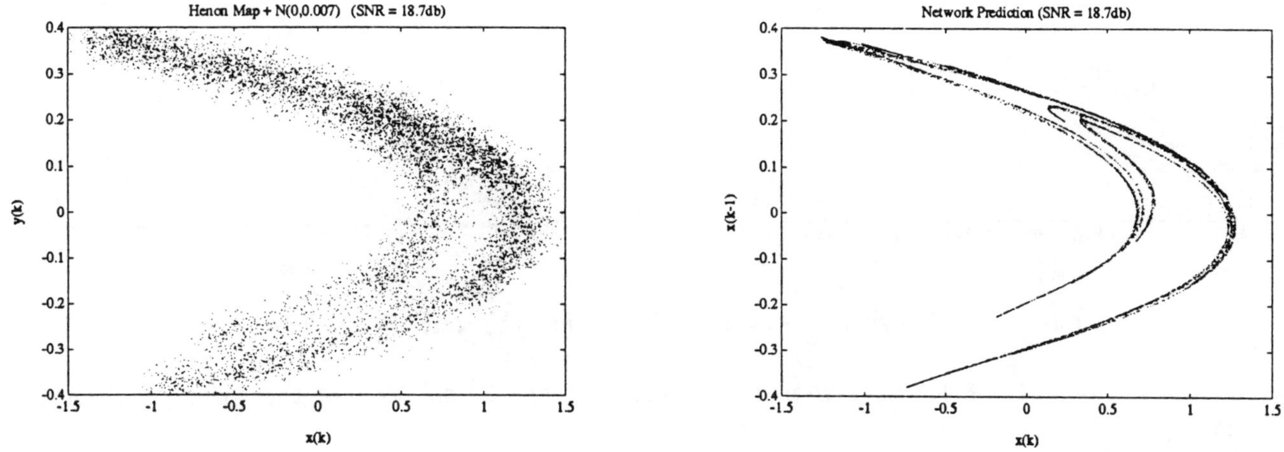

Figure 4: (a) Hénon Attractor plus Noise. SNR = 18.7db. (b) Resulting Network Attractor.

Ikeda Map: A more complicated equation corresponding the the plane-wave interactivity in an optical ring laser is the Ikeda map: $z_{n+1} = a + Rz_n exp\{i[\phi - p/(1 + |z_n|^2)]\}$, with $a = 1.0, R = 0.9, \phi = 0.4, p = 6$ [8]. The series corresponding to the imaginary part of z is shown in Fig. 5a. The phase-plot of real vs. imag. is shown in Fig. 5b. To make the problem even more difficult for the network, only the imaginary

[2]While a smaller network with lower embedding dimension would have worked for this problem, in general the actual dimensions of the system is not known in advance. Determining appropriate embedding dimensions for the network is rich topic of research and is beyond the scope of this paper.

sequence is used to train the network. Thus, in order to predict the next imaginary value, the network must implicitly extract the real part as well as learn the interrelation between the real and imaginary sequences. To visualize just the imaginary sequence, we show in Fig. 6 the plot of $Im[z_k]$ versus $Im[z_{k-1}]$. The image appears as a convoluted projection of the original attractor. After training, the network (dimensions: 1x13x13x1 nodes with 10:5:5 taps) was iterated forward in time to form the image shown in Fig. 6. Again it is clear that the network has captured much of the underlying dynamics.

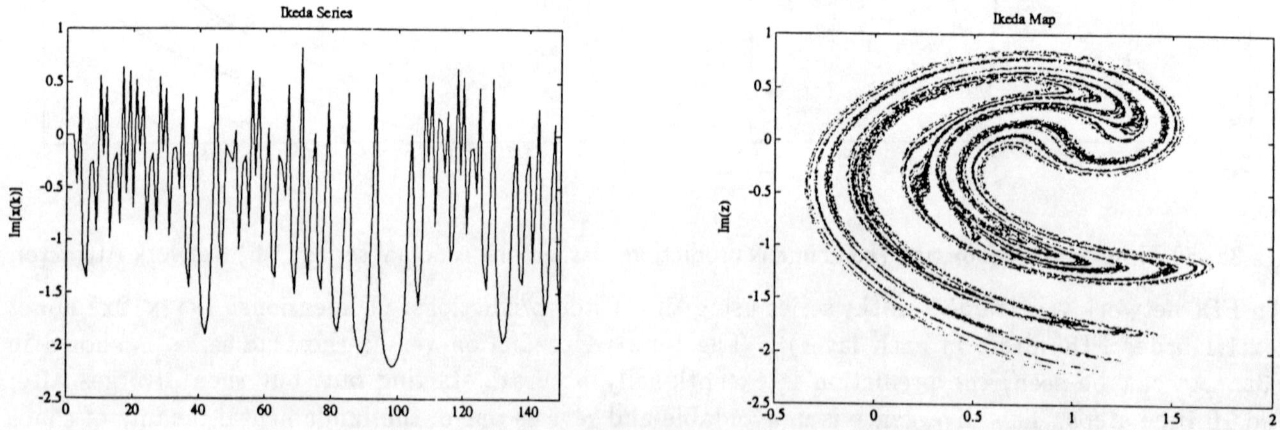

Figure 5: (a) Ikeda Time Series. (b) Ikeda Attractor.

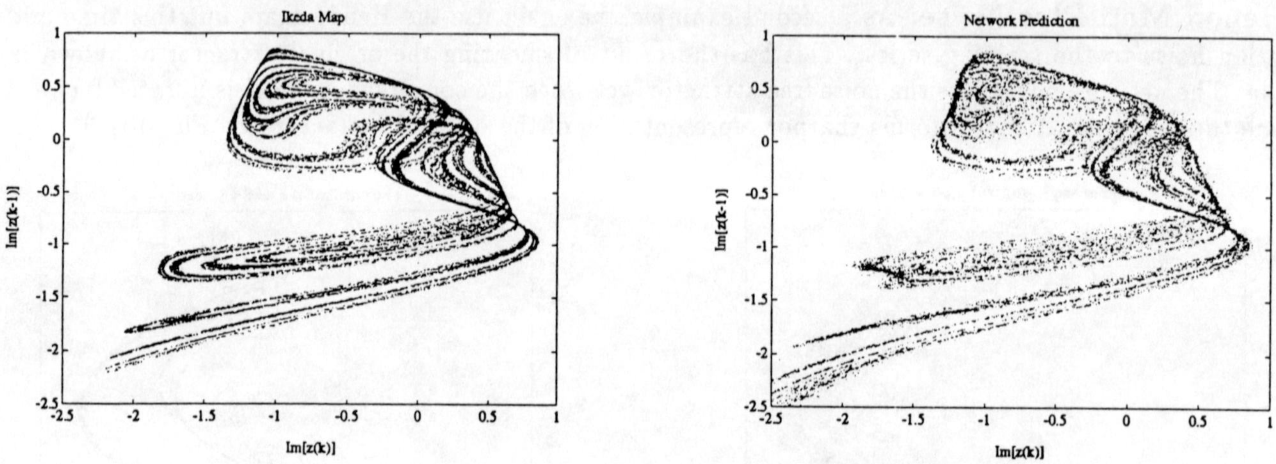

Figure 6: (a) Imaginary Delay Embedding Plot. (b) Network Prediction.

Lorenz Equations: A lorenz system [9] is described by the solution of the three simultaneous differential equations:

$$dx/dt = -\sigma x + \sigma y \quad (3)$$

$$dy/dt = -xz + rx - y \quad (4)$$

$$dz/dt = xy - bz \quad (5)$$

A projection of the trajectory in the $x - z$ plane for parameters values $\sigma = 10$, $r = 28$, and $b = 8/3$ is shown if Fig. 7a. A network (dimensions: 1x12x12x1 nodes with 2:5:5 taps) was trained on observations of only the x state sampled at a period of 0.05 seconds. In Fig. 7b a plot of $x(t)$ vs. $x(t - 0.1)$ is given. In Fig. 7c the iterated network prediction $\hat{x}(t)$ vs. $\hat{x}(t - 0.1)$ is shown for comparison.

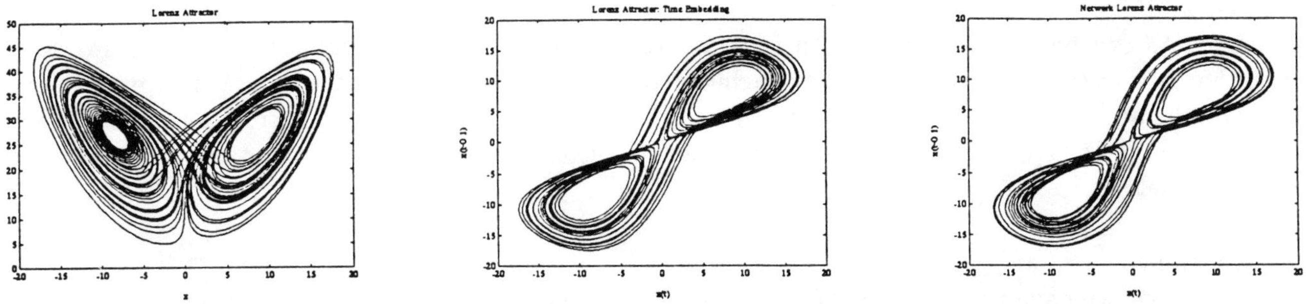

Figure 7: (a) Lorenz x-z phase-plot (b) x(t) vs. x(t-.1) (c) Network x(t) vs. x(t-0.1).

Laser Data: The plot in Fig. 8 shows the chaotic intensity pulsations of an NH_3 laser measured in a laboratory experiment.[3] This data was distributed as part of *The Santa Fe Institute Time Series Prediction and Analysis Competition* with the motivation of making rigorous comparisons between various techniques [11, 12]. For the laser data, 1000 samples of the sequence were provided. The goal was to predict the next 100 samples. During the course of the competition, the physical background of the data set, as well as the 100 point continuation, was withheld to avoid biasing the final prediction results.

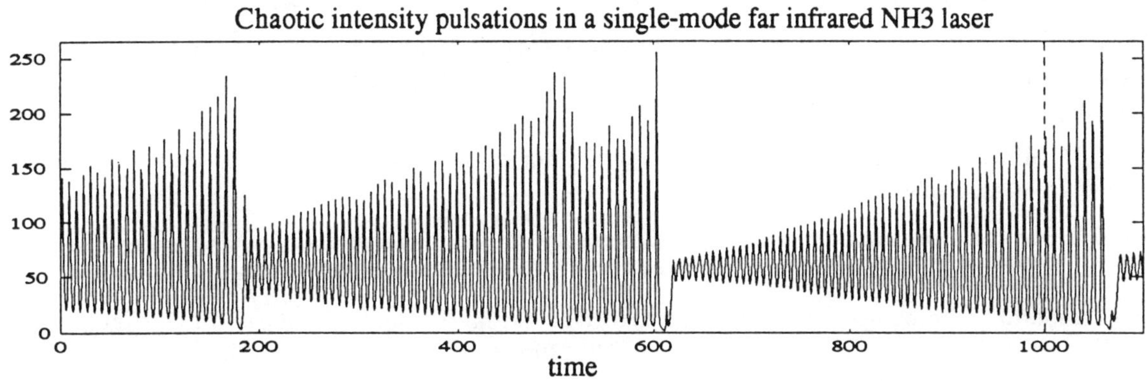

Figure 8: 1000 points of Laser Data.

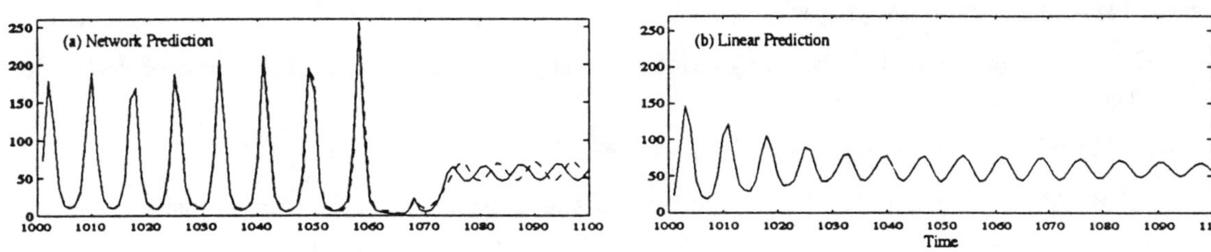

Figure 9: (a) Network Prediction and Series Continuation. (b) Linear AR Prediction.

A FIR network (1x12x12x1 nodes with 25:5:5 taps) was used to learn the training set. The 100 step prediction achieved using the FIR is shown in Fig. 9 along with the actual series continuation (the prediction achieved using a 25th order linear autoregression is also shown for further comparison). It is important to emphasize that this prediction was made based on only the past 1000 samples. True values of the series for time past 1000 were not provided to the network nor were they even available when the predictions were

[3]"Measurements were made on an 81.5-micron 14NH3 cw (FIR) laser, pumped optically by the P(13) line of an N20 laser via the vibrational aQ(8,7) NH3 transition" - U. Huebner [10].

submitted. As can be seen, the prediction is remarkably accurate with only some slight phase degradation. Other submissions to the competition included methods of k-d trees, piecewise linear interpolation, low-pass embedding, SVD, nearest neighbors, Wiener filters, as well as standard recurrent and feedforward neural networks. As reported in [11, 12], the FIR network described here outperformed all other methods on this data set.

Conclusion

In this paper we have provided several examples illustrating the potential of using neural networks for time series prediction and modeling. A more comprehensive paper which details the network structure, learning algorithm, and more practical issues of parameter selection, convergence, and testing, can be found in [6]

While we have focused on autonomous deterministic scalar series, it should be clear that the basic principles directly apply to many problem of interest to power engineers such as multivariate *load forecasting.* In addition, if control input are present, the exogenous inputs may easily be incorporated within the basic framework for problems of system identification and control.

References

[1] Ljung and Lenart. *System Identification: Theory for the User.* Prentice-Hall, Englewood Cliffs, NJ, 1987.

[2] F. Takens. Detecting strange attractors in fluid turbulence. In D.A. Rand and L.-S. Young, editors, *Dynamical Systems and Turbulenc*, volume 898 of *Lecture Notes in Mathematics*, pages 366–381, Berlin, 1981. (Warwick 1980), Springer-Verlag.

[3] K. Hornik, M. Stinchcombe, and H. White. Multilayer feedforward networks are universal approximators. *Neural Networks*, 2:359–366, 1989.

[4] B. Irie and S. Miyake. Capabilities of three-layered perceptrons. In *Proceedings of the IEEE Second International Conference on Neural Networks*, volume I, pages 641–647, San Diego, CA, July 1988.

[5] E. Wan. Temporal backpropagation: An efficient algorithm for finite impulse response. In G. Hinton D. Touretzky and T. Sejnowski, editors, *Proceedings of the 1990 Connectionist Models*, pages 131–140, San Mateo, CA, 1990. Morgan and Kaufmann.

[6] E. Wan. Finite impulse response neural networks for autoregressive time series prediction. In A. Weigend and N. Gershenfeld, editors, *To appear in Proceedings of the NATO Advanced Workshop on Time Series Prediction and Analysis.* Addison-Wesley, 1993.

[7] M. Hénon. A two-dimensional mapping with a strange attractor. In *Commun. Math. Phys. 50*, page 69, 1976.

[8] S. Hammel, C.K.R.T. Jones, and J.V. Moloney. Global dynamical systems, and bifurcations of vector fields. In *J. Opt. Soc. Am., B 2*, page 552, 1985.

[9] E.N. Lorenz. Deterministic nonperiodic flow. In *J. Atmos. Sci. 20*, pages 130–141, 1963.

[10] U. Huebner, N.B. Abrahm, and C.O. Weiss. Dimension and entropies of a chaotic intensity pulsations in a single-mode far-infrared nh3 laser. In *Physics Review*, volume A 40, page 6354, 1989.

[11] Proceedings of the NATO advanced research workshop on time series prediction and analysis, may 14-17 1992. Santa Fe, New Mexico, 1993. Addison-Wesley.

[12] A. Weigend and N. Gershenfeld. Results of the time series prediction competition at the santa fe institute. In *Neural Information Processing Systems*, Denver, CO, Dec 1992.

F. Control

Neural Control Systems

Bernard Widrow and Françoise Beaufays *

Department of Electrical Engineering
Stanford University
Stanford, CA 94305-4055

1 Introduction

In the past decade, the field of neural networks has grown from a mere handful of researchers to thousands spread throughout academics and industry worldwide. Applications have been developed in domains as varied as character recognition, pattern classification, dynamic system control, and stock market prediction. In this paper, we concentrate on the application of feedforward multilayer neural networks to nonlinear control. After a brief introduction on adaptive linear combiners and multilayer neural networks, we present one of the most famous algorithms used in neural control: the backpropagation-through-time algorithm. A series of applications developed at Stanford University are discussed to illustrate the applicability of the algorithm. We then briefly summarize some recent theoretical work related to the subject matter, and we conclude by listing a couple of examples of neural controllers actually used in the industry.

2 Adaptive Linear Combiner and LMS Algorithm

The basic building block of a layered neural network is the *adaptive linear combiner* shown in Fig.1. This element receives an input vector or *input pattern*, and outputs a weighted combination of its components. During the training process, the adaptive linear combiner is presented with input patterns and corresponding desired output responses. At each presentation, an error signal is defined as the difference between the actual and the desired outputs, and is used by a *training algorithm* to adjust the weights. The simplest and most popular training algorithm for the linear combiner is the LMS (least mean square) algorithm [1, 2], also called the Widrow-Hoff delta rule [3]. It proceeds by iteratively modifying the weights in such a way as to minimize the sum of the squared errors over the training set. In other words, it defines an error surface over the weight space, and searches it for its minimum using a stochastic steepest descent method. The resulting algorithm consists simply of adjusting the weights at each pattern presentation by an amount proportional to the product of their input and the error signal.

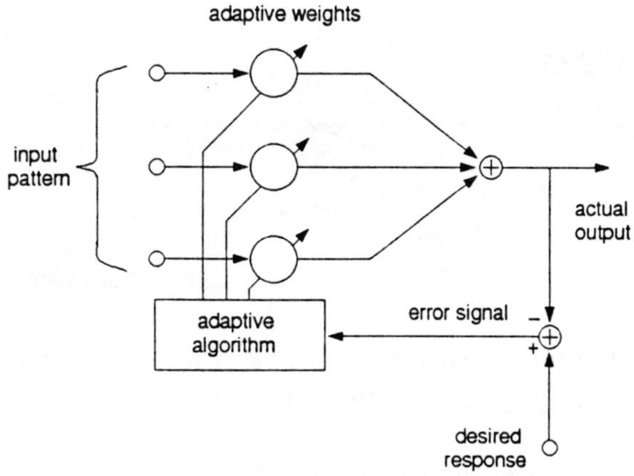

Figure 1: *Adaptive linear combiner.*

*This work was sponsored by NSF under grant NSF IRI 91-12531, and EPRI under contract RP8010-13.

The development of the LMS algorithm and its application to adaptive signal processing led to major commercial applications such as adaptive equalization in high-speed modems [4, 5] and echo cancellation for long-distance telephone and satellite circuits [6]. In the area of pattern classification, linear adaptive combiners were cascaded with hard-limiting thresholds to produce linear binary classifiers, and combinations of such elements were built to classify non-linearly separable patterns [7, 8]. Another family of adaptive structures based on the same principle was obtained by forming layered arrangements of linear combiners, each of them cascaded with an S-shape or *sigmoidal* nonlinear function. The resulting structure is the now famous feedforward multilayer neural network.

3 Feedforward Multilayer Neural Networks

A feedforward multilayer neural network is represented in Fig.2(a). It consists of successive layers of adaptive linear combiners or *neurons*, each of which is cascaded with a nonlinearity called a *sigmoid.* At each iteration of the training process, the neural network is presented with an input pattern and a corresponding desired response. Following the principle of LMS, an error surface defined over the weight space is iteratively minimized using a stochastic steepest descent method. Due to the layered structure of feedforward networks, steepest descent can be implemented in a very efficient way, the resulting algorithm is called *backpropagation* [9, 3].

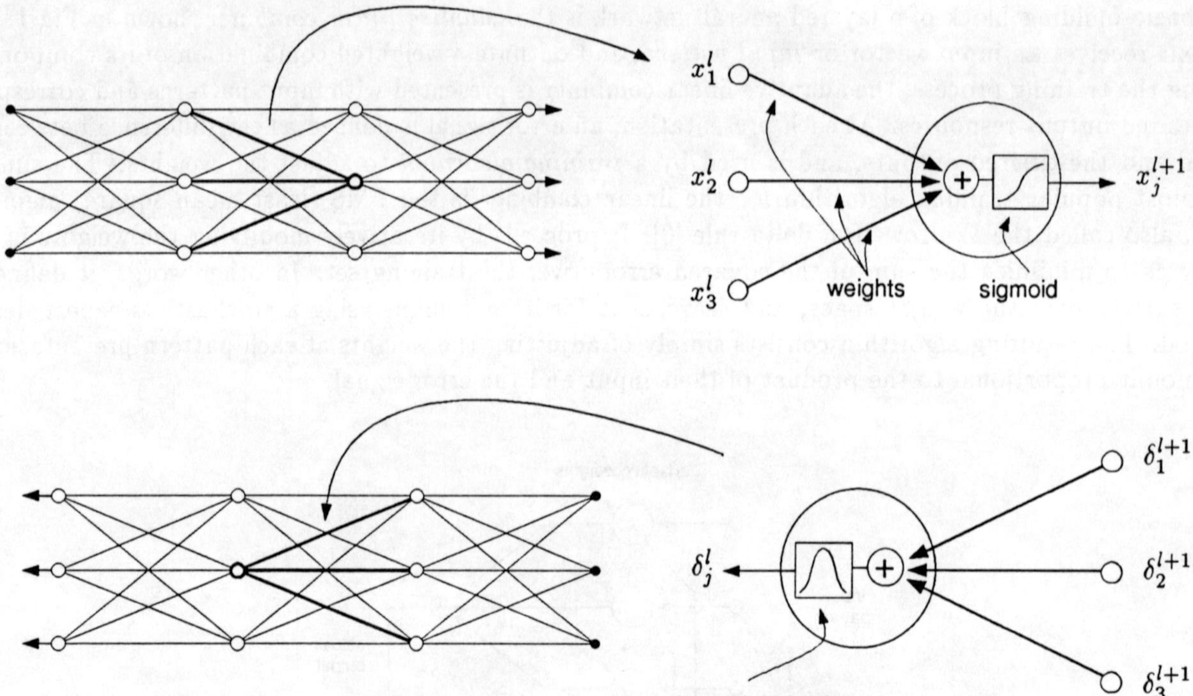

Figure 2: *Feedforward multilayer neural network and details of a specific neuron. (a) Forward propagation of the input vector* **x** *(b) Backward propagation of the error vector* δ.

Without proof, we will summarize the algorithm (a detailed derivation can be found in [3]). At each pattern presentation, the input vector is propagated forward through the network (see Fig.2(a)). The error vector is then propagated backward through a network identical to the original one but where the sigmoids have been replaced by their derivatives (see Fig.2(b)). The weights of each neuron are then updated by an amount proportional to the product of the inputs to that neuron and the error backpropagated to the output of the neuron. Note the similarity between this algorithm and the single-neuron LMS algorithm.

An alternative to backpropagation is the MRIII algorithm. Intuitively, backpropagating the error vector through a given layer of the network is equivalent to multiplying the errors by the derivatives of the outputs of the layer with respect to its inputs. MRIII estimates these derivatives by independently perturbing each neuron by a small amount and comparing the output of the perturbed network with the unperturbed one. The equivalence between the weight updates performed by MRIII and backpropagation is demonstrated in [10]. Although MRIII presents some hardware implementation advantages over backpropagation, it usually requires more computations and is therefore less often used in practice.

4 Neural Control Systems and Backpropagation-Through-Time

One of the most promising new areas of application for layered neural networks is the control of non-linear dynamic systems. Among the different algorithms dealing with the adaptation of neural networks embedded in dynamic structures, the most widely used is *backpropagation-through-time* [9, 3, 11]. In an early application to neural control, the algorithm was used by Nguyen and Widrow [12, 13] to teach a computer-simulated truck to backup to a loading dock. The control structure used to solve this problem is shown in Fig.3. The truck kinematics are modeled by a set of discrete time equations that constitute the plant model. At each time step, the neural controller takes as input the state of the truck (i.e. its position and orientation) and computes a steering angle. The truck is then backed up a fixed distance, with the steering angle defined by the neural controller. The new state of the truck is computed by the plant model, and the operation is repeated until the truck hits near the dock. An error vector is then defined as the difference between the state of the truck and its desired state (centered and perpendicular to the dock).

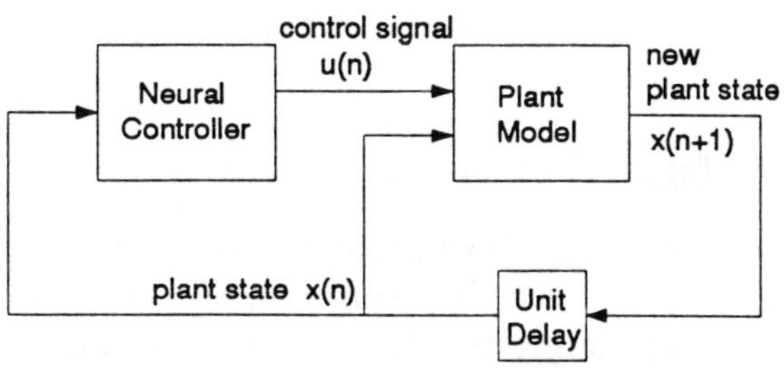

Figure 3: *Neural controller and plant model: block diagram.*

Note that no desired network outputs are made available to the learning algorithm. Only the desired final state of the truck is known. The approach used by Nguyen and Widrow to solve this problem is two-staged. First, a neural network is trained to emulate the kinematics of the truck, then the neural controller is adapted. Unravelling in time the feedback loop of Fig.3 and replacing the truck model by its neural emulator, we obtain a giant layered neural network (Fig.4) whose desired output is known. Backpropagation can be applied to the unravelled structure (hence the name backprop-through-time), and the weights of the controller can be adjusted accordingly.

After training, the truck was capable of backing up from almost any initial position, including initial positions never seen before. A sample trajectory is shown in Fig.5(a).

The same algorithm was then applied to teach a neural network to balance a pendulum mounted in equilibrium on top of a cart [14]. The neural network was constrained to output full-magnitude forces that were applied to the cart (bang-bang control). Again, a first neural network was trained to emulate the cart-broom discrete state-space equations, and a second network was trained to control the system. After adaptation, the neural controller was capable of bringing an initially-off-centered cart back to the center of the track while keeping the broom vertical.

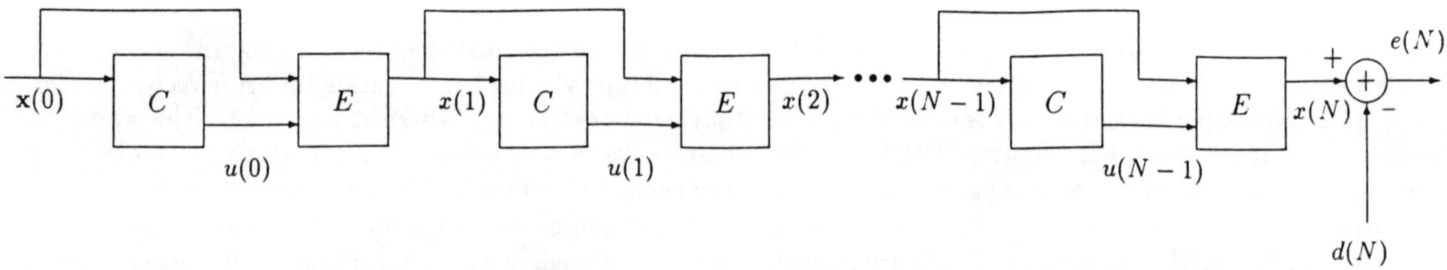

Figure 4: *Neural controller and plant model: block diagram unravelled in time. The C and E blocks represent the neural controller and emulator; $x(n), u(n)$ are the plant state and control signal at time n respectively; $d(N), e(N)$ are the desired response and the error signal at time N respectively.*

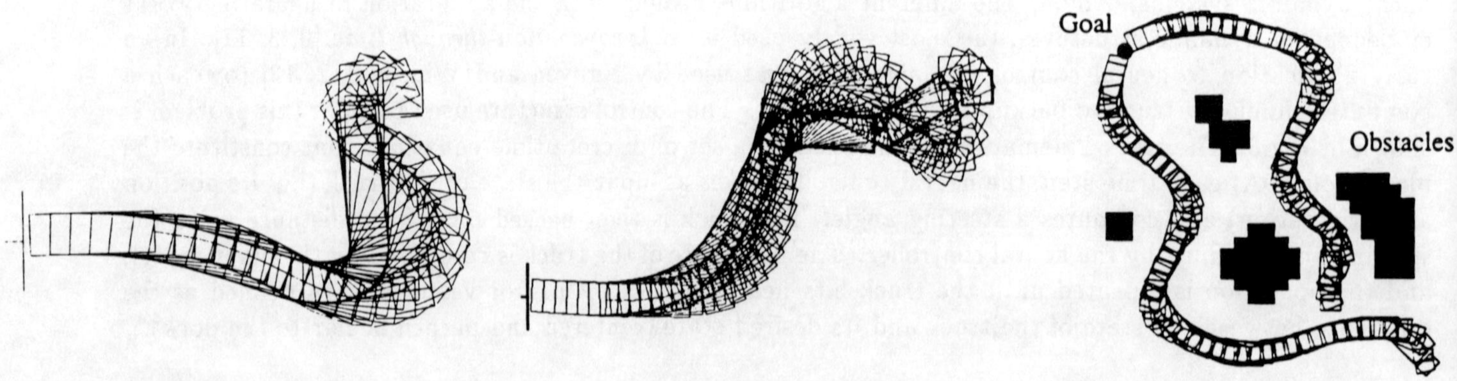

(a) Single-trailer truck (b) Double-trailer truck (c) Single-trailer truck around obstacles.

Figure 5: *Sample truck trajectories.*

An important theoretical step done by Widrow's group was to recognize that instead of backpropagating errors through a neural emulator of the plant, the errors can equivalently be multiplied by the Jacobian matrix of the plant, i.e. the matrix containing the derivatives of the outputs of the plant with respect to its inputs. Derivations of this result can be found in [15, 16]. Backpropagation directly through the plant model can be done whenever such a model is available, and results in a significant saving in development time. Based on this Jacobian method, a double-trailer truck was taught to backup to a loading dock (see sample trajectory in Fig.5(b)), and a double-broom-cart system was balanced.

As a further extension to the family of truck problems, a single-trailer truck was then taught to backup around obstacles [17, 18]. In solving this problem, a potential field is iteratively built around the obstacles (high values correspond to locations close to obstacles, low values to locations close to the dock). The truck then follows a steepest descent path in this potential field to reach the dock while avoiding the obstacles (see sample trajectory in Fig.5(c)).

In the area of power systems, backpropagation-through-time was used to solve an excitation control problem. In this application, a synchronous generator is connected to a load that can vary over time. As a result of load variations, the terminal voltage of the generator fluctuates. Proper action must be taken on the excitation or *field* current in order to bring the terminal voltage back to its nominal value after each load variation. Three neural networks were adapted to control the excitation current of the computer-simulated generator shown in Fig.6: one acting on the DC voltage of the excitor, one setting the position of a rheostat inserted in the excitor circuit, and one controlling both the excitor voltage and the rheostat. While the first control scheme could have been efficiently performed by a linear controller, the two neural controllers acting on the rheostat were solving nonlinear differential equations that would have

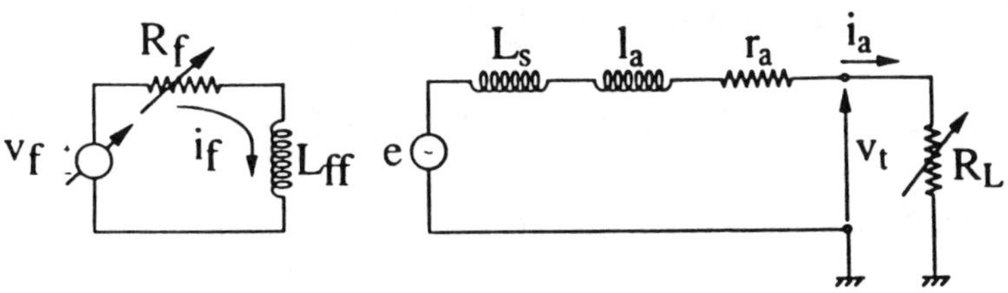

Figure 6: *Single generator driving a time-varying load: simplified model.* v_f, R_f, i_f, L_{ff} *are the DC excitation or* field *voltage, the excitation rheostat, the excitation current, and the rotor self-inductance respectively;* $e, L_s, l_a, r_a, i_a, v_t, R_L$ *are the stator induced emf, the armature reaction inductance, the armature leakage reactance, the armature resistance, the armature current, the terminal voltage, and the resistive time-varying load respectively.*

been difficult to solve by conventional methods. Among the three control schemes, simultaneous control of the DC voltage and rheostat has shown to allow the fastest terminal voltage recovery. Simulation results can be found in [19].

5 Recent Theoretical Work in Neural Control Systems

In terms of control theory, the control structure of Fig.3 can be seen as an extension of dynamic state feedback, where the feedback gains are replaced by a nonlinear neural network. Since the neural controller-plant system is nonlinear, notions such as deadbeat control can not be maintained. The number of time steps needed by the controlled system to reach its desired state is a parameter whose setting is left to the intuition of the control designer instead of being optimized by the neural controller. To overcome this problem, theory has recently been developed in our lab that shows how backpropagation-through-time can be rephrased as an extension of classical optimal control and how neural networks can be used to solve so-called "open final time problems" such as minimum-time control [20].

6 Neural Controllers in Industrial Applications

One could argue that all the examples presented so far are "toy-problems" and computer-simulated applications. We conclude this paper by listing a couple of examples of neural controllers actually implemented in industrial environments.

A first example, taken from the power systems industry, is a prototype system developed by BC Hydro in Vancouver. A neural network was taught to simultaneously control five synchronous condensors in such a way as to balance the imaginary power of an electric network without overstressing any particular machine. The set up successfully demonstrated the power of neural networks and illustrated the savings that they can bring in the power industry.

Another example is the Intelligent Arc Furnace Controller™, a neural controller developed by Neural Applications Corporation to position the electrodes of a steel melting furnace [21]. Replacing the rule-based system previously used, the neural controller was implemented at North Star Steel Co. It increased the furnace productivity, reduced the electrode consumption, and improved the electric power factor of the plant, resulting in major cost reductions.

Finaly, a more surprising application, the details of which are proprietary, is the use of backpropagation-type neural networks by M & M Mars to improve the productivity and reliability of candy manufacturing!

References

[1] B. Widrow and M. E. Hoff, Jr. Adaptive switching circuits. In *IRE WESCON Conv. Rec., pt. 4*, pages 96–104, 1960.

[2] B. Widrow and M. E. Hoff, Jr. Adaptive switching circuits. Technical Report 1553-1, Stanford Electron. Labs., Stanford, CA, June 30 1960.

[3] D. E. Rumelhart and J. L. McClelland, editors. *Parallel Distributed Processing.* The MIT Press, Cambridge, MA, 1986.

[4] R. W. Lucky. Automatic equalization for digital communication. *Bell Syst. Tech. J.*, 44:547–588, April 1965.

[5] R. W. Lucky, *et al. Principles of Data Communication.* McGraw-Hill, New York, 1968.

[6] M. M. Sondhi. An adaptive echo canceller. *Bell System Technical Journal*, 46:497–511, March 1967.

[7] B. Widrow. Adaline and madaline—1963, plenary speech. In *Proceedings of the IEEE First International Conference on Neural Networks*, volume I, pages 145–158, San Diego, CA, June 23 1987.

[8] B. Widrow and R. Winter. Neural nets for adaptive filtering and adaptive pattern recognition. *IEEE Computer*, pages 25–39, March 1988.

[9] P. Werbos. *Beyond Regression: New Tools for Prediction and Analysis in the Behavioral Sciences.* PhD thesis, Harvard University, Cambridge, MA, August 1974.

[10] D. Andes, B. Widrow, M. Lehr, and E. Wan. MRIII: A robust algorithm for training analog neural networks. In *International Joint Conference on Neural Networks*, volume I, pages 533–536, Washington, DC, January 1990.

[11] P. J. Werbos. Backpropagation through time: What it does and how to do it. *Proc. IEEE, special issue on neural networks*, 2:1550–1560, October 1990.

[12] D. Nguyen and B. Widrow. The truck backer-upper: An example of self-learning in neural networks. In *Proceedings of the International Joint Conference on Neural Networks*, volume II, pages 357–363, Washington, DC, June 1989.

[13] D. Nguyen and B. Widrow. Neural networks for self-learning control systems. *IEEE Control Systems Magazine*, April 1990.

[14] D. Nguyen. *Applications of Neural Networks in Adaptive Control.* PhD thesis, Information Systems Laboratory, Stanford University, Stanford, CA, 1991.

[15] S. W. Piche and B. Widrow. First-order gradient descent training of adaptive discrete-time dynamic networks. Technical Report RL-TR-91-62, Rome Laboratory, Air Force Systems Command, Griffiss Air Force Base, NY 13441-5700, May 1991.

[16] F. Beaufays, Y. Abdel-Magid, and B. Widrow. Application of neural networks to load-frequency control in power systems. *To be published in Neural Networks*, 1992.

[17] E. S. Plumer. Cascading a systolic array and a feedforward neural network for navigation and obstacle avoidance using potential fields. Technical Report CR-177575, National Aeronautics and Space Administration, Ames Research Center, Moffett Field, CA 94035-1000, February 1991.

[18] E. S. Plumer. Neural network structure for navigation using potential fields. In *Proceedings of the International Joint Conference on Neural Networks*, volume I, pages 327–332, Baltimore, MD, June 7-11 1992.

[19] F. Beaufays and B. Widrow. Application of neural networks to nonlinear control in power systems. *presented at the EPRI/NSF workshop on application of advanced mathematics to power systems*, September 1991.

[20] E. S. Plumer. Two neural network algorithms for designing optimal terminal controllers with open final-time. Technical Report CR-177599, National Aeronautics and Space Administration, Ames Research Center, Moffett Field, CA 94035-1000, July 1992.

[21] W. E. Staib and R. A. Staib. The Intelligent Arc Furnace^{TM} Controller: A neural network electrode position optimization system for the electric arc furnace. In *Proceedings of the International Joint Conference on Neural Networks*, Baltimore, June 1992.

POTENTIAL USES OF INTELLIGENT AND ADAPTIVE CONTROLS FOR ELECTRIC POWER SYSTEM OPERATIONS IN THE YEAR 2000 AND BEYOND

Roger L. King
P.O. Drawer EE
Mississippi State University
Mississippi State, MS 39762

Michael L. Oatts
P.O. Box 2625
Southern Company Services, Inc.
Birmingham, AL 35202

ABSTRACT - The problem with the present systems used for control and operation is that they employ at best, a loosely-coupled or, in the worst cases, a completely decoupled, hierarchial, control scheme. As new operating requirements and constraints are being placed upon the existing control systems, it is expected that they will become significantly less capable of satisfactorily managing the bulk power systems. Therefore, it is essential to begin to define the operating and technical requirements for the next generation of Energy Management Systems (EMS). This paper will present perspectives on the potential use of intelligent and adaptive controls in the next generation of EMS. The paper will address the power system of the future at two distinct time periods. The first will be a near term, evolutionary look at the power system of the year 2000. The second will be a more revolutionary look at one potential operating scenario for the year 2020. The research applications suggested for the near term are based upon existing problems and what we believe can be realized with intelligent and adaptive controls by the year 2000. The scenario for the year 2020 is more nebulous due to the problems of forecasting the future. Our engineering assessment of the future is complicated by the fact that we cannot "crystal ball" what new scientific discoveries or technologies will become economically practical that far into the future. However, we will make some assumptions about new technologies and comment upon the impact they possibly will have on real-time control in the year 2020.

INTRODUCTION - The classical view of power system control is depicted as three loosely coupled or decoupled hierarchial levels of control. The levels have been decoupled by choice; i.e. humans have decided that they would couple or coordinate the control between levels. The lowest level contains the automatic schemes programmed into the system, e.g. relays. The middle level contains functions such as supervisory control and data acquisition (SCADA), network security analysis, etc. And, the highest level considers such functions as the matching of load and generation (AGC). The lack of an integrated control hierarchy has been due to an unwillingness on the part of utilities to replace the human with an automated system. This unwillingness, however, is not unmerited. The degree to which a utility will be willing to couple the control layers will be dependent upon the degree of intelligence that can be put in place in the power system architecture.

Assuming a satisfactory degree of intelligence can be emplaced, issues about the degree of automation to be employed at a given time will be dependent upon the state of the system. For example, there is a major difference between automation in a normal state and an emergency state. Another issue of major impact is the power system computing architecture. Historically, utilities have used centralized computing with little distributed computing power. Any intelligent and adaptive control system will require major changes from centralized to localized control. From a practical standpoint, a change from centralized computation to localized computation is necessary to enable a change in the control philosophy.

Nevertheless, change must come. Today's utility was structured and planned "yesterday" as a company that generated power to meet its native load. In practice, utilities are providing bulk power transactions over long distances, drastic flow pattern changes are occurring, non-utility generators (NUGs) and independent power producers (IPPs) are becoming a fact of life, along with EPRI's flexible ac transmission system (FACTS). An ever widening gap is growing between the planner's vision of today and the operator's challenges of today. The only way to close this gap is with a new generation of power system controls. These system must be intelligent, adaptive, and capable of operating in real time.

INTELLIGENT AND ADAPTIVE CONTROL FUNCTION - Intelligence is as difficult to define as it is to measure and, therefore, we often envision intelligence in light of outward attributes. Intelligent beings can learn, adapt to changing situations, communicate, etc. In other words, an ability to learn from your environment/surroundings. Therefore, the following functional requirements for such an intelligent and adaptive control scheme for the electric power system can be stated:

1) Operates in real time.

2) Has the capacity to learn from its environment.
3) Must have accurate, detailed system information.
4) Must understand the decision-making process of a human operating the system.
5) Must be adaptable.

From a more pragmatic point of view, there are additional operating requirements that also must be considered. These systems must have the ability to recognize the state of the system (normal, alert, emergency, restorative). The gaps in communication between hierarchial control levels must be closed. Sensors must be available for providing accurate system information. The speed of operation - real time or near real time - must be defined. And, conscious decisions must be made as to what controls the human operator will give up to the intelligent system.

THE FUTURE - What is the power system architecture of interest for the year 2000 and beyond? It obviously will encompass much of the existing infrastructure. Generation, transmission, and control operations owned and operated by a utility are not going to be phased out arbitrarily. It is reasonable to expect that the actual devices in use today will still be in operation in the year 2000. However, changes to the power system architecture will evolve due to two considerations. The first consideration, is what we describe as business drivers. These encompass changes in the architecture due to non-utility generators (NUGs), independent power producers (IPPs), and load control. The second consideration is new technology or systems. This would encompass such items as energy storage and flexible ac transmission systems (FACTS).

We would now like to turn to the area of some potential research projects. The paper will address the applications to two distinct time periods. The first will be a near term, evolutionary look at the power system of the year 2000. The second will be a more revolutionary look at one potential operating scenario for the year 2020. The research applications suggested for the near term are based upon existing problems and what we believe can be realized with intelligent and adaptive controls by the year 2000. The scenario for the year 2020 is more nebulous due to the problems of forecasting the future. Our engineering assessment of the future is complicated by the fact that we cannot "crystal ball" what new scientific discoveries or technologies will become economically practical that far into the future. However, we will make some assumptions about new technologies and comment upon the impact they possibly will have on real-time control in the year 2020.

The Power System in the Year 2000: We do not foresee any major breakthroughs in fusion, superconductivity, or similar technologies which will be made by this time that will warrant immediate implementation from an economic perspective. We believe that the basic infrastructure of transformers, generators, transmission lines, etc. that are in place today will still exist. The major changes will take place in the control and protective systems. We believe that there is a tendency towards decentralizing the computing resources of the electric utility. Induction disk relays are being replaced with microprocessor based ones, remote terminal units (RTUs) are becoming more intelligent, and computer workstations are becoming more common place. We believe that a typical system of the year 2000 will look very much like it does today. However, it will have a larger degree of distributed computing power at the power plants and substations. We also believe that better communication between the buses of a system will be possible with the advances of fiber optics.

An Intelligent System for Emergencies: Some research has been conducted on how intelligent and adaptive controls can be used to recognize emergency situations on the system and to recommend actions to alleviate the problem. Reference 1 illustrates how intelligent systems can be used to recognize overloaded lines and recommend corrective actions on the IEEE 30 bus system. The recognition of the overload condition is accomplished using a feed forward neural network. The training sets for the network have been generated via numerous loadflow simulations. The corrective actions are recommendations in the scheduling of available generation or, in the worst case, load shedding. The knowledge for the corrective actions is captured with a hybrid neural network and expert system. This operation and controls problem has shown a potential for success from the initial studies and could serve as an excellent operations aid. The fielded system would continuously monitor the state of the system's lines from the SCADA data, recognize overload conditions, check for generator availability, compare this with its knowledge about how to respond to various operating scenarios, and make recommended actions to the operator.

An Intelligent System for Detection of Loop Flows: The traditional idea of always rescheduling internal generators or line status when an overloaded internal line is detected may not be the preferred alternative when the cause of the overload is externally created loop flows. It is recognized that some loop flows are inevitable and are an expected part of interconnected operation. However, these inadvertent flows do not normally interfere with a utility's economics or ability to use its own network.

The current trend to long term, bulk sales of power has disrupted the normal cyclic swapping of power flow directions which historically balanced or shared the effects of loop flows among neighboring utilities. Therefore, the issue becomes one of identification of the source of the loop flows and recommended corrective action. Although there is an increasing amount of data exchange between utilities, this can not be expected to provide a reliable source of loop flow identification without the knowledge for interpretation. Therefore, it is becoming more important to develop intelligent control systems that can identify the cause of major inadvertent loop flows on a utility's network. Again, the intelligent system would have to be interfaced with the SCADA system, off line training sets, and simulators, as well as with external data exchange systems. Once the system determines the source of the loop flows, it could then identify the source(s) and make recommendations to the operator to minimize the problem.

To accomplish such a system, it is believed that an intelligent system consisting of a neural network and an expert system must be used. The neural network will be used for pattern recognition and the expert system for procedural expertise. The neural network would monitor area generation and transaction schedules and tie line power flows for any abnormal patterns. This would have to be coupled with knowledge contained within the expert system about the state of the surrounding control areas and how changes in their generation affect power flows within the internal control area. If the system identifies an inadvertent loop flow, then the operators of the affected system would be notified. The recommended actions provided by the system to the operator is another area of research.

Near Real Time Reliability Index: This is a more difficult problem because it forces a utility to create its own definitions for normal, alert, and emergency states. This is not trivial because the definition includes static security, steady-state stability, and transient stability issues. Also, there are degrees for each of the three states. For example, suppose a utility has only one contingency that could move the system from alert to emergency. This utility is at a "safer" level of reliability than if there were 40 contingencies each of which could move the system from alert to emergency. Typically, off line studies identify states which are vulnerable to contingencies that would result in less reliable conditions.

The concept is to use the system description (as available from state estimation or real time data) to determine a real time reliability index and, practically, what to do to move the index to a predetermined level. The reliability index would not have to be a "global" index. There could be reliability indices for different critical areas in a large system. It is believed that practical training sets (alluded to in the previous paragraph) are available for this type of study.

Advanced AGC Concepts: Due to timing constraints and data availability, AGC typically uses extremely simple generator models. Whether this is by convention or by practicality is not important at this time. AGC traditionally has had difficulty in predicting expected unit response based upon control requests because the actual response is a function of:

1) how many continuous requests have been made to the unit,
2) have there been reversals (i.e., increase output then decrease output),
3) how long has it been since unit maintenance,
4) fuel quality,
5) plant operator on duty,
6) linkage problem in the plant control system (e.g., number of mills or other auxiliary equipment operating for a coal unit).

What is needed is an ability to intelligently select and predict response. To accomplish this it is recommended that use be made of the existing real time records of actual unit response for given control requests. Coupling this with short term load forecasts (in minutes) and knowledge concerning ramp rates required to meet daily load shapes, this type of information could be used to make a more intelligent AGC without expensive unit modeling.

The Power System in the Year 2020: Again, we do not believe that any major technological breakthroughs are going to be made by this time that can be economically implemented. We still believe that the basic infrastructure of transformers, generators, transmission lines, etc. that are in place today will still be there. However, we believe that it is possible to visualize breakthroughs in photovoltaic systems, fuel cells, and their accompanying energy storage devices that would make localized (even on a residential scale) power production feasible. We also can visualize that EPRI's FACTS could also be cost effective by this time period. Also, by this time the evolution to distributed computing should be complete and we are assuming that communications to handle the massive data transfers between all buses will be cost effective.

Smart Demand Side Options: With the use of photovoltaic systems at residences and commercial sites the number of energy sources will skyrocket into the 1000's. This will require more sophisticated control options since load patterns will be dynamic. However, demand side options will now require a degree of intelligence. No longer will the distribution automation systems just turn off the water heaters/air conditioners of arbitrary houses or businesses. Now, a decision will need to be made to permit the system to respond intelligently to problems such as extended cloud cover in particular areas. Extended cloud cover will reduce the localized production capacity within the affected area and the customers will desire to purchase power from the utility. To be able to accomplish this, the distribution automation system will need to locate areas unaffected and affected by the cloud cover, and through a combination of intelligent automatic generation control, transaction scheduling and FACTS utilization, redirect power to support shifting load patterns in a secure manner.

Intelligent Power Brokering Systems: The power brokering systems in place today are hardly more than "match makers" between utilities which wish to buy and sell energy. With the advent of transmission access, exploding numbers of small power producers, and implementation of FACTS on power systems with reduced operating margins, the potential for more intelligent power brokering systems will be enormous. In addition to more intelligently matching buyer and sellers, these systems could consider and recommend actions which involve unit commitments and both dynamic and static security issues. The intelligent broker could also be integrated with FACTS to bring about energy exchanges which might not have otherwise occurred.

An Intelligent Power System: Assume that the power system achieves its goals of distributed computing power with local intelligence and excellent communication between buses. What could such a system be expected to accomplish? Coordination of the system could take place by simply training a neural network off line that would recognize fault conditions. Then, by loading the weights of this network at each processing element (bus) the system would be coordinated. If the transmission network changes (e.g. line out), then a new set of weights could be loaded at all of the buses to account for the change. Again, the system would be coordinated. Also, are not the conventional setpoints now obsolete? Since the system is now adaptable, it should be able to use its distributed computing power and intelligence to ascertain changes in the system and make the appropriate changes in the protective system, i.e. real-time settings. Also, sensors could now detect abnormal or dangerous situations and adjust the system to minimize customer outages if the problem does manifest itself. For example, connections to the existing lightning detection system would permit the system to track a storm across its lines and adjust generation scheduling and/or network configurations (à la FACTS) to direct critical power transfers away from the potentially adversely affected areas.

ACKNOWLEDGEMENT - Portions of this paper are a result of research conducted under Contract No. SJ547V for the Oak Ridge National Laboratory managed by Martin Marietta Energy Systems, Inc. for the U.S. Department of Energy under Contract No. DE-AC05-84OR21400 [2].

CONCLUSIONS - Several potential applications for intelligent and adaptive control have been described for what we viewed as the evolutionary change to the year 2000 and for the revolutionary change to the year 2020. We believe that all of the projects discussed for the year 2000 are feasible and cost effective. The year 2020 system is visionary. Perhaps it will stimulate discussion as to where we can really go with intelligent real-time control of the electric power system. It is hoped that this paper will stimulate readers in their planning and prioritizing of a research plan for the real-time control and operation of electric power systems.

REFERENCES

1) Damir Novosel and Roger L. King, "Identification of Power System Emergency Actions Using Neural Networks," Mohamed A. El-Sharkawi and Robert J. Marks II, eds., Proceedings of First International Forum on Applications of Neural Networks to Power Systems, University of Washington, July 23-26, 1991, IEEE 91TH0374-9, pp 205-209.

2) Roger L. King and Michael L. Oatts, "Perspectives on the Potential Use of Intelligent and Adaptive Controls for Electric Power System Operations," Workshop on Real-Time Control and Operation of Electric Power Systems, November 19-21, 1991, Denver, CO, pp 113-122.

Load-Frequency Control Using Neural Networks.

Françoise Beaufays and Bernard Widrow *

Abstract

A single generator unit feeds a power line to various users whose power demand can vary over time. As a consequence of load variations, the frequency of the generator changes over time. A feedforward neural network is trained to control the steam admission valve of the turbine that drives the generator, thereby restoring the frequency to its nominal value. Frequency transients are minimized and zero steady-state error is obtained. The same neural net technique is then applied to control a two-area system. Electric load variations can happen independently in both areas. Both neural controllers are trained with the backpropagation-through-time algorithm.

1 Introduction

We first consider a single isolated generator unit connected to an electric bus serving different users. Variations in users power demand cause the electric load on the bus, and hence the generator frequency, to change over time. In conventional systems, an integral controller acts on the steam admission valve of the unit turbine to bring the frequency back to its nominal value after load variations. This type of controller is slow and does not allow the control designer to take into account possible nonlinearities in the generator unit.

We here propose a neural net load-frequency controller. The neural controller makes use of a piece of information that is not used in conventional controllers: an estimate of the electric load perturbation. This estimate could be obtained by a linear estimator or by a nonlinear neural net estimator. We show by simulation that when a load estimate is available, the neural network can achieve extremely good dynamic response.

The same neural net technique is then extended to control a two-area system where electric load perturbations can happen independently in both areas.

Both neural networks are feedforward multilayer structures whose weights are adapted using the backpropagation-through-time algorithm [1, 2, 3], a generalization of the well-know backpropagation algorithm [4, 5, 6]. The dynamic plant to be controlled is represented by its discrete state space equations. Models, equations, and orders of magnitude of parameters are chosen in accord with examples in the classic book by O. Elgerd, "Electric Energy Systems Theory" [7], chapter 9, sections 9.3 and 9.4.

2 Plant models and control by conventional means

2.1 Automatic load-frequency control in single-area systems

In a single-area system, mechanical power is produced by a turbine and delivered to a synchronous generator serving different users. The frequency of the current and voltage waveforms at the output of the generator is mainly determined by the turbine steam flow. It is also affected by changes in user power demands which appear, therefore, as electric "perturbations". If, for example, the electric load on the bus suddenly increases, the generator shaft slows down, and the frequency of the generator decreases. The control system must immediately detect the load variation and command the steam admission valve to open more so that the turbine increases its mechanical power production, counteracts the load increase, and brings the shaft speed and hence the generator frequency back to its nominal value.

---

*The authors are with the department of Electrical Engineering, Stanford University, Stanford, CA 94305-4055. This work was sponsored by EPRI under contract RP8010-13.

A detailed description of the system to be controlled can be found in [8]. In this paper, we will just briefly introduce a model of the plant and pursue from there. Since most devices in power systems are extremely nonlinear, one usually likes to linearize the plant and think of different variables in terms of their fluctuations about given operating points. Nonlinearities are then modeled by making the parameters of the linearized system functions of the operating point. The resulting small-signal models consist of linear operators having variable parameters whose values depend upon the state of the system. Such a model is given in Fig.1. The model includes a generator, a speed regulator and an integral controller, which by observing the frequency fluctuations* $\Delta F(s)$ produce a control signal, a hydraulic amplifier to amplify the control signal, and a turbine feeding the generator. The load fluctuation $\Delta P_E(s)$ is modeled as an input perturbation to the generator [7].

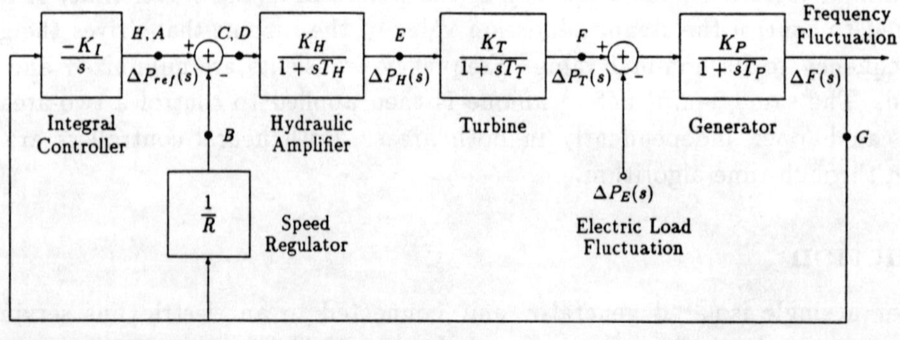

Figure 1: Conventional single-area system: small-signal model, simplified block diagram.

For any non-negative value of the integrator gain K_I, and assuming that the electric perturbation $\Delta P_E(s)$ is a step function, the controlled plant is stable; that is, its state vector converges to a finite steady-state value. In particular, the frequency fluctuation $\Delta F(s)$ converges to zero. In practice, K_I will be chosen to be the *critical gain*, i.e. the highest gain that yields no overshoot.

2.2 Automatic load-frequency control in two-area systems

A two-area system consists of two single-area systems connected through a power line called the *tie-line*: each area feeds its user pool, and the tie-line allows electric power to flow between the areas. Since both areas are tied together, a load perturbation in one area affects the output frequencies of both areas as well as the power flow on the tie-line. For the same reason, the control system of each area needs information about the transient situation in both areas in order to bring the local frequency back to its steady-state value. Information about the local area is found in the output frequency fluctuation of that area. Information about the other area is found in the tie-line power fluctuations. Therefore, the tie-line power is sensed, and the resulting tie-line power signal is fed back into the controllers of both areas. This basic scheme is illustrated in Fig.2(a). A more complete diagram is given in Fig.2(b).

In conventional systems, the position of the turbine steam valve of each area is set by an integral controller whose input is a linear combination of the frequency fluctuations in both areas. For any non-negative integrator gains $K_{I,1}$, $K_{I,2}$, when a step-load perturbation has occured in area 1 and/or in area 2, and after transients have died out, the frequency variations in both areas converge to zero, as well as the tie-line power.

Note that the plant models used so far were linearized models. We tacitly assumed that the operating point of the plant did not change much when a step-load perturbation occured on the bus, and that, therefore, all the plant parameters could be kept constant. In practice though, several devices, such as the speed regulator, are characterized by parameters that depend in a highly nonlinear way upon the turbine power level. The presence of such nonlinearities and the slowness of integral controllers is what motivates our substituting a neural controller for the integral control scheme.

*$\Delta F(s)$ represents the Laplace transform of the frequency fluctuation $\Delta f(t)$, i.e. the variation of the frequency about its nominal value (60 Hz in the U.S.). $\Delta f(n)$ represents the discretized frequency fluctuation.

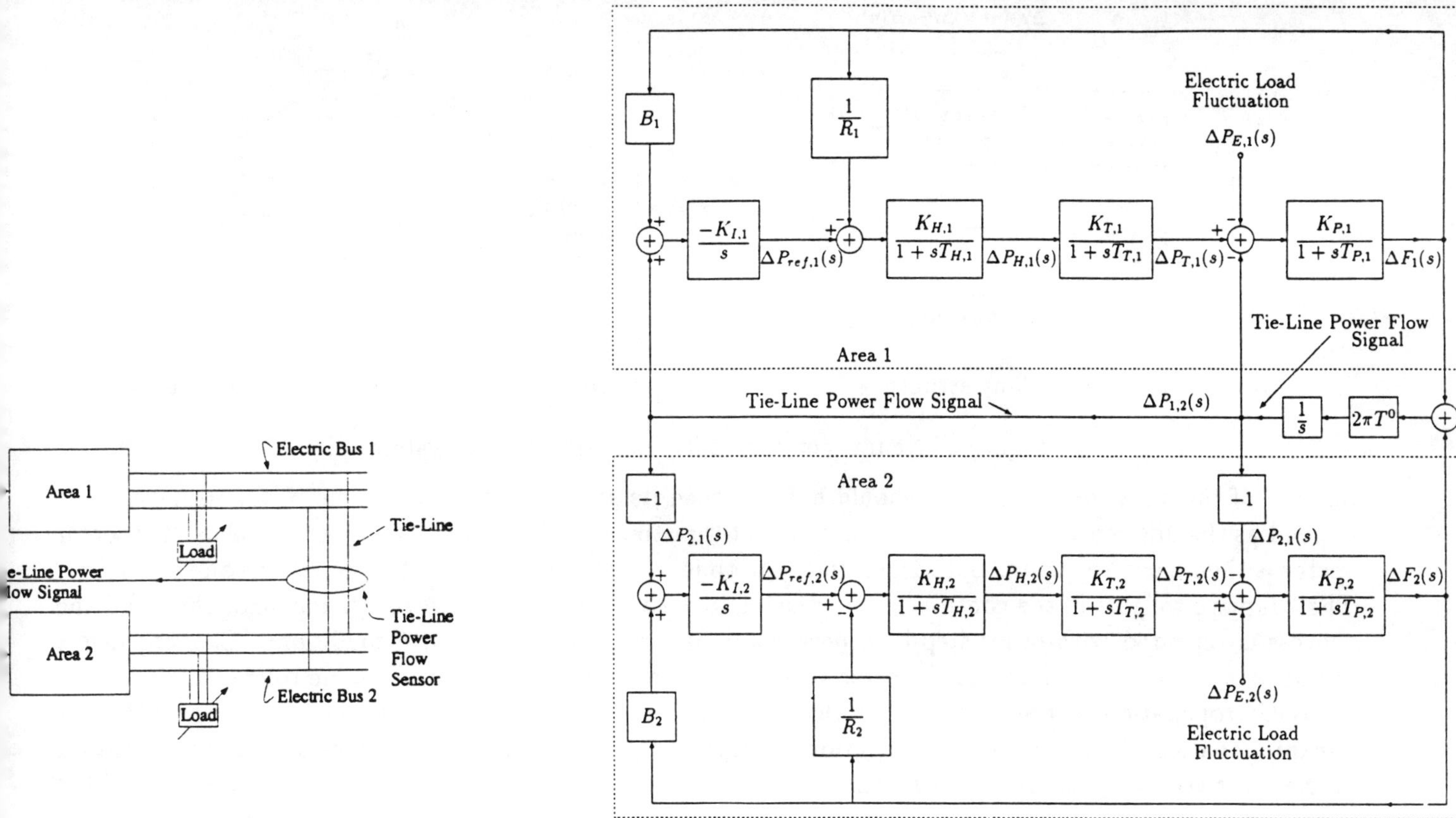

(a) Basic block diagram (b) Small-signal model

Figure 2: Conventional two-area system.

3 Nonlinear control by means of neural networks

3.1 Neural net control of a single-area system

Let us first consider a single-area system initially in steady-state. The plant is suddenly hit by a step-load perturbation. Transients ensue. The neural controller that replaces the integral controller should make the plant converge to the same steady state while limiting the duration and magnitude of the transients.

The dynamic controller-plant structure used here is shown in Fig.3(a). The plant model (dashed box) contains a set of discrete state-space equations describing the dynamics of the plant. Such equations are easily derived from the plant small-signal model of Fig.1. The internal structure of the neural controller is detailed in Fig.3(b). It contains two layers of neurons: a hidden layer and a single neuron output layer. No sigmoid is used at the output of the network. Bias terms (i.e. constant inputs equal to 1) are included in all the hidden units and in the output unit.

The neural controller takes as input the discrete states of the plant $\mathbf{x}(n)$ and makes use of a piece of information that is not used in conventional control: an estimate of the load perturbation, $\Delta \hat{p_E}(n)$. At each instant of time, it produces an output signal, the turbine reference power $\Delta p_{ref}(n)$, which is used to drive the plant (i.e. to position its steam admission valve).

In general, the load perturbation of a large system is not directly measurable. It must therefore be estimated by a linear estimator or by a nonlinear neural net estimator if the nonlinearities in the system justify it. Such an estimator takes as input a series of K samples of the frequency fluctuation at the output of the generator, and estimates the instantaneous value of the load perturbation $\Delta p_E(n)$ based on this input vector. The estimate $\Delta \hat{p_E}(n)$ is then used to drive the plant controller. The implementation of such an estimator is beyond the scope of this paper. Here, we assume that the load estimate $\Delta \hat{p_E}(n)$ is available and that, for adaptation purposes, this estimate is perfect, that is, $\Delta \hat{p_E}(n) \equiv \Delta p_E(n)$. We also assume that the electric perturbation is a step function of amplitude ΔP_E: $\Delta p_E(n) = \Delta P_E \, . \, H(n)$ where $H(n) = 1$ for $n \geq 0$ and 0 for $n < 0$.

The adaptive weights of the neural controller are adjusted as follows. A step perturbation is applied to the system. As shown in Fig.3(a), consecutive control signals and state vectors are evaluated. When, in the

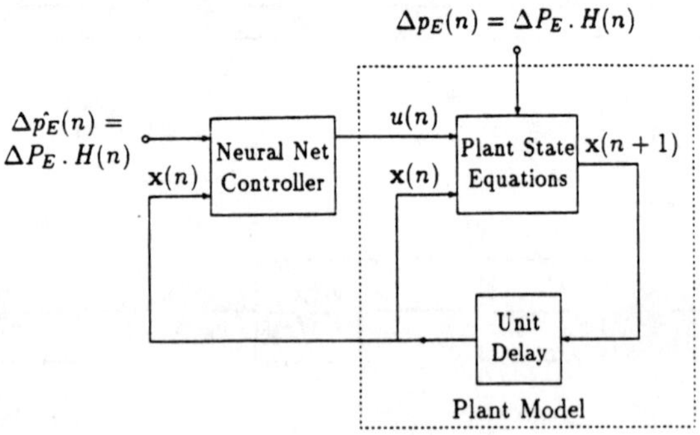

(a) Neural net controller-plant structure

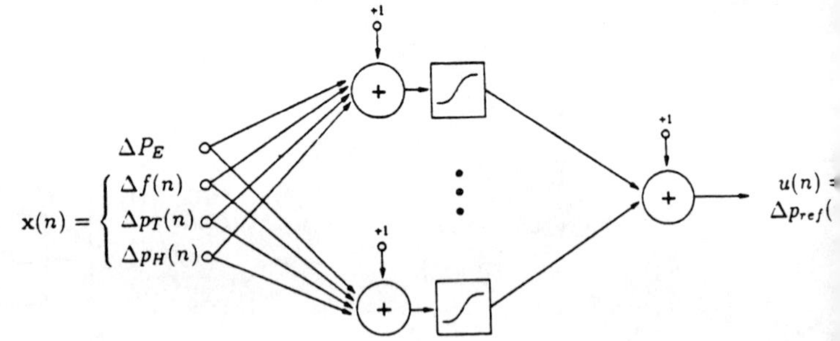

(b) Neural controller: internal structure

Figure 3: Neural net controller for a single-area system.

opinion of the designer, the system should have reached its steady-state, say at time N, an error vector is defined as the difference between the state vector at that time and its desired steady-state value. This error vector is then backpropagated back from time N to time 0, through the plant state equations and controller. The adaptive weights of the controller are adjusted in accord with the backpropagation algorithm, and the process is repeated with other step-load perturbations and plant initial state conditions. Convergence is obtained when the magnitude of the error vectors are consistently smaller than some tolerance.

Backpropagation of the error vector through the neural controller is performed in accord with the backpropagation algorithm, backpropagation through the plant is performed by means of a multiplication of the error vector by the Jacobian matrix of the plant (i.e. the matrix that contains the derivatives of the output state variables with respect to the input one) [9]. This approach is an alternative to the approach suggested by Nguyen in his original presentation of backpropagation-through-time [1]. In Nguyen's method, a neural emulator of the plant is built prior to the adaptation of the neural controller, and is used as a support for backpropagating error gradients during the controller adaptation. Our technique avoids the introduction and training of such an emulator, resulting in a substantial saving in development time. The equivalence of both methods is proven in [8].

3.2 Neural net control of a two-area system

The neural control scheme used for the two-area system is basically the same as for the one-area system. The inputs to the neural net are the state variables of the plant model and the load perturbation estimates in both areas. The outputs are the turbine reference powers in both areas.

4 Simulations

Computer simulations have been conducted to illustrate the behavior of single-area and two-area systems subject to step load perturbations. To illustrate the capacity of the neural controllers to deal with plant nonlinearities, we made the speed regulator parameter (R in Fig.1) be a nonlinear function of the power level p_T of the plant: $R(p_T) = 2.4(1 + e^{-0.2(p_T/10-5)^2}(1 - 0.4(p_T/10 - 5)^2))$ (see [8] for more details).

4.1 Single-area system simulations

A feedforward neural network containing one hidden layer with 20 hidden units was adapted to control this nonlinear plant. Fig.4 shows the frequency fluctuations at the output of the power unit when a 10% step-load increase occurs on the bus. Integral control with critical gain and neural control are compared. The neural network clearly achieves faster transient recovery while still limiting the overshoot. This result is not unexpected. The neural net controller makes use of the estimate of the load-perturbation that is provided to it as an input signal to immediately counteract the effect of the perturbation hitting the system. The integral controller, on the other hand, proceeds by accumulating the frequency error and using it to drive the plant. Therefore, it has to wait untill the error has built up before significantly effecting the system response. Most of the improvement observed in Fig.4 is thus due to the introduction of the perturbation estimate in the control scheme. However, in a highly nonlinear system such as the one used for our simulation, proper use of this estimate could not be achieved without a nonlinear controller. This, a posteriori, justifies the use of a nonlinear neural controller. A subject of future research will be a sensitivity analysis of the output of the neural controller to imprecisions in the load perturbation estimate.

4.2 Two-area system simulations

A one-hidden layer neural net controller with 20 hidden units was then trained to control a nonlinear two-area system. Fig.5 compares the dynamic response of a two-area system subject to a 1% step-load increase when controlled (a) by a neural network, (b) by two conventional integral controllers with critical gains. As explained previously, the neural controller takes as inputs the estimates of the load-perturbations in both areas. These estimates are evaluated by a special device which needs a short, but non-zero, lapse of time to output accurate estimates of the perturbations. For this reason, a 500 msec delay was allowed before the neural net started controlling the plant. During these first 500 msec, the control signals in both areas were set to zero. The use of a neural net controller has greatly reduced the frequency transient recovery period. Zero steady-state error is achieved. Fig.6 shows the frequency recovery of a two-area system subject to a 1% step-load increase in area 1 and, at the same time, a 2% step-load decrease in area 2. Dynamic responses obtained with a neural network and with integral controllers are compared. Again, the neural network achieves fast and steady transient recovery.

5 Conclusion

Layered neural networks have been successfully applied to control the turbine reference power of a computer-simulated generator unit. The same principle was then applied to a simluated two-area system. Both neural networks were adapted using backpropagation-through-time. In this paper, the frequency variations in both areas of the two-area system were inputted into the neural net controller. The neural network thus implements a global control scheme. A subject of future research consists in building independent neural net controllers for the different areas of a multi-area system. Each neural net controller receives only local information about the system (frequency in that specific area and powers in the tie-lines connected to that area). Such an architecture decentralizes the control of the overall system and reduces the amount of information to be exchanged between different nodes of the power grid. Performance with local control only will be compared with that obtained with global control.

References

[1] D. Nguyen and B. Widrow. The truck backer-upper: An example of self-learning in neural networks. In *Proceedings of the International Joint Conference on Neural Networks*, volume II, pages 357–363, Washington, DC, June 1989.

[2] Derrick Nguyen and Bernard Widrow. Neural networks for self-learning control systems. *IEEE Control Systems Magazine*, April 1990.

[3] Paul J. Werbos. Backpropagation through time: What it does and how to do it. *Proc. IEEE, special issue on neural networks*, 2:1550–1560, October 1990.

[4] Paul Werbos. *Beyond Regression: New Tools for Prediction and Analysis in the Behavioral Sciences.* PhD thesis, Harvard University, Cambridge, MA, August 1974.

[5] David E. Rumelhart and James L. McClelland, editors. *Parallel Distributed Processing.* The MIT Press, Cambridge, MA, 1986.

[6] B. Widrow and M. A. Lehr. 30 years of adaptive neural networks: Perceptron, madaline, and backpropagation. *Proc. IEEE*, 78(9):1415–1442, September 1990.

[7] Olle I. Elgerd. *Electric Energy Systems Theory.* McGraw-Hill, Inc., second edition, 1982.

[8] F. Beaufays, Y. Abdel-Magid, and B. Widrow. Application of neural networks to load-frequency control in power systems. *To be published in Neural Networks*, 1992.

[9] S. W. Piche and B. Widrow. First-order gradient descent training of adaptive discrete-time dynamic networks. Technical Report RL-TR-91-62, Rome Laboratory, Air Force Systems Command, Griffiss Air Force Base, NY 13441-5700, May 1991.

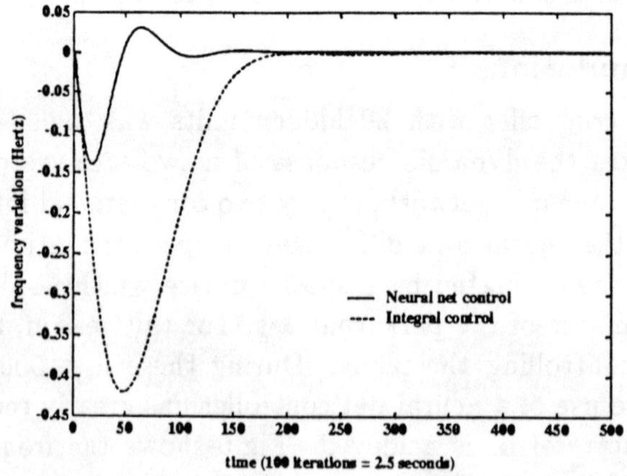

Figure 4: Step response of a single-area system subject to a 10% load increase.

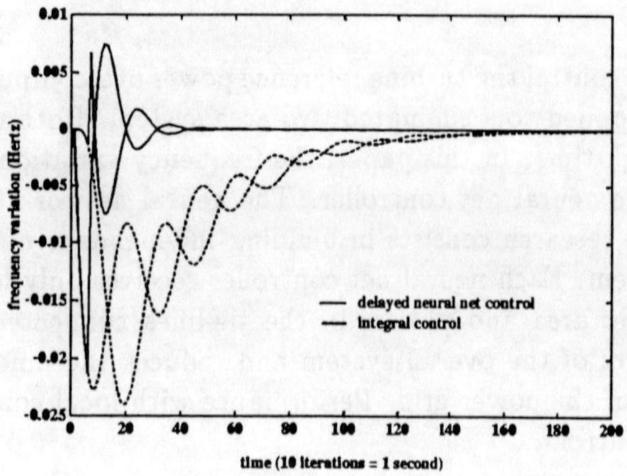

(a) Frequency variations in both areas.

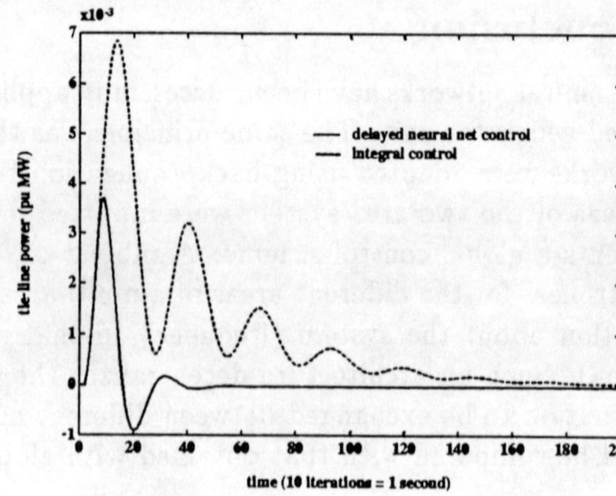

(b) Tie-line power variations.

Figure 5: Dynamic response of a two-area system subject to a 1% step-load increase in area 2.

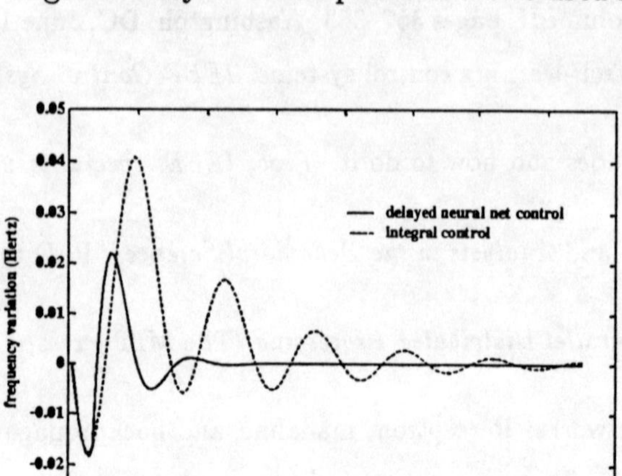

(a) Frequency variations in area 1.

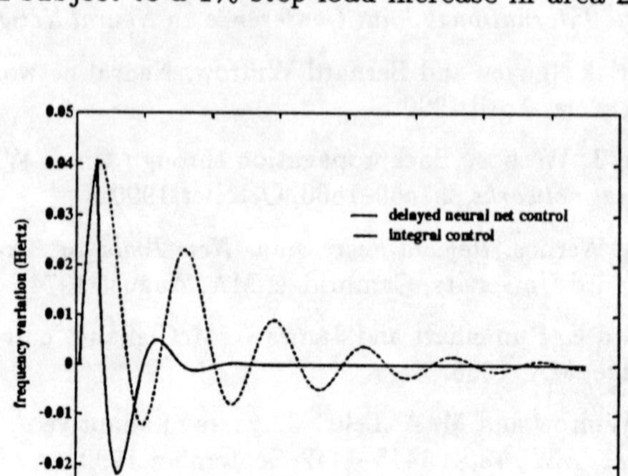

(b) Frequency variations in area 2.

Figure 6: Dynamic response of a two-area system subject to a 1% step-load increase in area 1 and a 2% step-load decrease in area 2.

REINFORCEMENT LEARNING FOR ADAPTIVE CONTROL

Larry L. Adams
Woodward Governor Company
1000 E. Drake Road
Fort Collins, Colorado 80525

Abstract

An adaptive control problem is solved by combining a reinforcement learning neural network with elements of conventional PD control. The control task is the second-order problem from the 1988 American Control Conference "Showcase of Adaptive Controller Designs." A neural network simulation is shown that learns the gain mapping required to control the second-order dynamic system. Results are presented showing system performance with command and disturbance input changes and wide variations in plant parameters.

1. INTRODUCTION

Contemporary adaptive control strategies [1], are based on the assumption that parameters of the system model are known. Mathematical functions describing a dynamic system model can be complex and exact form is often unknown. Multilayer neural networks have the ability to learn to estimate continuous mathematical functions without specifying the input-output dependence. This model-free property makes the possibility of using neural networks to identify and control unknown dynamic systems an attractive alternative.

Reinforcement learning algorithms [2-4] are an extension to supervised learning algorithms that provide a potential solution when *a priori* training data is not available. In contrast to supervised learning, the learning unit itself has the task of discovering correct response. These learning algorithms utilize a measure of actual real-time system performance feedback to enable the network to develop an initial control strategy and continue to improve in performance over time.

The described control task chosen to evaluate reinforcement learning is the second-order problem from the 1988 American Control Conference "Showcase of Adaptive Controller Designs" [5]. In the showcase, various experts in adaptive control technology presented their design methods and results from a set of standard example problems. The second-order ACC Showcase problem provides a "benchmark" for comparing these contemporary adaptive control techniques with the reinforcement learning method and results reported in this paper.

2. THE CONTROL PROBLEM

The transfer function of the second-order plant from Figure 1 is defined by $G(s) = K / (s^2 + a_2 s + a_1)$ where nominal plant parameters and model response are defined as $K = 1.0$, $a_2 = 1.4$, and $a_1 = 1.0$. Variation of plant parameters adapted for use in these experiments lie in the intervals $0.5 \le K \le 5.0$, $-1.0 \le a_2 \le 10.0$, and $-5.0 \le a_1 \le 10.0$. Plant disturbance and command inputs specified for the ACC Showcase problem are shown with an overlay of the model response in Figure 2. Controller performance is evaluated by its ability to force plant output to follow the model response with specified command and disturbance inputs over an interval of 20 seconds. After each interval, a different set of plant parameters is selected and the simulation rerun. It is assumed that the controller does not have direct access to plant parameters or disturbance signal.

3. IMPLEMENTATION

The reinforcement learning unit (RLU) shown in Figures 1 and 3 is an adaptation of the adaptive heuristic critic pole-balancing network reported by Anderson [4]. The system contains two neural network components which must simultaneously learn to predict and control. The evaluation network "predicts" the eventual outcome of a control action. An action network maps the input state vector to a real-valued output that predicts the control action to be performed.

The input state vector elements to the RLU are derived from the current and delayed values of the error between plant response $y(t)$, model response $mr(t)$, and a constant bias input. The state inputs are $x[0] = e(t-2) \cdot 0.25$, $x[1] = e(t-1) \cdot 0.5$, $x[2] = e(t)$, $x[1] = e(t) + \dot{e}(t) \cdot 10.0$, and $x[0] = 1.0$, where error is $e(t) = y(t) - mr(t)$ and the change in error is $\dot{e}(t) = e(t) - e(t-1)$. The error state parameter $x[2]$, corresponds to the proportional term (P) and the error plus error rate parameter $x[1]$, the derivative (D) of a PD controller.

An RLU internal reinforcement signal, $\hat{r}(t+1) = r(t+1) + \gamma\sigma(t+1) - \sigma(t)$, is computed from the change in state evaluation from time t to $t+1$, along with the performance feedback signal, $r = -e^2$. The value of $\hat{r}$ indicates the degree to which the state of the system changed as a result of a control action $u(t)$. If the action resulted in better/worse performance (less/more error) the weights in the action network are modified to increase/decrease the probability of performing that action in the future. A non-zero value of $\hat{r}(t+1)$ also indicates a result different than predicted and is used to modify the evaluation network to improve the evaluation function. The weighting parameter, γ, $0 \leq \gamma < 1$, discounts the current evaluation [6], thus preventing an infinite sum and allowing convergence of the algorithm.

A real-valued output from the RLU is determined stochastically [7] by $u(t) = \Psi[\mu(t), \sigma^2(t)]$, where the output of the action network μ is the mean and output from the evaluation network σ (squared), the standard deviation of a normal distribution function, Ψ. This algorithm automatically provides a variable amplitude excitation source for system identification. When expected reinforcement is high, indicating a large predicted error in plant response, the random component is high, thus increasing the search space. As performance improves, the output noise component is automatically reduced.

4. SIMULATION RESULTS

Figure 4 shows results of closed-loop plant output response and control signal for the first 20 seconds of the experiment. The initial run was made with nominal plant parameters after starting from random network weights. A successful control strategy was developed in approximately 6 seconds (simulated), or 600 iterations of the learning unit. Time required to learn depends on initial random values of network weights. Learning time was observed to vary from 3 to 15 seconds.

Figure 5 shows results of initial learning over the next 4 runs where the gain K, is stepped through each specified value with a_2 and a_1 held at their nominal values. Plant response and performance measure in each case is nearly identical. Sensitivity to disturbance inputs and reference changes is low in all cases. The greatest magnitude control signal corresponds to the lowest gain case in all cases.

Figure 6 shows results from the second set of runs. Stochastic noise can be observed on the control signal in the case of lower gain parameters. This, along with the performance measure and plant response curves, show learning continues at an observable level for these cases. Results from a run with each specified parameter set are shown in Figures 7-9 including cases where the plant is open loop unstable.

5. CONCLUSIONS

The key difficulty is steady state error indicated by slope in performance measure curves. It was expected that learning would continue to reduce steady state error, but this was not found to be the case. The evaluation function is shown to be the problem. If $\sigma(t) = \sigma(t+1)$, which occurs at steady state after initial learning, then $\hat{r}(t+1) = 0$ when $r(t+1) = (1+\gamma)\sigma$ at (t) or $(t+1)$. When the evaluation is 0, updating of network weights does not occur.

The control function can be equated to a simple PD control. The error plus error rate in state feedback provides derivative function and the RLU learns to set proportional and derivative gains encoded in network weights. A constant gain PD controller was designed by application of Routh stability criteria to system characteristic equations which give a set of inequality constraints which must be satisfied for stability. Values of P and D (250 and 0.1 sec) were picked which satisfy the criteria for all cases of plant parameters used in these experiments. Simulations run with this fixed gain PD controller produced essentially identical results to those obtained with the reinforcement learning system.

The reinforcement learning solution to the second-order problem resulted in much more stable performance and covered a wider range of plant parameters than the contemporary adaptive control methods reported in the ACC Showcase.

References

[1] K.J. Åström, B. Wittenmark, *Adaptive Control*, Addison-Wesley: Reading, MA, 1989.

[2] A.G. Barto, R.S. Sutton, and C.W. Anderson, "Neuronlike adaptive elements that can solve difficult learning control problems," *IEEE Trans. on Systems, Man, and Cybernetics.*, Vol. SMC-13, pp. 834-846, Sept./Oct. 1983.

[3] C.W. Anderson, "Strategy learning with multilayer connectionist representations," Tech. Report TR87-509.3, GTE Laboratories, Waltham, MA, 1989.

[4] C.W. Anderson, "Learning to control an inverted pendulum using neural networks." *IEEE Control Systems*, Vol. 9, No. 3, 1989.

[5] M.K. Masten and H.E. Cohen (Eds), "Special issue on adaptive controller designs." *International Journal of Adaptive Control and Signal Processing*, Vol. 3, No. 2, 1989.

[6] R.S. Sutton, "Learning to predict by the methods of temporal differences," *Machine Learning 3*, pp. 9-44, Boston, MA: Kluwer Academic Publishers, 1988.

[7] V. Gullapalli, "A stochastic algorithm for learning real-valued functions via reinforcement feedback," Tech. Report 88-91, University of Massachusetts, Amherst, MA, 1988.

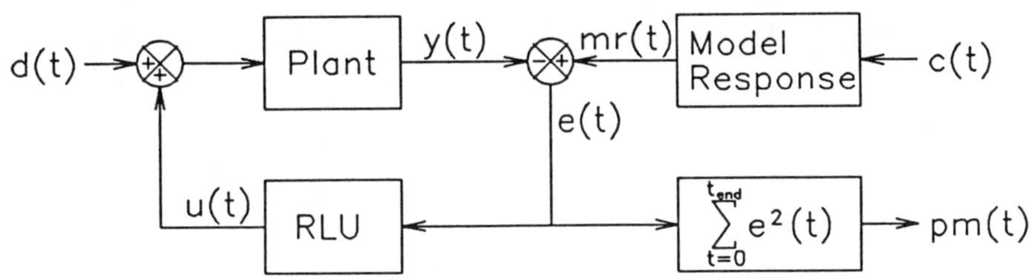

Figure 1. Block diagram of system configuration.

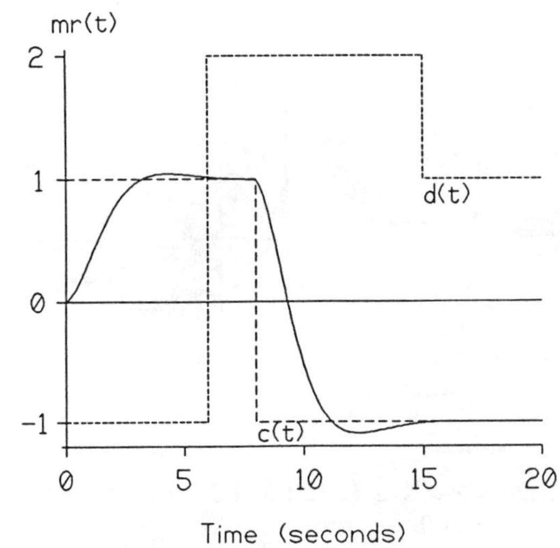

Figure 2. Command and disturbance inputs with model response curve for the undisturbed plant at nominal parameters.

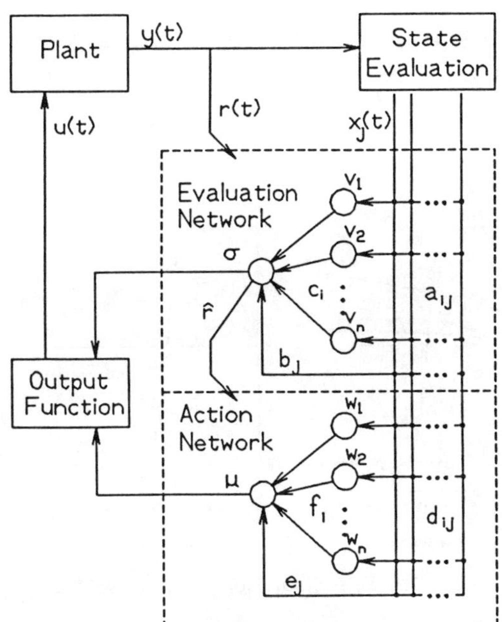

Figure 3. Block diagram of the reinforcement learning unit and plant interface.

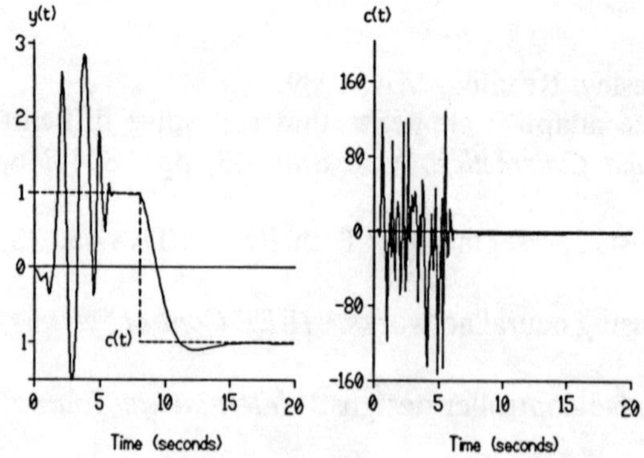

Figure 4. Plant response and control signal during the first 20 seconds of learning. Case K = 1.0, a2 = 1.4, a1 = 1.0.

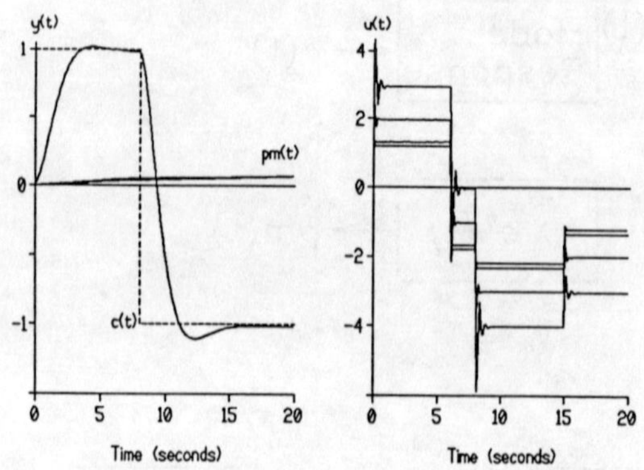

Figure 5. Case K ∈ {0.5, 1.0, 3.0, 5.0} with a2 = 1.4, a1 = 1.0.

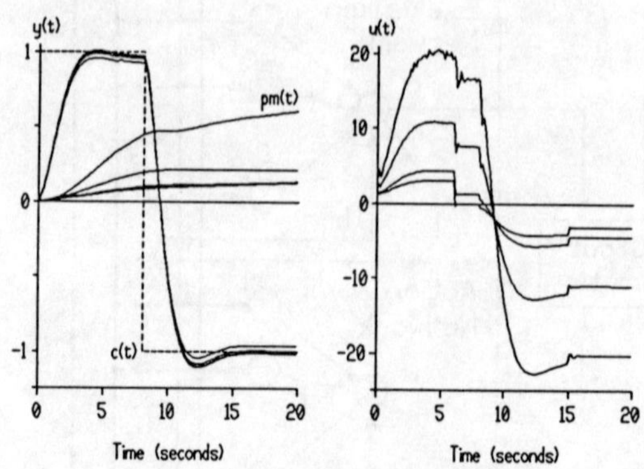

Figure 6. Case K ∈ {0.5, 1.0, 3.0, 5.0} with a2 = 1.4, a1 = 10.0.

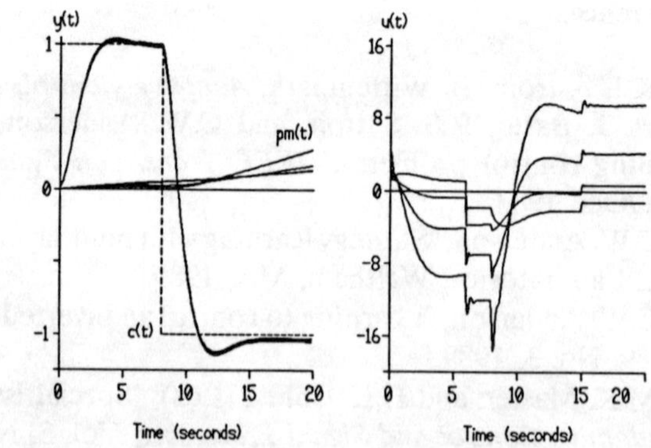

Figure 7. Case K ∈ {0.5, 1.0, 3.0, 5.0} with a2 = 1.4, a1 = -5.0.

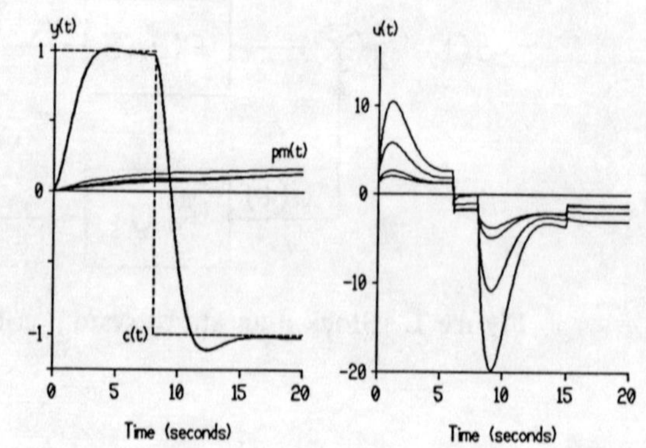

Figure 8. Case K ∈ {0.5, 1.0, 3.0, 5.0} with a2 = 10.0, a1 = 1.0.

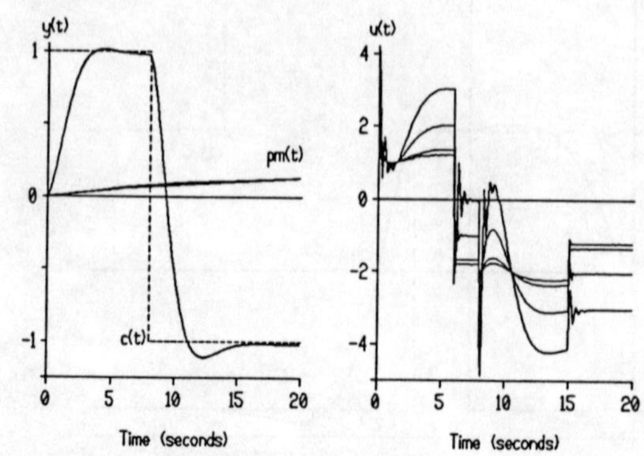

Figure 9. Case K ∈ {0.5, 1.0, 3.0, 5.0} with a2 = -1.0, a1 = 1.0.

G. Load Forecasting

APPLICATION OF ARTIFICIAL NEURAL NETWORKS TO LOAD FORECASTING

Alain J. Germond **Nicolas Macabrey** * **Thomas Baumann**

Laboratoire de Réseaux d'Energie Electrique
Ecole Polytechnique Fédérale de Lausanne
CH-1015 Lausanne (Switzerland)

Abstract

After a brief review of neural network models and references to the application of supervised learning for short-term load forecasting, this paper analyses the application of Kohonen's self-organizing feature map to short-term forecasting of peak electrical load. The Kohonen self-organizing feature map is used for the classification of electrical loads. The network not only learns similarities of load patterns in an unsupervised manner, but it uses the information stored in the weight vectors of the Kohonen network to forecast the future load. The results are evaluated by using one year of hourly load data of a real system to train the network, and forecasting the daily peak hour loads for the next year.

1. Introduction

The improvement of system operational security and economy are the motivations for forecasting the load of an electric power system. It is quoted in reference [1] that the impact upon operating costs of an increase of 1% in the short-term load forecasting error for the British power system was evaluated as 10 million £/year in 1985.

A survey [1] of the research efforts devoted to short-term load forecasting lists more than 70 papers. The principal methods developed so far for short-term load forecasting of peak electrical load are extrapolation, correlation, temporal series and regression. The idea of using pattern recognition technique for load forecasting was proposed in 1975 [2]. This 1975 paper can be considered as the first application of an artificial neural network to a power system problem. Expert system [3] and fuzzy techniques [4] were also proposed.

Already a number of papers on artificial neural networks for load forecasting were published in the last two years [5-18]. A comparison of forecasting accuracy of neural networks and traditional techniques can be found in reference [19].

Section 2 reviews applications of supervised neural network models to load forecasting and presents the Kohonen network. Section 3 discusses its application to load forecasting, and section 4 the evaluation. More details are given in [20].

2. Neural Network Models

Figure one displays the major neural network models, according to the learning paradigm.

Neural networks with *supervised learning* are very familiar to power engineers, and most of the applications of neural networks to load forecasting use multilayer feedforward networks with *supervised learning* [5-14]. Improvements such as nonfully connected model [15] have been proposed to improve the learning efficiency. A strategy based on the concept of minimum distance to historical patterns was proposed for selecting the training cases [16].

The first use of *unsupervised learning* for the electric load forecasting problem dates back to 1975 [2]. Unsupervised learning with Kohonen's self-organizing feature maps [21] was used recently for classifying the type of load [17], whereas the prediction itself was made with a multilayer perceptron [18]. Our new approach of the load forecasting problem is to use a Kohonen map to classify data representing load patterns and other parameters, such as meteorological data, and to use directly the information stored in the weight vectors of the Kohonen map to predict the load [20].

---

* Thomas Baumann is now with : Siemens AG Oesterreich
Gudrunstraße 11
A- 1100 Wien (Austria)

Artificial neural network models

Networks with supervised learning

- Perceptron
- Adaline
- Hopfield
- Multi-layer perceptron
- LVQ and LVQ2
- Functional Link Net

Networks with unsupervised learning

- Adaptive Resonance Theory
- Kohonen network
- Hérault and Jutten net

- Counterpropagation

Figure 1: Taxonomy of neural network models

2.1 The Kohonen network

The Kohonen network [21, 22] is an array of neurons, usually arranged on a square grid in 2-dimensional space, and connected according to a specified neighborhood order: for example, all 8 neurons adjacent to a given neuron can be defined as its neighbors of the first order, then the neurons adjacent to a first order neighbor, excluding those already considered, as neighbors of the second order etc. The dimension of the vectors which will be presented as inputs to train the network is N.

Figure 2 represents a 2-dimensional Kohonen network of size 8 x 8, with 6-dimensional inputs. Each component of the input vector is connected to every individual neuron in the Kohonen network (although only the connections to the first column of the Kohonen network are represented in figure 2). Thus there is a connection between each neuron i of the Kohonen network and every component of the N-dimensional input vector. The weights of these connections form a synaptic weight vector w_i of dimension N for each neuron i.

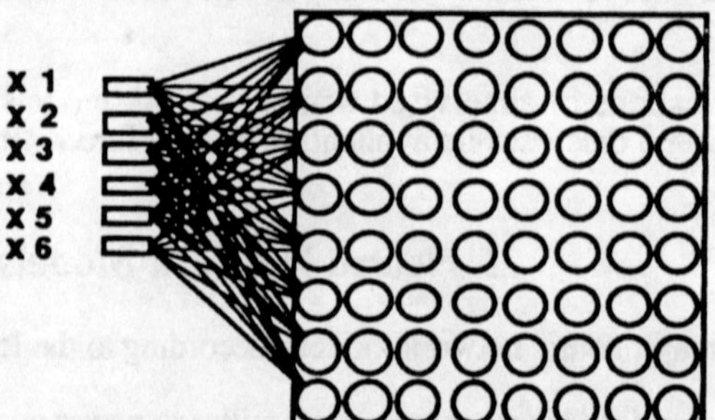

Figure 2: A 8 x 8 Kohonen network with 6-dimensional input

Learning starts with the weight vectors w_i initialized randomly. At each learning step k, an input vector x(k) is drawn from the training set. The Euclidean distance from this input vector to each weight vector is calculated and the weight vector w_c with the minimal Euclidean distance to x(k) is designated as the winner. The components of the weight vectors w_i are adapted according to the rule

$$w_i(k+1) = w_i(k) + \alpha\,(c, k) * (x(k) - w_i(k))$$

where $0 < \alpha\,(c, k) < 1$ is the learning gain, which is maximal for the winning neuron, and decreases with the neighborhood order c and with the learning steps. After several thousand learning steps, the network is organized, and the N-dimensional input vectors have been mapped on the 2-dimensional Kohonen map.

Figure 3 is a graphical representation of the weight vectors stored in neurons Nr 8 to 54 of a 8 x 8 Kohonen network, trained with 6-dimensional input vectors . The neurons of the Kohonen map are numbered 0 to 63.

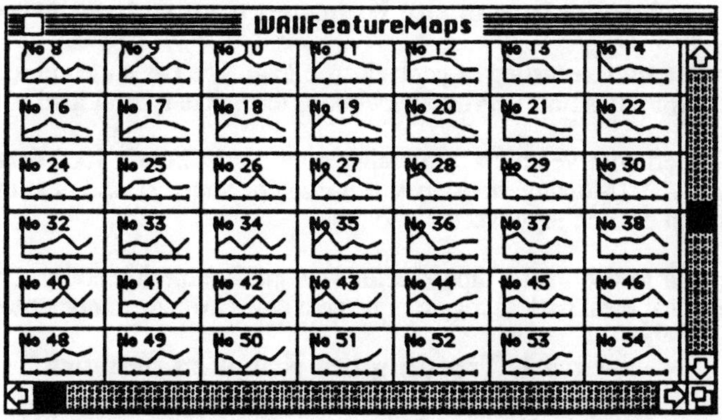

Figure 3: Example of Kohonen feature map : organization of 6-dimension weight vectors on a 8 x 8 map

Let us assume that the 6 components of each input vector in this example represent the amplitudes of a function at 6 consecutive instants in time : the weight vectors stored in the Kohonen network represent then *prototypes* of the input vectors. The number of prototypes is at most equal to the number of neurons, and the Kohonen network divides the input vector space into sub-spaces, according to the density of the input vectors in each subspace. Furthermore, the prototypes of neighboring subspaces are represented by the weight vectors of neighboring neurons in the Kohonen map. These properties are illustrated in Figure 4. The mathematical foundations and the convergence properties of the Kohonen network are studied in references [21] to [26].

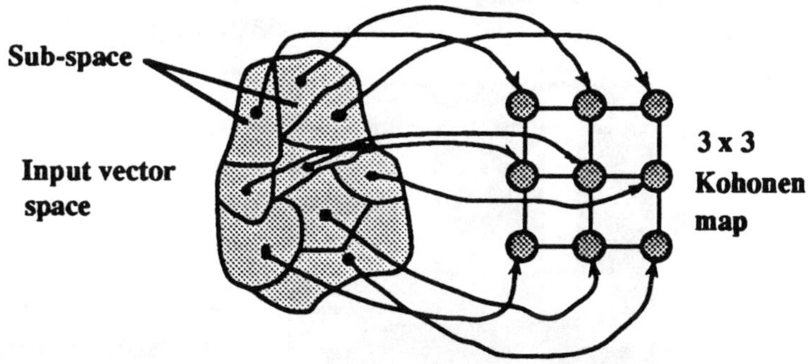

Figure 4: Conservation of the topological neighborhood of the input vector space

The weight vectors represented in Figure 3 exhibit an organization, which can be outlined in our example by observing the amplitudes of a specific component , e. g. component Nr 0, as displayed in Figure 5. The area of the squares in this figure are proportional to the amplitude of the corresponding component of the weight vector.

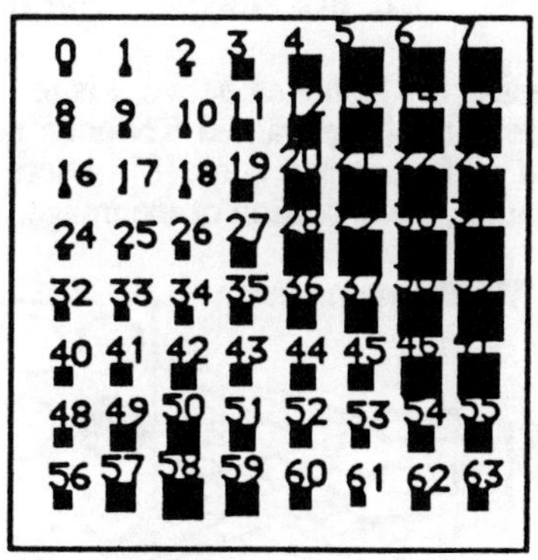

Figure 5: Representation of component Nr 0 of the weight vectors of figure 3.

The Kohonen network has been used in power systems for diagnosis and classification problems [27, 28]. In these applications, the input vectors are samples of measurements on an apparatus or vectors representing the state of a power system. The Kohonen network is used on-line to recognize whether the result of a test, or the present state of the system, is similar to any situation already met in the past.

3. Application of the Kohonen map to load forecasting

The objective is short term forecasting of peak electric load. The basic idea in using the Kohonen network for load forecasting is to train this unsupervised network to recognize typical load patterns.

A training set is formed with vectors x containing each the 24 hourly load data of a day, 4 indices (0 or 1) indicating the type of day (Monday, mid-week, Saturday or Sunday), and next day's load at the peak hour. In that case the dimension of each input vector is 29.

The choice of the data to use in the input vector was made experimentally. Other cases were treated with additional meteorological data, such as for example next day's average temperature and cloudiness.

3.1. Training the Kohonen network

A Kohonen network is trained with vectors formed from a year of a real power system data, drawn randomly. After training, the organization of the map can be interpreted, by presenting again each of the input vectors, and observing where the nearest weight vector to a given input vector is located on the map.

Figure 6 represents such a map obtained after training a 10 x 10 Kohonen network. It can be noticed that the neurons, or rather their weight vectors, are grouped into clusters representing input vectors with similar characteristics, such as season and day of the week. The unshaded squares represent neurons for which the weight vector was never selected as the nearest to an input vector. These neurons have the capability to generalize.

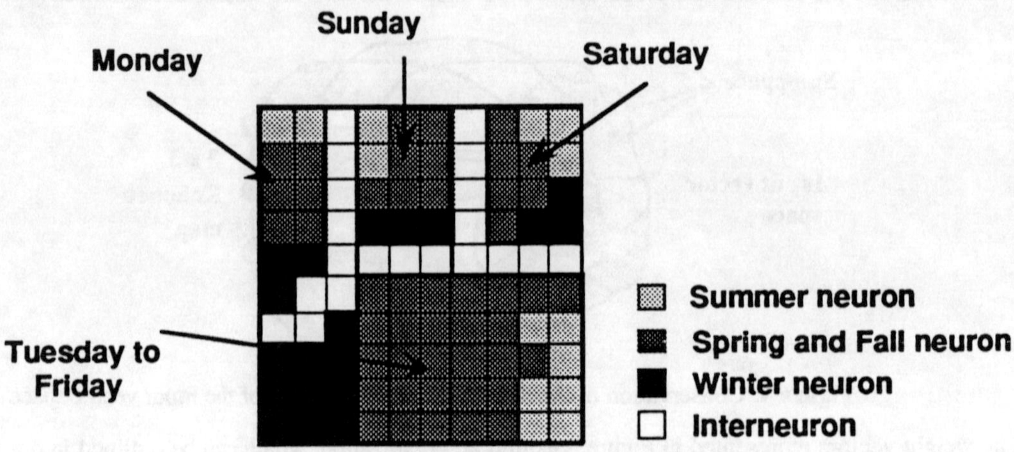

Figure 6: Organization of a Kohonen network after learning power system load profiles

3.2. Forecasting the load

In order to be able to do any forecast, the Kohonen network is used as an autoassociator: the peak load of the next day is extracted from the weights of the trained Kohonen network by autoassociation. The process is similar to experiments described in Kohonen's book [8], where digitized images are learned and can be reconstructed from partial information, i. e. a portion of the image. This is illustrated in Figure 7.

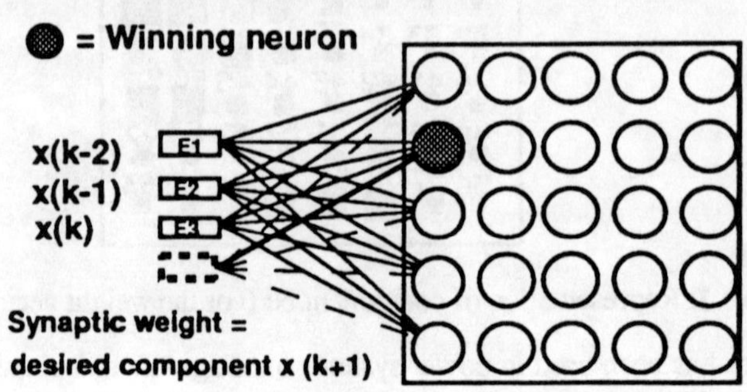

Figure 7: Autoassociation

An incomplete input vector is formed with the components representing today's load, the type of day, and optionally the meteorological forecast for the next day. The Euclidean distance to each of the weight vectors is calculated, using only the components present in the input vector. The winning neuron is found and the component of its weight vector corresponding to the next day's peak load is the forecasted load.

4. Evaluation

An evaluation was made, using the hourly loads from the EOS power system of the Western part of Switzerland, from October 1984 to September 1986. The meteorological data supplied for the same period were the average daily temperature and the daily cloudiness for Geneva. The data were split into a training set (the data from October 1984 to September 1985) and a test set (the data from October 1985 to September 1986).

After training the network with the first set of data, which required 25000 steps and several minutes on a workstation, the second set of data was used to forecast the load at peak hour for every day of the test year, and the results of the forecast were compared to the historical load data. The experiment was made for different Kohonen network sizes, and did not include meteorological data. Table 1 displays in each case the mean absolute percent error, the mean percent error, and the standard deviation of the percent error over a one year forecast. The percent error is defined as 100 * [(actual load - forecasted load) / actual load].

Size of the Kohonen net	Mean absolute percent error	Mean percent error	Standard deviation of the percent error
7*7	3.366	1.93	4.00
10*10	3.067	1.58	3.72
15*15	2.998	1.51	3.61
20*20	2.938	1.44	3.58
25*25	2.838	1.40	3.45
30*30	2.814	1.37	3.43

Table 1: Mean absolute percent error, mean percent error, and standard deviation of the percent error, in function of the Kohonen network size.

The 3 error measures get smaller with increasing network size. The positive value of the mean percent error seems to reflect the annual increase of the load between the year used for training, and the year used for the forecast.

The frequency of the percent error in the case of a 30 x 30 size Kohonen network is shown in Figure 8.

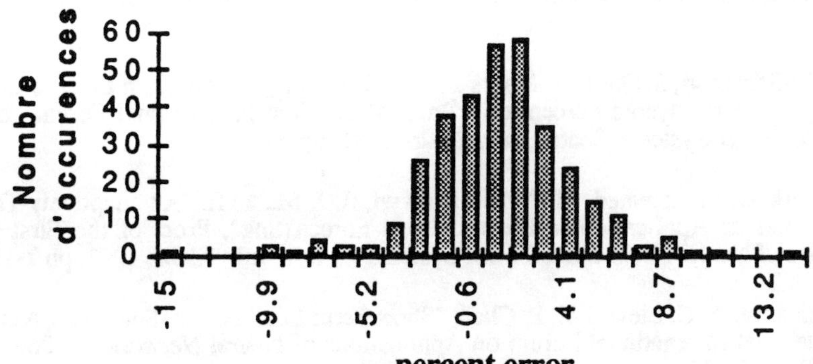

Figure 8: frequency of the errors in the case of a forecast with a 30 x 30 network, without meteorological data.

Adding the average temperature to the input vector, and using the forecasted temperature improves only slightly the accuracy, essentially by eliminating the few cases with the largest errors. The explanation for that phenomenon is that the information about the temperature is already implicitly included in the load profile, and therefore the explicit use of the temperature in the forecast only improves the results in the case where there are major temperature changes from one day to the next.

5. Conclusion

There is a major economical motivation for improving the accuracy of electrical load forecasting. Classical methods have been developed for this purpose over the past 30 years, but artificial neural networks are considered as very promising for this kind of task, since they can be used for modelling the behavior of a complex system without the need to assume predetermined relationships, for example linear or exponential, between variables.

There are several neural network models, and multilayer networks with supervised learning have been used most often in the power system applications. The technique developed in the paper applies the Kohonen network, which tries to classify data with unsupervised learning, according to their distance relationship.

The quality of the results need to be improved by the selection and pretreatment of the variables, and the optimal size of the network need to be investigated. The computation effort to train a Kohonen network, simulated on a workstation, with one year of load data is a matter of minutes.

6. Acknowledgements

The support of the Swiss National Science Foundation, under project Nr 20.26557.89, for the research described in this paper is gratefully acknowledged. We also would like to thank MM. S Birnstein and Y. Jaccard, S. A. l'Energie Ouest Suisse (EOS), Lausanne, for their cooperation.

7. References

[1] G. Gross, F. D. Galiana, "Short-term Load Forecasting ", Proceedings of the IEEE, vol 75, no 12, Dec. 1987, pp 1558-1572.

[2] T. S. Dillon, K. Morsztyn, K. Phua, "Short Term Load Forecasting Using Adaptive Pattern Recognition and Self-Organizing Techniques", Proc. of Fifth Power Systems Computation Conference, Cambridge, Paper 2.4/3, pp. 1-16, 1975.

[3] S. Rahman, R. Bhatnagar, "An Expert-System Based Algorithm for Short-Term Load Forecast,"IEEE Transactions on Power Systems, Vol 3, No 2, pp 392-399, May 1988.

[4] G. Lambert-Torres, C. O. Traore, F. G. Mandolesi, D. Mukhedkar, "Short-Term Load Forecasting Using A Fuzzy Engineering Tool", Proc. of the First International Forum on Applications of Neural Networks to Power Systems, Seattle, July 23-26, 1991, pp 36-40.

[5] D. C.Park, M. A. El-Sharkawi, R. J. Marks II, L. E. Atlas, M. J. Damborg, "Electric Load Forecasting Using An Artificial Neural Network", IEEE Transactions on Power Systems, Vol. 6, No 2, pp 442-448, May 1991.

[6] M. A. El-Sharkawi, S. Oh, R. J. Marks II, M. J. Damborg, "Short Term Electric Load Forecasting Using an Adaptively Trained Layered Perceptron", Proc. of the First International Forum on Applications of Neural Networks to Power Systems, Seattle, July 23-26, 1991, pp 3-6.

[7] D. C. Park, O. Mohammed, M. A. El-Sharkawi, R. J. Marks II, "An adaptively Trainable Neural Network Algorithm and its Application to Electric Load Forecasting", Proc. of the First International Forum on Applications of Neural Networks to Power Systems, Seattle, July 23-26, 1991, pp 7-11.

[8] D. Srinivasan, A. C. Liew, J. S. P. Chen, "Short Term Load Forecasting Using Neural Network Approach", Proc. of the First International Forum on Applications of Neural Networks to Power Systems, Seattle, July 23-26, 1991, pp 12-16.

[9] T. S. Dillon, S. Sistito, S. Leung, "An adaptive Neural Network Approach in Load Forecasting in a Power System", Proc. of the First International Forum on Applications of Neural Networks to Power Systems, Seattle, July 23-26, 1991, pp 17-21.

[10] J. T. Conner, L. E. Atlas, D. Martin, "Recurrent Neural Networks and Load Forecasting", Proc. of the First International Forum on Applications of Neural Networks to Power Systems, Seattle, July 23-26, 1991, pp 22-25.

[11] K. Y. Lee, Y. T. Cha, C. C. Ku, "A Study of Neural Networks for Short-term Load Forecasting", Proc. of the First International Forum on Applications of Neural Networks to Power Systems, Seattle, July 23-26, 1991, pp 26-30.

[12] J. N. Hwang, S. Moon, "Temporal Difference Method for Multi-Step Prediction: Application to Load Forecasting", Proc. of the First International Forum on Applications of Neural Networks to Power Systems, Seattle, July 23-26, 1991, pp 41-45.

[13] K. Y. Lee, Y. T. Cha, J. H. Park, "Short-Term Load Forecasting using an Artificial Neural Network", IEEE Transactions on Power Systems, Vol 7, No 1, pp 124-130, Feb 1992.

[14] K.-L. Ho, Y.-Y Hsu, C.-C. Yang, "Short term load Forecasting Using a Multilayer Neural Network with an Adaptive Learning Algorithm", IEEE Transactions on Power Systems, Vol 7, No 1, pp 141-149, Feb 1992.

[15] S.-T. Chen, D. C, Yu, A. R. Moghaddamo, "Weather Sensitive Short-Term Load Forecasting Using Nonfully Connected Artificial Neural Network", IEEE Transactions on Power Systems, Vol 7, No 3, pp 1098-1105, Aug. 1992.

[16] T. M. Peng, N. F. Hubele, G. G. Karady, "Advancement in the Application of Neural Networks for Short-term Load Forecasting", IEEE Transactions on Power Systems, Vol 7, No 1, pp 250-257, Feb. 1992.

[17] Y.-Y. Hsu, C.-C. Yang, "Design of artificial neural networks for short-term load forecasting. Part I: Self-organising feature maps for day type identification", IEE Proceedings-C, Vol. 138, No 5, pp 407-413, September 1991.

[18] Y.-Y. Hsu, C.-C. Yang, "Design of artificial neural networks for short-term load forecasting. Part II: Multilayer feedforward networks for peak load and valley load forecasting", IEE Proceedings-C, Vol. 138, No 5, pp 414-418, September 1991.

[19] M. C. Brace, J. Schmidt, M. Hadlin, "Comparison of the Forecasting Accuracy of Neural Networks with Other Established Techniques", Proc. of the First International Forum on Applications of Neural Networks to Power Systems, Seattle, July 23-26, 1991, pp 31-35.

[20] N. Macabrey, T. Baumann, A. Germond, "Prévision de charge dans un réseau électrique à l'aide du réseau de neurones de Kohonen", Bulletin ASE/UCS, 5/1992, pp 13-19.

[21] T. Kohonen., "Self-Organized Formation of Topologically Correct Feature Map", Biological Cybernetics, vol. 43, pp. 59-69, 1982.

[22] T. Kohonen, "Self-Organization and Associative Memory", 3rd edition, Springer-Verlag, Berlin, 1989.

[23] M. Cottrell, J. Fort, "A Model of Retinotopy: A Self-Organizing Process", Biol. Cybernetics no. 53, 1986, pp 405-411.

[24] H. Ritter, K. Schulten, "Convergence Properties of Kohonen's Topology Conserving Maps: Fluctuations, Stability and Dimension Selection", Biol. Cybernetics no. 60, 1988, pp 59-71.

[25] H. Yang, T. S. Dillon, "Convergence of Self-Organizing Neural Algorithms", Neural Networks, Vol. 5, No 3, 1992, pp 485-493.

[26] H.Ritter, T. Martinetz, K. Schulten, "Neural Computation and Self-Organizing Maps, An Introduction", Addison-Wesley, Reading, 1992.

[27] T. Baumann, A. Germond, D. Tschudi, "Impulse Test Fault Diagnosis on Power Transformers Using Kohonen's Self-Organizing Neural Network", Proceedings of the 3rd Symposium on Expert System Application to Power Systems, Tokyo, April 1-5, 1991, pp 642-647.

[28] D. Niebur, A. Germond, "Power System Static Security Assessment Using the Kohonen Neural Network Classifier", IEEE Transactions on Power Systems, Vol 7, No 2, pp 865-872, May 1992.

SHORT-TERM ELECTRIC LOAD FORECASTING USING NEURAL NETWORKS

Mostafa Khadem
Alex Lago
ABB Systems Control
2550 Walsh Avenue
Santa Clara, CA 95051

Edward Dobrowolski
Philadelphia Electric Company
2301 Market Street
Philadelphia, PA 19103

Abstract

This paper presents the application of neural networks to forecast the electrical load for the next 168 hours (1 week). A multi-layer neural network is proposed to provide a non-linear mapping between weather parameters and electric load. Using historical weather parameters (such as dry bulb temperature, relative humidity, and sky cover), and historical hourly loads, a neural network is trained for each day type and each weather-defined season. Using the forecast of weather parameters for the next 168 hours, the trained neural networks are then used to forecast the hourly electric load for the same forecast period. The proposed neural network is a fully connected feedforward network. The input parameters to each network include the hour of the day, the previous hourly loads, the previous hourly weather variables, the combination of previous hourly weather variables, and the forecast of weather variables. Each season has a different set of input parameters. The learning algorithm is based on backpropagation with a momentum term included in the weight adjustment equation. The results have shown that absolute relative errors of 2.5% may be attainable with proper selection of the input parameters for each season. The results for winter and summer are presented using the historical hourly weather and load data provided by the Philadelphia Electric Company (PECo). A comparison is made between the actual and forecasted load data.

INTRODUCTION

Energy Management Systems (EMS) in electric utilities require a load forecasting subsystem to provide forecasted load for use by other subsystems, such as unit commitment and network power flow. The accuracy of the forecast is essential to properly dispatch generating units. Conventional approaches to load forecasting have been based on time series and regression techniques . Model development and parameter estimation using these approaches is a tedious task, and requires extensive customization for each electric utility.

An alternative approach is the use of neural networks technique. A number of cases of using neural networks in electric load forecasting have been reported [1-6]. The neural network uses the most recent historical data, and does not require any assumption of the functional relationship between load and weather parameters. When the neural network is completely trained, the non-linear mapping between load and weather data is obtained. Neural network model development and customization is based on analysis of historical data, resulting in proper selection of the input parameters of the neural network. This process is less cumbersome and less time consuming than the conventional approaches. Another feature of the neural network approach is the ease of updating the model parameters when new load/weather patterns become available.

In this paper the results of using neural networks to forecast the hourly load for the next 168 hours is presented. Forecasts are made for a week in summer and two consecutive weeks in winter using the most recent historical data as training cases for each forecast.

NEURAL NETWORK ALGORITHM

Backpropagation algorithm is used to train each of the neural networks for each day type. This algorithm is based on the Generalized Delta Rule [7] which defines the changing rule of the neural network weights upon presentation of an input/output pair "p", and is given by:

$$\Delta_p w_{ji} = \eta \, \delta_{pj} \, o_{pi} \qquad (1)$$

where o_{pi} is the output value of node i obtained from presentation of pattern "p", δ_{pj} is a measure of prediction error at node j, and $\Delta_p w_{ji}$ is the change in weights from ith to jth node to be made upon the presentation of input/output pair "p". Parameter η is the learning rate which is selected to be a number less than one. The procedure of updating the weights is as follows:

a) A pair of input/output "p" is presented to the network. At this phase the input propagates forward through the network, and the output value for each node j, o_{pj} , is computed. If the jth node is in the output layer, then o_{pj} is compared with the corresponding target value in that node, which provides the error signal δ_{pj} for that node, given by :

$$\delta_{pj} = (t_{pj} - o_{pj}) \, f'_j \, (net_{pj}) \qquad (2)$$

for the output node

b) Since there is no target value for the hidden nodes, then the error signal δ_{pj} for the output nodes is passed backward through the network, to provide an equivalent error signal for the hidden nodes. This pass calculates δ_{pj} for the hidden nodes by :

$$\delta_{pj} = f'_j \, (net_{pj}) \sum_k \delta_{pk} \, w_{kj} \qquad (3)$$

for the hidden nodes

c) Then the weight changes for all the nodes can be calculated from Equation 1.

The activation function f is selected to be a hyperbolic tangent function.

It is shown [7] that adding a momentum term to Equation 1 can speed up the learning process. The momentum term includes the effect of previous change of w_{ji} in the current changes of w_{ji}. Then Equation 1 is modified to include the momentum term as:

$$\Delta_p w^{(n)}{}_{ji} = \eta \, \delta_{pj} \, o_{pi} + \alpha \, \Delta_p w^{(n-1)}{}_{ji} \qquad (4)$$

Parameter α is the momentum rate which is selected to be a number less than 0.9.

The above procedure is called the "pattern learning" [8] where the weights are adjusted after presentation of each input/output pattern, which is more common to use than "batch learning" where weights are adjusted after all the input/output patterns are presented.

After all of the input/output patterns are observed with weights adjusted after each observation, then process is repeated for another iteration pass. The number of maximum iteration can be set to a large number and the program will stop when a test of convergence is passed

NEURAL NETWORK STRUCTURE

The structure of the neural network used for this study is shown in Fig. 1. The input parameters for the winter and summer seasons are as shown in Tables 1 and 2 respectively. The selection of these parameters were partly based on the analysis of historical weather and load data, and partly by performing various training tests. The human comfort temperature of 60 Deg F, where heating or cooling in most cases is not required has also been considered in selection of input parameters.

Each neural network has four hidden nodes in one layer. The output parameter of the network is the difference between the load corresponding to the hour of the day "t", and the load at "t-168". A bias node is also included in each network.

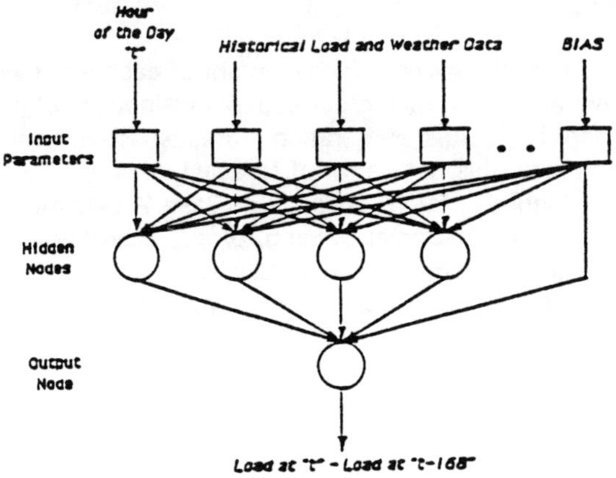

Fig. 1. Neural Network Structure

1. Hour of the day "t"
2. Load at "t-168" hour
3. Load at "t-169" hour
4. Load at "t-170" hour
5. Maximum load at "t-168"
6. Minimum load at "t-168"
7. Maximum temperature at "t-168"
8. Minimum temperature at "t-168"
9. Difference of temperature at "t" and "t-24"
10. Difference of temperature at "t" and "t-168"
11. Difference of temperature at "t" and 60 degree F
12. Difference of temperature at "t" and average of last four hours

Table 1. Neural Network Input Parameters for Winter

1. Hour of the day "t"
2. Load at "t-168" hour
3. Load at "t-169" hour
4. Load at "t-170" hour
5. Difference of temperature at "t" and "t-24"
6. Difference of temperature at "t" and "t-48"
7. Difference of temperature at "t-168" and 60 degree F
8. Difference of temperature at "t" and "t-168"
9. Difference of temperature at "t" and 60 degree F
10. Difference of THI at "t" and "t-168"
11. Difference of temperature at "t" and average of last four hours
12. Square of difference of temperature at "t" and 60 degree F
13. Difference of relative humidity at "t" and "t-168"
14. Relative humidity at "t"
15. Difference of cloud cover at "t" and "t-168"
16. Cloud cover at "t"
17. Square of difference of temperature at "t-168" and 60 degree F
18. Difference of temperature at "t" and same hour average of last seven days
19. Minimum load at "t-168"

Table 2. Neural Network Input Parameters for Summer

TEST CASES

Historical data was provided by PECo for 1991 and the first half of 1992. Table 3 provides the training and forecasting dates for each of the three test cases. For each of the test cases, one neural network is trained for each day of the week excluding holidays.

Figures 2, 3, and 4 present the hourly temperature and load variation from Monday through Sunday for each of the cases respectively. For Case 3, as shown in Figure 4, there is a large temperature drop from Tuesday to Wednesday. The Load for Wednesday has reflected this temperature change.

FORECASTING RESULTS

As mentioned earlier, each day type was trained by a separate neural network. Each element of each input vector of the neural network was normalized by subtracting the average value of the corresponding input vector, and dividing by its standard deviation. During training, after each 10 rounds of iteration, forecast RMS error was calculated. It was observed that for some selection of learning rate, this RMS error of forecast does not decrease monotonically as the training error decreases. This is shown in Figure 5, for Wednesday of Case 1. One may have to stop training before the increase in the forecast error happens to avoid over training, or decrease the learning rate. This over training happens usually for learning rates of 0.02 and larger. Very smaller values of learning rate (i.e. 1.0E-4) did not create such problem. Learning rates of 9.0E-5 to 0.01, and momentum rates of 0.0 to 0.9 are used for the training tests.

The hourly absolute relative error is defined as:

$$\text{Error} = 100 \times (|\text{actual load-forecasted load}|)/\text{actual Load}$$

The daily average absolute relative error for each case and for each day type is presented in Table 4. Figures 6, 7, and 8 present the one week forecast for the above three cases.

The most comprehensive comparison of load forecasting using neural networks with other established techniques is made by M.C. Brace [9] at Puget Sound Power & Light Company, in the form of a forecasting contest. The 1990-1991 contest included neural networks models, regression models, and a judgmental forecast. The forecast is made for 24 hours of next day. The evaluation of each method was based on the forecast results of 137 different forecast days. The lowest average absolute relative error for weekdays forecast was obtained by a regression model at about 2.3%. The neural networks models resulted in error of about 5%. The judgmental forecast has shown an error of about 4%.

The work presented in this paper forecasts the hourly load with a lead time of one full week (168 hours). The authors are not aware of any comprehensive comparison for a forecast with such a lead time. Chen et. al [3], have compared the Auto Regression Integrated Moving Average (ARIMA) with a neural network model, using a lead time of one week. No average value of error is presented for the ARIMA model. A few figures for various days within the forecast period show a much less accurate forecast results for the ARIMA model.

CONCLUSIONS

The 168 hours-ahead electric load forecasting results using neural networks technique is presented. The backpropagation algorithm is used for training of the fully connected feedforward neural networks. Preliminary test results show averages of 2.81, 2.44 and 2.58 % for the daily absolute error for each of the three test cases. These results indicate that with proper selection of input parameters and training data, daily average absolute relative error of 2.5% and lower may be attainable. The forecast is very satisfactory for the case when there was a cold front.

Forecast error should be calculated during training process to avoid the over training which can be caused by improper selection of the learning rate.

The training time for each day type of the test cases vary from 1 to 2 minutes of CPU time using the 6610 DEC VAX computer with CPU speed of 32 MIPS, for 800 iterations where convergence is acceptable. Currently, work is in progress to model transitional seasons between winter and summer. Use of Kohonen neural networks is being investigated for classification of weather-defined seasons. Modeling of holidays and special days is also in progress.

ACKNOWLEDGEMENTS

The authors would like to thank Mr. Farrokh Albuyeh, Mr. Aru Arunasalam, and Mr. Art Cohen from ABB for their valuable suggestions.

REFERENCES

[1] T. M. Peng, et. al, "An Adaptive Neural Network Approach to One-week ahead Load Forecasting", 92SM 407-7 PWRS.

[2] D. C. Park, et. al,"Electric Load Forecasting Using an Artificial Neural Network", IEEE Transaction on Power Systems, Vol. 6, No. 2, May 1991.

[3] S. Chen, et. al,"Weather Sensitive Short-Term Load Forecasting Using Nonfully Connected Artificial Neural Network", IEEE/PES 1991 Meeting, San Diego.

[4] T. M. Peng, et. al,"Advancement in the Application of Neural Networks for Short-Term Load Forecasting", IEEE/PES 1991 Meeting, San Diego.

[5] M. A. El-Sharkawi, et. al, "Short Term Electric Load Forecasting Using an Adaptively Trained Layered Perceptron", Proceedings of the First Int. Forum on Applications of Neural Networks to Power Systems, Seattle, July 1991.

[6] D. C. Park, et. al, "An Adaptively Trainable Neural Network Algorithm and its Application to Electric Load Forecasting", Proceedings of the First Int. Forum on Applications of Neural Networks to Power Systems, Seattle, July 1991.

[7] D. E. Rumelhart, G. E. Hinton, and R. J. Williams, "Learning Internal Representation by Error Propagation", Parallel Distributed Processing, Volume 1, MIT Press, 1989.

[8] P. J. Werbos, "Backpropagation Through Time: What It Does and How to Do It", Proceedings of the IEEE, Vol. 78, No. 10, October 1990.

[9] M. C. Brace, et. al, "Comparison of the Forecasting Accuracy of Neural Networks with Other Established Techniques", Proceedings of the First Int. Forum on Applications of Neural Networks to Power Systems, Seattle, July 1991.

Cases	Training Dates	Forecasting Dates
Case 1	05/13/91 to 08/11/91	08/12/91 to 08/18/91
Case 2	01/01/92 to 03/01/92	03/02/92 to 03/08/92
Case 3	01/01/92 to 03/08/92	03/09/92 to 03/15/92

Table 3. Test Cases Data

Day Type	Case 1	Case 2	Case 3
Monday	2.95	2.00	2.12
Tuesday	3.30	2.43	2.59
Wednesday	3.51	2.65	2.29
Thursday	2.55	1.73	2.55
Friday	2.54	2.40	3.27
Saturday	1.97	3.59	3.57
Sunday	2.82	2.27	1.69
Average	2.81	2.44	2.58

Table 4. Daily Average Absolute Relative Error (%) of Hourly Load Forecast

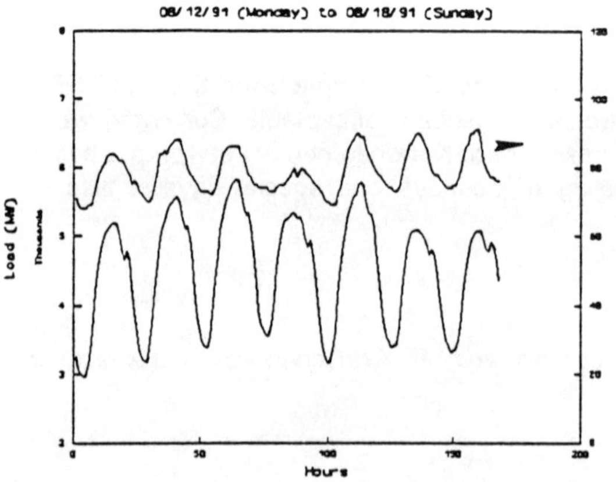

Fig. 2. Hourly Load and Temperature for Case 1

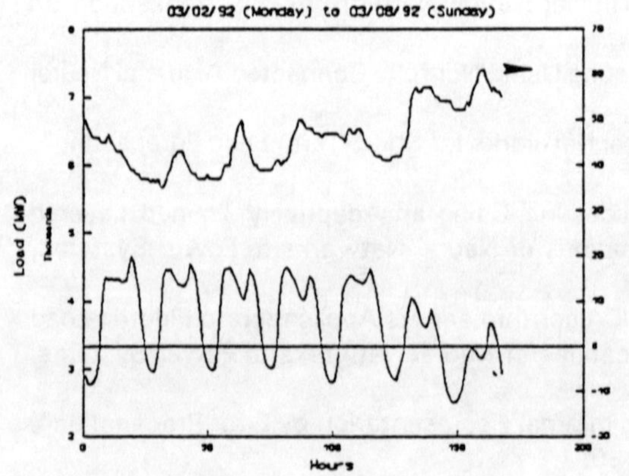

Fig. 3. Hourly Load and Temperature for Case 2

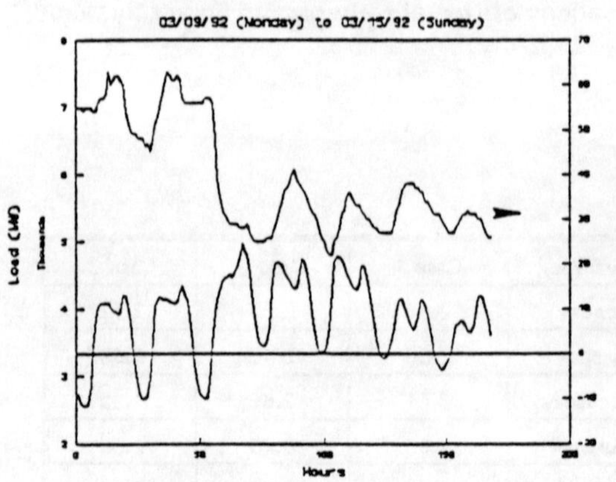

Fig. 4. Hourly Load and Temperature for Case 3

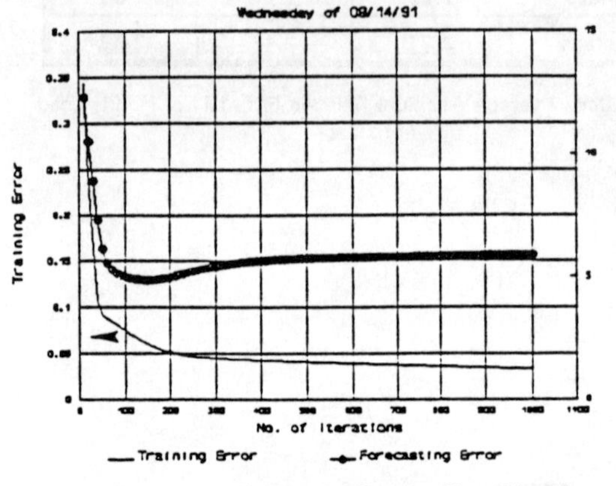

Fig. 5. Training and Forecasting Error vs. No. of Iterations (learning rate = 0.02, momentum rate = 0.9)

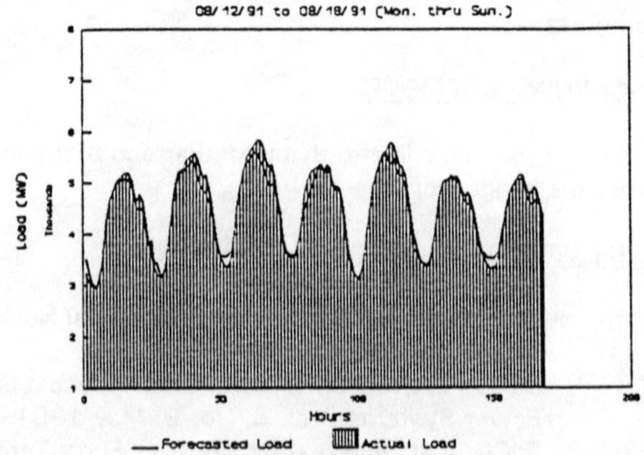

Fig. 6. Forecast Results for Case 1

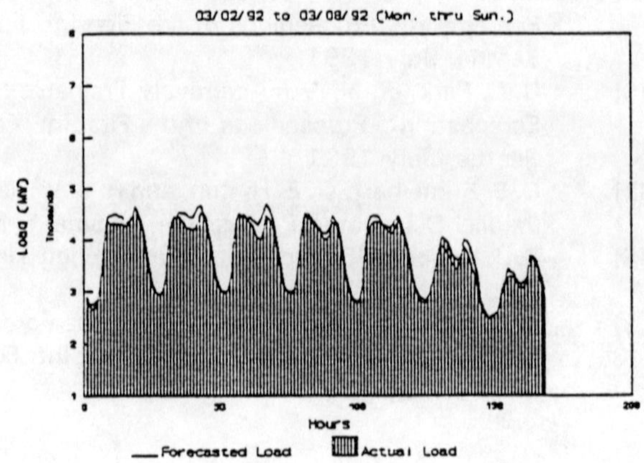

Fig. 7. Forecast Results for Case 2

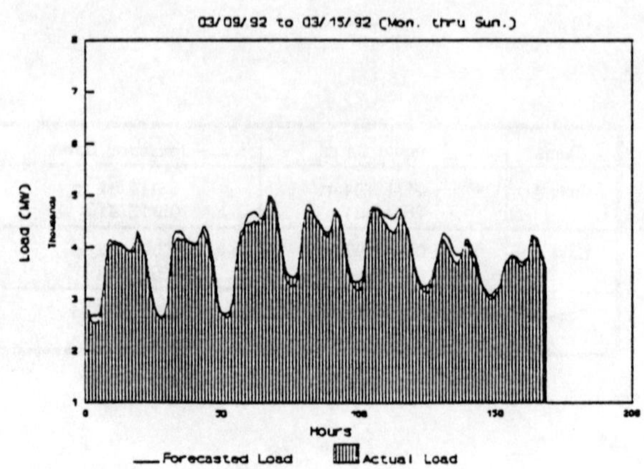

Fig. 8. Forecast Results for Case 3

LOAD FORECASTING BY HIERARCHICAL NEURAL NETWORKS THAT INCORPORATE KNOWN LOAD CHARACTERISTICS

John Y. Cheung, David C. Chance*, John Fagan
School of Electrical Engineering, University of Oklahoma,
*Department of Psychology, University of Central Oklahoma.

ABSTRACT

Forecasting of electrical load is an important aspect of planning and control of the electric utility industry. Proper prediction provides the much needed information to ensure anticipated load is met while minimizing the cost of electric power production resulting in savings to the customers. In this paper, we examine the use of hierarchical neural networks as a means for load forecasting. While the use of neural networks for prediction has been attempted by many researchers in diverse fields, the present effort differs from previous attempts in that the modeling and prediction by the neural net is controlled and influenced by what is known about the load characteristics.

INTRODUCTION

Accurate load forecasting is crucial in short term utility operation and long term energy management. Hourly, daily, and weekly forecast is necessary for short term utility operation to ensure that the anticipated load can be met with the least cost. On the other hand, seasonal, annual, 5-year, and 10-year forecast is essential for long term energy management such as decisions concerning facility maintenance, upgrade, and construction.

Short term forecasting is exceptionally difficult because the instantaneous power is dependent on many variables. Accurate forecasting depends very much on adequate modelling of all intervening parameters. These include the hour of the day, the day of the week, and meteorological data such as temperature, humidity, wind speed, wind chill factor, etc. Weather parameters has been studied by a number of researchers [1-3]. While the industrial load is more or less predictable as a function of the hour of the day and the day of the week, the residential load is less predictable because it not only depend on the hour of the day and the day of the week, but it also depends on the meteorological parameters as well. When the temperature is high, the air conditioning units are used more frequently for most of the day and the evenings. On the other hand, when the temperature is low, those homes with electric heaters also demands more energy. Hence, it is crucial to monitor the meteorological data as important parameters in predicting load.

Long term forecasting also depends on meteorological data though in a more subtle way. The total load may fluctuate from year to year depending on whether the year is hotter than usual or colder than usual. The dependence of meteorological data then must be removed before adequate forecasting of the annual load trend can be made. Furthermore, prediction of load for a particular year must also be modified according to the anticipated meteorological conditions.

In this paper, we examine the use of hierarchical neural networks as a means for load forecasting. While the use of neural networks for prediction has been attempted by many researchers in diverse fields, the present effort differs from previous attempts in that the modeling and prediction by the neural net is controlled and influenced by what is known about the load characteristics.
In other words, the neural net is not used merely used as a black box to predict the data blindly given the past data. Rather, the known load characteristics such as the daily cycles, the weekly periodicity, the seasonal changes, and the effects of temperature are all incorporated into the neural network in an effort to improve the performance of the neural network. What is known is used to constrain the network as much as possible. What is not known is left for the neural network to find out. A second feature of the present work is the use of a network of small neural networks. This is the "divide and conquer" strategy. Each subnetwork is used to handle the effects of one constraining factor. For example, there is a network for producing the daily cycle. There is another for producing the weekly cycles, etc. This results in a hierarchical neural network

structure. To test the network, the past ten years of data from PSO was used partly for training and partly for testing. We will compare the simulation results of the proposed network with traditional Box and Jenkins type forecasting and with other different types of neural network strategies.

SHORT TERM FORECASTING

Load data from the PSO, a electric utility company, has been collected between 1981 and 1988. At the same time, meteorological data, mainly temperature data for this paper, has also collected for the same time period. Since PSO serves six primary regions, an equivalent temperature index is defined as the weighted temperature or composite temperature weighted according to the percentage of service of each region [1]. When the equivalent temperature index is plotted against the hourly load for the entire year, it is evident that there is a direct and a high degree of correlation between the load and the temperature as shown in Figure 1. The relationship, however is a non-linear one as summer cooling and winter heating both put a demand on the electric load consumption. The temperature-load relationship can be modelled by a piecewise linear approximation using two segments, one for the summer, i.e. temperature above a predetermined threshold, and another one for the winter, i.e. temperature below the same predetermined threshold. It is observed that the temperature-load relationship for the winter is much weaker than that for the summer. However, both relationships are very nearly linear. Hence, temperature below the predetermined threshold can be represented by an equivalent temperature above the predetermined threshold since both temperatures demand the same increase in load. In this way, by converting temperature below the predetermined threshold to an equivalent temperature above the threshold, it is possible to define an equivalent temperature index that represents the modification on the load required due to temperature effectsas shown in Figure 2.

By modelling this temperature-load relationship, we can then accurately predict the instantaneous hour by hour load. A neural network is used to predict the load modification based on the temperature parameter. The network chosen closely resembles the first order piecewise approximation of the temperature-load relationship observed before. Five years of training data from 1981 to 1985 is used to train the network to accurately predict the load adjustment based on the equivalent temperature index as shown in Figure 3. Using the data from 1981, we can predict the load for 1981 and 1982. Correlation analysis shows that the predicted load for 1981 is 0.9 and for 1982 is 0.8 .

LONG TERM FORECASTING

For long term forecasting, it is desirable to know the trend of the load growth. In this case, it is desirable to remove the dependence of the load on the temperature of the year in question. In other words, a year with many hot days should not be interpreted as a year with strong growth, or a year with moderate temperatures should not be interpreted with a year with poor growth either. Hence, a meaningful load growth curve can be obtained only after temperature variation has been removed. Figure 4 shows the load growth curve before temperature adjustment, Figure 5 shows the equivalent temperature index for the same years, and Figure 6 shows the load growth curve after temperature adjustment. It is obvious that the temperature adjusted curve is much smoother and more representative of a growth curve than the non-temperature adjusted curve.

As in the case of short term forecasting, the long term prediction is also handled by the neural network approach. Instead of hourly load as in the case of short term forecasting, we use weekly load data for long term forecast. An equivalent temperature index for the week is first obtained and is then used to predict the load for the corresponding week. The same data is applied to the neural network for training. After the network is trained, it is used to predict the load for the next two years, from 1986 to 1988. The correlation coefficient for the long term forecasting is 0.8 for load forecast without temperature adjustment and 0.9 with temperature adjustment.

NEURAL NET MODELING

In the previous section, we have shown how meteorological data can be used to more accurately predict the load profiles for short term and long term forecasting. While the form of the temperature-load dependency is known, the parameters are evaluated by a neural network approach. Forecasting by neural networks has been attempted by a number of researchers in the past year [2]. A series of neural networks are used in the prediction. The lowest level of networks provides the short term forecast while the highest level of networks provides the long term forecast.

A number of forecasting has been reported in the literature based on the neural network technique. For short term forecasting, the simplest case is to take a moving window of the time series as input and use the windowed data to predict the value of the next data point. This has been done for the present data set. The neural network used has twenty input points representing the last twenty sample values, ten hidden nodes in the hidden layer and a single output node representing the predicted value. Simulation results showed that a correlation of 0.89 can be obtained for short term forecasting. When the equivalent temperature index was also included in the predicted resulting in a network of 40 input nodes, 20 hidden nodes, and one output node, simulation results showed that a correlation of 0.92 can be obtained for short term forecasting. Approaching the problem from another angle, if the temperature-load dependence can be modeled using the known functional form, then the neural network can be trained to find out the parameters of the functional forms. This results in a much smaller neural network. Simulation results showed that the simpler network that incorporates the known load characteristics gives a higher correlation than the straightforward implementation.

For long term forecasting, it is not enough just to predict the load based on the equivalent temperature index, it is also necessary to predict the growth rate of the load. A straight forward neural network can be constructed in similar fashion as the short term forecasting case resulting in a correlation of 0.91 . Using what is known in the temperature-load dependency and a known functional form of the growth rate, a much simpler neural network can be formed that yields a higher correlation than the straight forward neural network.

SUMMARY

Examination of load curve shows an obvious dependence of load demand on meteorological data, specifically temperature. In this paper we define a composite temperature index that can be used to predict the instantaneous load profile. While the form of the functional dependence is known, the parameters of the non-linear functional dependence is found by the neural network approach. Models for both short term and long term forecasting is proposed. These models incorporates the meteorological data for load forecast. Using the load data and the equivalent temperature index, several neural networks have been constructed. The straight forward networks that uses past values as inputs require a large number of nodes but perform worse than the networks that are constructed to approximate the known functional dependency. The latter networks that incorporates the known load characteristics are smaller and perform better than the straight forward networks.

REFERENCE

[1] D. Zhai, "Short Term Load Forecasting", MS Thesis, School of Electrical Engineering and Computer Science, University of Oklahoma, 1991.

[2] D.A. Fajkus, "Forecast Method Removes Weather Bias", Electric World, Feb. 1, 1975.

[3] R. Campo and R. Bhatnagar, "Adaptive Weather-Sensitive Short Term Load Forecast", IEEE Trans. on Power Systems, vol.2, no.3, pp.592-600, 1987.

[4] M.T. Hagan and S.M. Behr, "The Time Series Approach to Short Term Load Forecasting," IEEE Trans. on Power Systems, vol.2, no.3, pp.785-791, 1987.

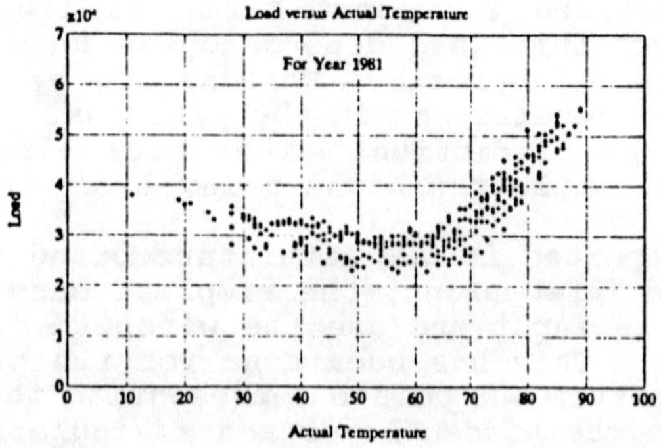

Fig 1. The load-temperature curve.

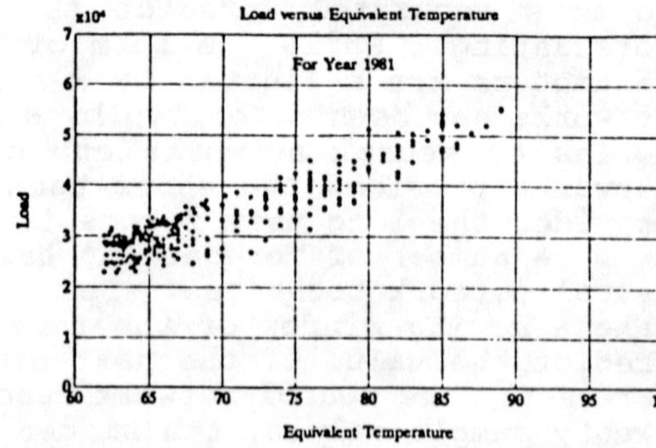

Fig 2. The equivalent temperature index.

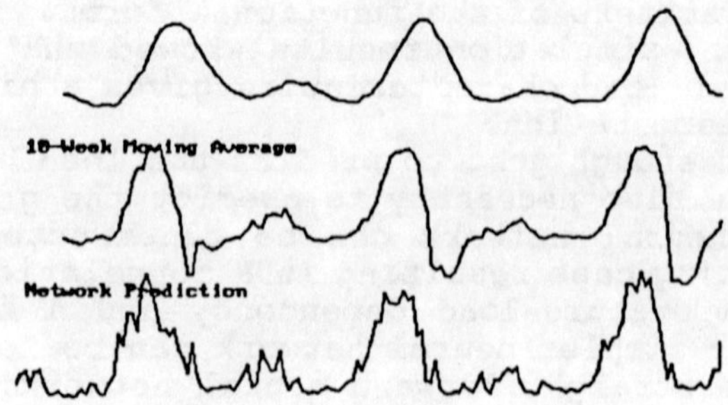

Fig 3. Short term forecasting by neural network.

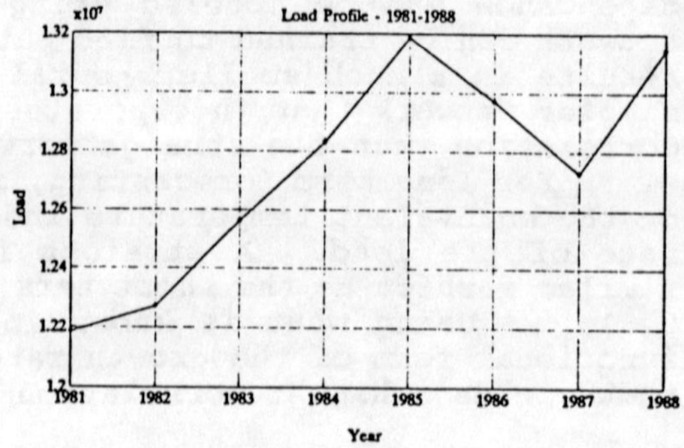

Fig 4. Non-temperature adjusted growth curve.

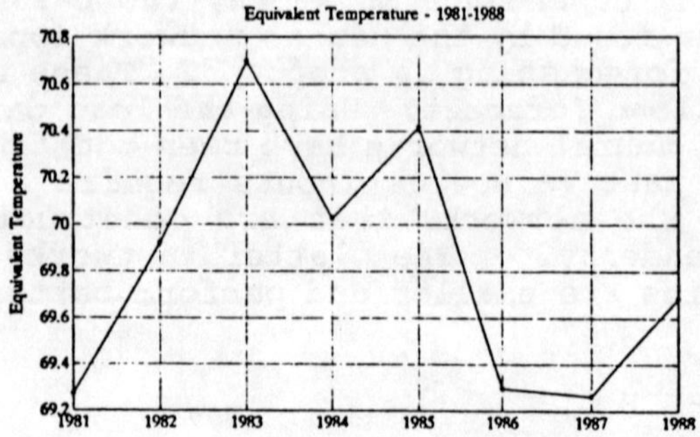

Fig 5. Equivalent temperature index.

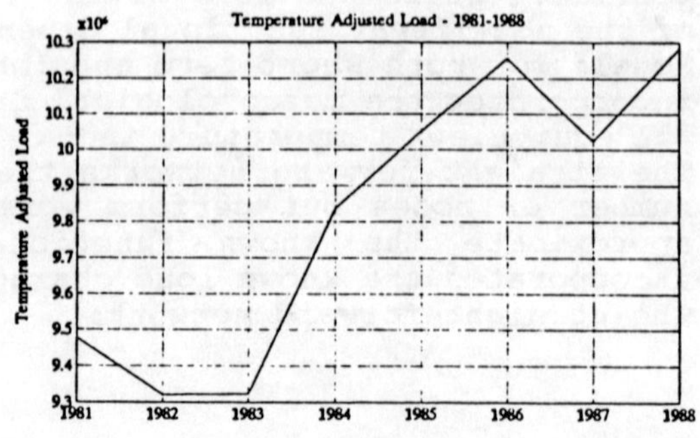

Fig 6. Temperature adjusted growth curve.

H. Scheduling and Optimization

A Solution Method for Maintenance Scheduling of Thermal Units by Artificial Neural Networks

H.Sasaki Y. Takiuchi J. Kubokawa
Dept. of Electrical Engineering Hiroshima University
Higashi - Hiroshima, 724, Japan

1. INTRODUCTION

Thermal units in power systems must legally undergo maintenance works once a year to prevent their deterioration. Since the capacities of thermal units are large and the maintenance work takes a long time, the maintenance scheduling greatly affects the reliability and economy of power systems. In determining maintenance scheduling, reliability should have priority over economy. In general, this problem has been formulated as a 0–1 integer programming problem. Major solution methods proposed so far may be classified as:

(1) Integer programming [1]
(2) Dynamic programming [2]
(3) Branch and bound method [3]
(4) Simulated annealing method
(5) Expert system based methods [4]

Although methods (1) – (3) give an optimal solution to a small scale problem, it is very difficult to obtain an optimal solution for a large scale problem since the number of combinations increases exponentially. Simulated annealing method may be applicable to a large scale problem, in which some trouble happens in a cooling schedule of thermal-parameter. An expert system is an efficient tool to economize most of combinatorial approaches by making use of domain experts' knowledge. It should be noted the knowledge of experts is not necessarily optimal as to maintenance scheduling problems. Moreover, almost all expert system assisted methods must sequentially determine the order of maintenance work. This feature makes a clear contrast with the neural network approach in determining the optimal combinations.

In this study we have noticed Hopfield type neural networks which are very effective for combinatorial optimization problem, thus being able to apply it to the maintenance scheduling of thermal units [5]-[8].

A solution given by the artificial neural networks is not necessarily the optimal one since it is often stuck at one of local minima. Therefore, we investigate the probabilistic property of solutions by the neural networks and utilize the best solution selected from a certain number of trials as the current best solution and proceed to the branch and bound method.

2. FORMULATION OF MAINTENANCE SCHEDULING OF THERMAL UNITS

In this study, we define this problem as "to determine the maintenance scheduling of thermal units that minimizes the sum of maintenance costs and fuel costs during a period of study, while satisfying the demand of power system."

The objective function and constraints in this problem has been defined as follows:

[Objective function]
(Maintenance cost) + (Fuel cost)

[Constraints]

(1) Constraint on starting period of maintenance

Starting period of maintenance must be legally selected to complete the maintenance work within one year ± one month after the latest work.

(2) Spinning reserve margin

The spinning reserve larger than the specified value must be secured over a period of study.

(3) Not repaired at the same sub period

This is the constraint on units which the maintenance work can not be done simultaneously in a thermal station.

(4) Constraint on the order of the maintenance work

This constraint applies to the units for which the order of maintenance work is specified in advance.

The cost of unit commitment is not taken into consideration to the objective function since it has no appreciable effect. Generators are successively commissioned according a prespecified merit order so that they meet the demand at each subperiod. It is assumed that load duration curve at each period is given. Since variables are represented as discrete value, the maintenance scheduling can be formulated as a 0-1 integer programming.

3. NEURAL NETWORK FOR 0-1 INTEGER PROGRAMMING

The neural network proposed by Hopfield and Tank can obtain a solution of combinatorial optimization problems, but this network can not handle problems with equality constraints. Therefore we have proposed[1] the network which can obtain a solution for such problems effectively.

A 0-1 integer programming which is constrained by inequality constraints is generally formulated as:

$$\text{minmize} \ : f(V) \qquad V = (V_1, \dots, V_N) \qquad (1)$$

$$\text{subject to} \ : hj(V) \geq 0 \qquad (j = 1, \dots, M) \qquad (2)$$

where **V** denotes the N+1 variable vector. The network for this problem is shown in Fig.1.

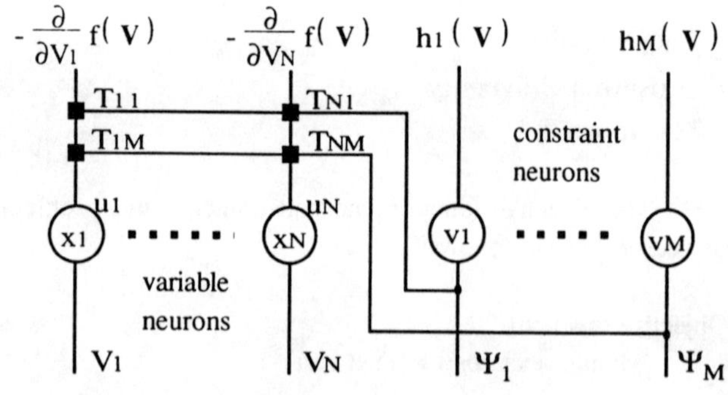

Fig.1 The neural network for a 0-1 integer programming

In this network, the input-output relationships of neurons are expressed as follows:

< Variable neuron >

$$V_i = x_i(u_i) = \frac{1}{2}(1 + \tanh(u_i)) \quad (i = 1, \cdots, N) \tag{3}$$

< Constraint neuron >

$$y_j(z) = \begin{cases} 0 & (z \geq 0) \\ Kz & (z < 0) \end{cases} \tag{4}$$

where K is a positive constant.

It should be noted that a constraint neuron must have a mach faster response time than that of variable neurons. Then the network converges to one of local minima to which energy function E reduces with time.

$$E = f(V) + \sum_{i=1}^{N} \int \sum_{j=1}^{M} T_{ij} y_j(h_j(V)) dV \tag{5}$$

If inputs from constraint neurons to variable neurons are appropriate, a solution which satisfies inequality constraints is obtained.

Next, let us consider the case where the problem includes equality constraints.

$$\text{subject to: } g_k(V) = 0 \tag{6}$$

So far, equality constraints have been treated by incorporating into the objective function. However, it is difficult to decide the coefficients or weights of equality constraint terms when the number of constraints increases because the convergence of the network is greatly affected by the selection of coefficients. Therefore, we have noticed that an equality constraint is equivalent to the two inequality constraints, i.e.;

$$g_k(V) \geq 0, \quad g_k(V) \leq 0 \tag{7}$$

4. MAPPING OF MAINTENANCE SCHEDULING ONTO NEURAL NETWORK

4.1 Establishment of Variable Neurons

At first, the number of subperiods under study is denoted by N, and the number of units by M. For each thermal unit, a subperiod at which the maintenance work is able to start is restricted by the latest maintenance. We refer this subperiod as the start up period of maintenance, and assign one variable neuron to each period. Neuron V_{kt} (the state of unit k at period t) takes on 0 or 1. If $V_{kt} = 1$, then the work for unit k should be started at period t. Under these assumptions, the maintenance scheduling may be mapped onto a neural network. The construction of the network is illustrated in Fig.2

4.2 Mapping of Objective Function

By making use of the variable neurons, an objective function is formulated as follows:

$$f(V) = (\text{maintenance cost}) + (\text{fuel cost}) = \sum_{k=1}^{M} \left(\sum_{t=1}^{T(k)} \alpha_{kt} V_{kt} + \sum_{w=1}^{N} \beta_{kw} Q_{kw} \right) \tag{8}$$

where

$T(k)$: the number of possible start up subperiods of unit k

α_{kt} : the maintenance cost of unit k when the maintenance work starts up at period t

β_{kw} : the fuel cost of unit k at period w

Q_{kw} : the electric energy produced by unit k at period w

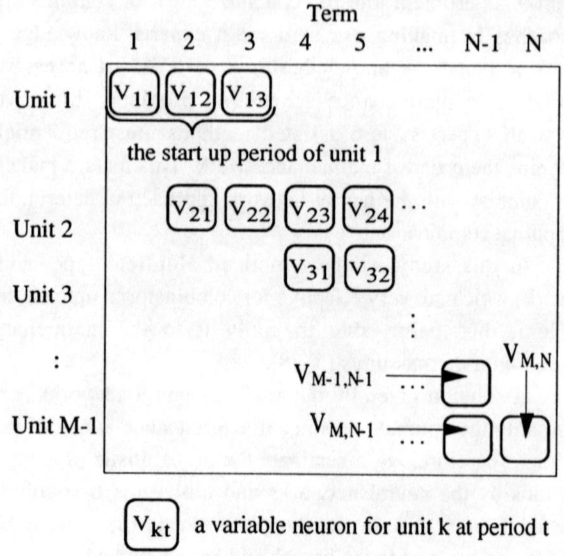

Fig.2 The construction of the network

The inverse function of the load duration curve at period w is shown in Fig.3.

$$Q_{kw} = \int L_w(x) dx \tag{9}$$

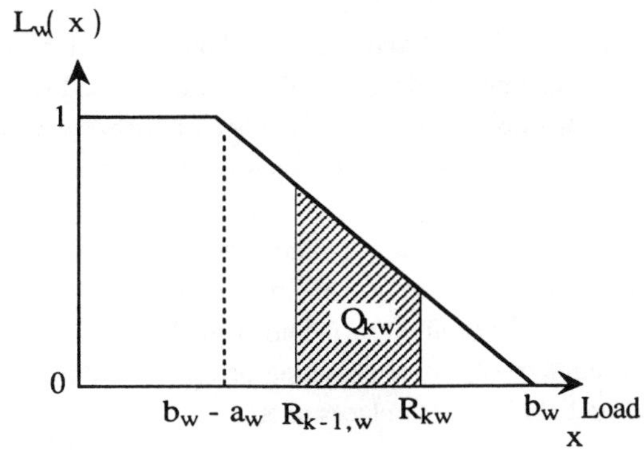

Fig.3 The inverse function of the load duration curve

The Partial differentiation of the objective function with respect to V_{ki} yields inputs from the variable neurons to the variable neurons.

4.3 Mapping of the Constraints

(1) Constraint on the starting period of maintenance work

In this study, we have used the same number of variable neurons as the number of start up periods of maintenance work. These variable neurons are subject to a condition that one and only one neuron must fire for each unit. i.e.;

$$1 \geq \sum_{i=1}^{T(k)} V_{ki} \geq 1 \quad (k = 1, \ldots, M) \tag{10}$$

Thus the outputs of constraint neurons are expressed as

$$\begin{cases} \psi_{1k} = y_1\left(\sum_{i=1}^{T(k)} V_{ki} - 1\right) \\ \psi_{2k} = y_2\left(-\sum_{i=1}^{T(k)} V_{ki} + 1\right) \end{cases} \tag{11}$$

, and the connections from y_{1k}, y_{2k} to V_{kt} are

$$D_{ki,1j} = \begin{cases} -1 \ (k = j) \\ 0 \ (k \neq j) \end{cases} \quad (i=1, \ldots, T(k)) \tag{12}$$

$$D_{ki,2j} = \begin{cases} 1 \ (k = j) \\ 0 \ (k \neq j) \end{cases} \quad (i=1, \ldots, T(k)) \tag{13}$$

The number of the constraint neurons is 2M, and the number of connections requires twice the number of variable neurons.

(2) Spinning reserve margin

This constraint must satisfy that spinning reserve should be secured more than specified value over a period of study:

$$\sum_{k=1}^{M} P_k - \sum_{k=1}^{M} \sum_{i=1}^{T(k)} V_{ki} P_k \geq P_{R_w} \tag{14}$$

where

P_{RW} : (maximum load at period t) + (spinning reserve)

For the lack of space, omit the derivation of the connections.

(3) Constraint on the number of units at the same time

For the units which can not be undertaken maintenance work repaired simultaneously, the neurons corresponding to start up periods are restricted.

$$V_{ki} + \sum_{m(\neq k) \in \xi} \sum_{j \in \eta(k,i)} V_{mj} \leq 1 \quad (k \in \xi) \tag{15}$$

where

ξ : the set of units for which works are not started simultaneously.

$\eta(k,i)$: the set of neurons in period k

(4) Constraint on the order of maintenance work

Suppose that the maintenance work of unit k1 must be performed before unit k2. Then this constraint is expressed as follows.

$$V_{k_1 i_1} + V_{k_2 i_2} \leq 1 \tag{16}$$
$$(i_1 = 1, \ldots, T(k_1),\ i_2 = 1, \ldots, T(k_2),\ k_1, k_2 \in \zeta)$$

where

ζ : the set of units which are restricted about the order of maintenance work.

Under the above conditions, the network for maintenance scheduling of thermal units can be constructed.

5. SIMULATION RESULTS AND DISCUSSIONS

In order to demonstrate the effectiveness of the proposed method, the neural network has been implemented by the neural network simulator developed in our research laboratory. The input to output responses of constraint neurons have to be much quicker than those of variable neurons. We therefore assume in the simulation that the time constants of variable neurons are infinitely long compared with those of constraint neurons. To realize this, we have adopted a strategy that the update of any one of the variable neurons is followed by the simultaneous updates of all the constraint neurons. Unfortunately, as already pointed out, a solution obtained by a neural network is not necessarily the optimal one. Therefore we investigate the probabilistic property of solutions to yield the best possible starting value to the branch and bound method.

Maintenance scheduling of thermal units by means of the neural network is solved by following the procedure shown in Fig.4.

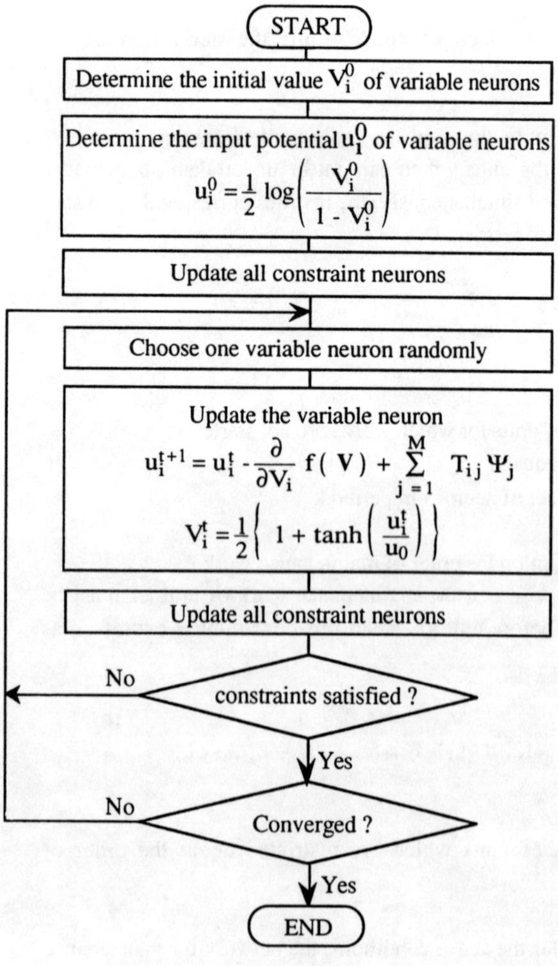

Fig. 4 The flow chart

In the first place, as a simple example we have solved the case of 15 units over 26 periods. Assume that spinning reserve is fixed at 10 percent of peak value and the start-up periods are 5 periods. We have simulated on the above data. The histogram shown in Fig.5 illustrates the distribution of feasible solutions obtained by the neural network simulator. The simulator has been run 1000 times from 1000 different initial conditions supplied by a uniform random variable generator.

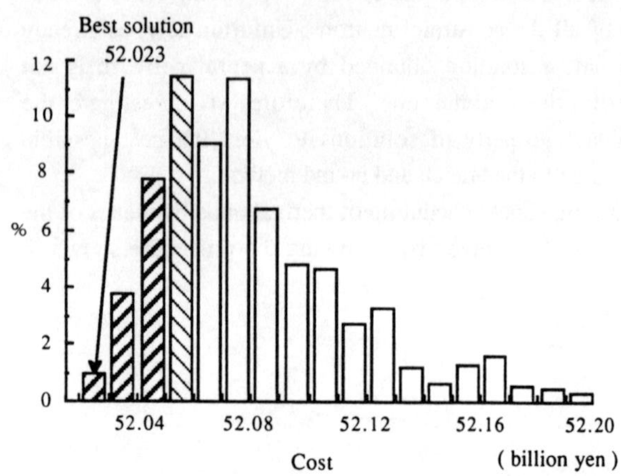

Fig. 5 Solutions for the case of 15 units over 26 periods

We regard solutions of under 52.05 billions as quasi-best solutions and investigate their probabilistic properties. Fig.6 shows the relationship between the number of times of the simulations and the probability that the quasi-best solutions are obtained. We can infer from this graph that either the best or quasi-best solutions can be obtained with the probability of more than 90 percent within 20 times simulations. If solutions under 52.06 (thin hatched portion) are included in the quasi-best solutions , the number of run of obtaining one of the solutions with the same probability level reduces to 10.

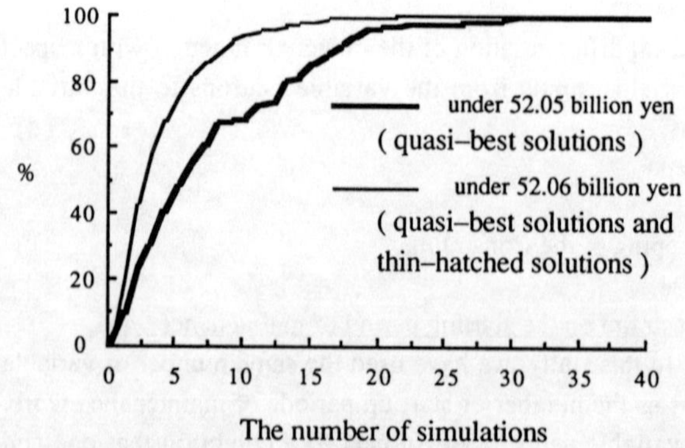

Fig. 6 The probabilistic property

In the next, the case of 20 units over 52 periods has been simulated. Assume that spinning reserve is fixed at 10 percent as in the previous case, the start up periods are 9 periods. Simulation results are depicted as a histogram in Fig.7. Let us regard solutions of under 120 billions as quasi-best solutions. The probability is shown in Fig.8. One quasi-best solution appears at the rate of 90 percent after 20 simulations.

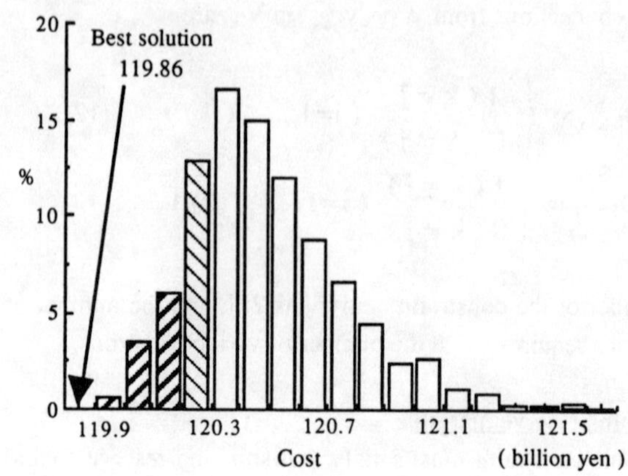

Fig .7 Solutions for the case of 20 units over 52 periods

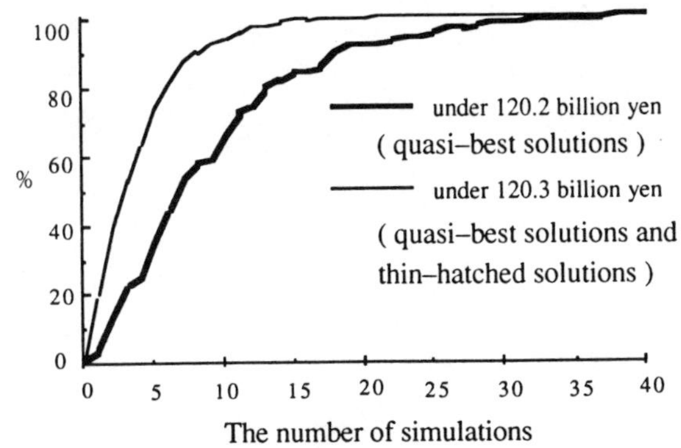

Fig .8 The probabilistic property

We have proposed to adopt the best of feasible solutions obtained in 20 simulations as the current best solution in the branch and bound method. Table.1 shows the search times of solutions and computation time in both the case where only the branch and bound method has been utilized and the case where the proposed method has been utilized. The application of neural network reduced search times of solutions by 10 percent to the problem of 15 units over 26 periods, by 15 percent to the problem of 20 units over 52 periods. In this simulation, we have applied only total cost to the brunch and bound methods. The authors expect that if tentative maintenance conditions (the outputs of variable neurons) which are obtained by the neural networks can be utilized for bounding operation, then the optimal solution is achieved more quickly.

Table 1. Simulation result

15 units over 26 periods (best solution 52.03 billion yen)

	initial value (billion yen)	search times	time (second)
branch and bound method	52.082	6331	7.4
the proposed method	52.050	5713	6.3

20 units over 52 periods (best solution 119.85 billion yen)

	initial value (billion yen)	search times	time
branch and bound method	120.64	439426	1'08"53
the proposed method	119.92	375374	0'59"12

6. CONCLUSION

In this paper, maintenance scheduling of thermal units has been mapped onto a neural network which has incorporated neural networks for enforcing inequality constraints, and the best solution selected within a certain number of trials has been adopted in the branch and bound method as the current best solution. There are little differences between the result obtained by a neural network and the result obtained by branch and bound method. This arises from that the area where feasible solutions can be obtained is narrow for these small scale problems. However, it must be difficult for the large scale problems to obtain the best solution by branch and bound method owing to the exponential increase in computation time. We believe, therefore, that if one wants to solve a large scale problem, the method described in this paper is a fruitful one.

REFERENCES

[1] J.F. Dopazo, H.M. Merrill : "Optimal Generator Maintenance Scheduling Using Integer Programming", IEEE Transactions Power Apparatus and Systems, Vol. 5, Sept/Oct 1975.

[2] H.H. Zurn, V.H. Quintana : " Generator Maintenance Scheduling via Successive Approximation Dynamic Programming", IEEE Transactions Power Apparatus and Systems, Vol. PAS-94, No. 2 March/April 1975.

[3] Cohen and Yoshimura : " A Branch-And-Bound Algorithm For Unit Commitment", IEEE Trans. On PAS, Vol PAS-102, No. 2, February 1983.

[4] B.Y. Choueiry, Y. Sekine : " Knowledge based method for power generators maintenance scheduling", Procs. of the First ESAP, Stockholm-Helsinki, Aug. 22-26, pp.9.7-9.12.

[5] J.J. Hopfield : " Neural Computation of Decisions in Optimization Problems ", Biol. Cybern. 52, 141-152 (1985)

[6] D.W. Tank : " Simple "Neural" Optimization Networks : An A/D Converter, Signal Decision Circuit, and a Linear Programming Circuit, and a Linear Programming Circuit ", IEEE Trans. Circuits & Systems, Vol. CAS-33, No.5 (1986)

[7] M.P. Kennedy : " Neural Networks for Nonlinear Programming ", IEEE Trans. Circuit & Systems, Vol. 35, No.5 (1988)

[8] H. Sasaki : " A Solution Method of Unit Commitment by Artificial Neural Networks ", IEEE 91 SM 437-4 PWRS

GENERATION DISPATCH ALGORITHM COORDINATING ECONOMY AND STABILTY BY USING ARTIFICIAL NEURAL NETWORK

H.SAITOH Y.SHIMOTORI J.TOYODA

Department of Electrical Engineering, Tohoku University
Aramaki Aza-Aoba, Sendai 980, JAPAN

Abstract : This paper proposes a new generation dispatch algorithm in which the stability evaluation of power systems is embedded by applying artificial neural network (ANN). The purpose of the algorithm is to determine the generation distribution which can improve the damping on electromechanical modes and does not extremely increase the total fuel cost of generation from the most economical operating point. In the algorithm, first, the most economical generation distribution is obtained using the equal incremental cost criterion. Next, a number of generation distributions are created which are close to the most economical distribution and do not cause the large increment of fuel cost. ANN is applied for evaluating the effect of those created generation distributions on the system damping. The training of ANN is done in off-line with the error-back-propagation learning algorithm in such a way that a stability index is outputted when the corresponding generation distribution is inputted. The stability index is equivalent to the energy of the mode, and can be extracted from the fluctuations of line power flows. Finally, the generation distribution with the smallest value of the index is selected as the final solution coordinating the system economy and stability. From the viewpoint of a multiple objective problem, the selected distribution corresponds to one of the noninferior solutions because the minimization of fuel cost and the enhancement of stability are the competitive objectives of generation dispatch problem. The effectiveness of the proposed algorithm is confirmed through the simulation of a simplified power system model.

INTRODUCTION

Economical and secure operation of modern power systems becomes more difficult because the increasing concern on environmental problems decelerates the construction of new power plants and the expansion of transmission network in spite of the continuous growth of electric demand. In this situation, more sophistication of Energy Management System (EMS) is expected for achieving the coordinated operation between the competitive objectives : system economy and security. On one hand, the recent progress of artificial neural network (ANN) stimulates the applications of it to many kinds of power system problems such as load forecasting, security assessment and system planning [1]. Since ANN has potential for the identification of unknown system and the adaptive control, it will be an effective tool for developing the advanced EMS functions.

This paper proposes a new generation dispatch algorithm in which the stability evaluation of power systems is embedded by using ANN. The purpose of the proposed algorithm is to determine the generation distribution which can improve the damping on electromechanical modes compared with the most economical distribution obtained from the conventional economic dispatch algorithm, and which does not cause the large increment of total fuel cost of generation from the most economical operating point. The conventional algorithm which takes the static security constraints into account, provides the generation distribution which is the most economical and does not violate the limits of bus voltage and transmission line flow. However, such distribution does not necessarily make the power system be in the secure state in the sense of stability. Under any circumstances, the damping on the modes is weakened and the system may move into the dynamically alert state even if the system is statically secure. Therefore, in order to avoid such state transition, the stability evaluation is embedded in the proposed algorithm.

In the algorithm proposed in this paper, first, the most economical generation distribution is obtained using the conventional economic dispatch algorithm based on the equal incremental cost criterion. Next, a number of generation distributions are created which are close to the most economical distribution and do not cause the large increment of fuel cost. ANN is applied for evaluating the effect of those created generation distributions on the system damping. The ANN is trained in off-line with the error-back-propagation learning algorithm in such a way that it outputs a stability index when the corresponding generation distribution is inputted. Finally, the generation distribution with the smallest value of the index is selected as the final solution coordinating the system economy and stability. From the viewpoint of a multiple objective problem, the selected distribution corresponds to one of the noninferior solutions because the minimization of fuel cost and the enhancement of stability are the competitive objectives of generation dispatch problem.

In order to confirm the effectiveness of the proposed algorithm, a simulation study is done using a simplified power system model. The result shows that the algorithm is useful for coordinating the system economy and stability in the phase of generation dispatch scheduling.

GENERATION DISPATCH COORDINATING ECONOMY AND STABILITY

The generation dispatch algorithm proposed in this paper determines the most effective generation distribution for improving the system damping without the large increment of total fuel cost from the most economical operating point. The algorithm is described as follows. Here, it is assumed that the algorithm is cyclically applied to a sequence of forecasted load data every Δt minutes.

STEP-1 : The following conventional economic dispatch problem, which is to determine the generation distribution minimizing the total fuel cost of generation under the constraint of energy balance between generation and demand, is solved by using the equal incremental cost criterion.

$$\min_{x(k)} c(\,x(k)\,), \quad \text{subject to} \quad u0^T \cdot x(k) = L(k) \tag{1}$$

where, $x(k) = [x1(k)\ x2(k) \ldots . xn(k)]^T$: generation distribution at time tk,
$xi(k)$: electrical output of generator #i,
$c(*)$: total fuel cost function of generation,
$L(k)$: forecasted load,
$u0^T = [1\ 1 \ldots . 1]$: n dimensional vector.

The solution $x^*(k)$ obtained in this step corresponds to the bottom point on the parabolic manifold in Fig.1 (a).

STEP-2 : The perturbation of generation distribution from the most economical one, which results in the minimum increment of total fuel cost, is obtained by solving the following minimization problem.

$$\min_{\Delta x(k)} \Delta c(\,\Delta x(k)), \quad \text{subject to} \quad u0^T \cdot \Delta x(k) = 0, \quad \exists\, \Delta xi(k) \neq 0 \tag{2}$$

where, $\Delta x(k) = x(k) - x^*(k)$: perturbation of generation distribution;
$x^*(k)$: the most economical generation distribution obtained in STEP-1;
$\Delta c(\Delta x(k)) = c(x^*(k)+\Delta x(k)) - c(x^*(k))$: the increment of total fuel cost caused by the perturbation.

The solution of this problem can be expressed as follows.

$$\Delta x(k) = b \cdot U \cdot v_{min} \tag{3}$$

where, $U = [u_1\ u_2 \ldots\ldots u_{n-1}]$: $n*(n-1)$ matrix,

u_i : n dimensional normalized orthogonal vector which is also orthogonal to u_0,

v_{min} : eigenvector of matrix $(U^T AU)$ corresponding to the minimum eigenvalue,

$A = \mathrm{diag}[\ \partial^2 c/\partial x^2{}_i(k) \mid x^*{}_i(k)\ ]$: diagonal matrix of which the element is the differential coefficients of the second order of total fuel cost function $c(x(k))$,

b : arbitrary real coefficient.

Eq.(3) means that the perturbation $\Delta x(k)$ in the direction of vector $(U \cdot v_{min})$ does not cause the larger increment of fuel cost than that in the other directions. In Fig.1 (a), the direction is denoted by the arrow on the $(x_1,\ldots,x_n)$ - plane.

<u>STEP-3</u> : By using Eq.(3), a number of feasible generation distributions satisfying the following inequality constraints are created.

feasible generation distribution : $x_i(k) = x^*(k) + b_i \cdot U \cdot v_{min} \quad (i = 1,2,\ldots m)$ (4)

constraints : $| x(k) - x_f(k-1) | \leq x_r \cdot \Delta t$: load pick up capability

$x_L \leq x(k) \leq x_U$: generation output limits

$\Delta c(\Delta x(k)) \leq \Delta c_{LIM}$: acceptable limit of fuel cost increment

where, $x_f(k-1)$: generation distribution which is finally determined at the previous time t_{k-1},

x_r : maximum rate of change in electrical output of generator,

x_L, x_U : lower and upper limits of electrical output.

A number of feasible generation distributions $x_i(k)$ are obtained by changing the parameter b_i uniformly in such a way that the all inequality constraints are not violated. The feasible distributions created in this step correspond to the points indicated by cross marks in Fig.1 (a). The value of Δc_{LIM} is given by a system dispatcher according to the system condition. For instance, if the system tends to be unstable, then the large value may be set, and the solution space where the feasible distributions can be created is expanded so that many number of generation distributions effective for stability improvement are included in that space.

<u>STEP-4</u> : Each feasible generation distribution $x_i(k)$ is mapped into the corresponding stability index by ANN.

$$S_i \longleftarrow \mathrm{ANN}[\ x_i(k), L(k)\ ] \tag{5}$$

where, $S_i = [\ s_{1i}\ s_{2i} \ldots s_{pi}]$: stability index vector.

The stability index s_{ji} is equivalent to the energy of a oscillatory mode of power system, and can be extracted from the fluctuations on line power flows by using the developed technique in reference [2]. The smaller value of stability index means the stronger damping on the mode, that is, the power system tends to be more stable. The process in this step corresponds to mapping the feasible generation distributions on the $(x_1,\ldots,x_n)$ - plane in Fig.1 (a) into the points on the (fuel cost, stability index) - space in Fig.1 (b).

<u>STEP-5</u> : The generation distribution $x_f(k)$ corresponding to the stability index vector with the smallest norm is selected among the created feasible distributions in STEP-3. This process means the selection of the noninferior solution indicated by the rectangular mark in Fig.1 (b).

Since minimizing the fuel cost and enhancing the stability are generally the competitive objectives of generation dispatch scheduling, the generation dispatch problem treated in this paper is inherently a multiple objective problem ,that is, there does not exist the solution which can optimize the both objectives simultaneously. Therefore, the proposed algorithm finds out the noninferior solutions of the multiple objective problem, and then selects one of those solutions as the generation distribution coordinating economy and stability.

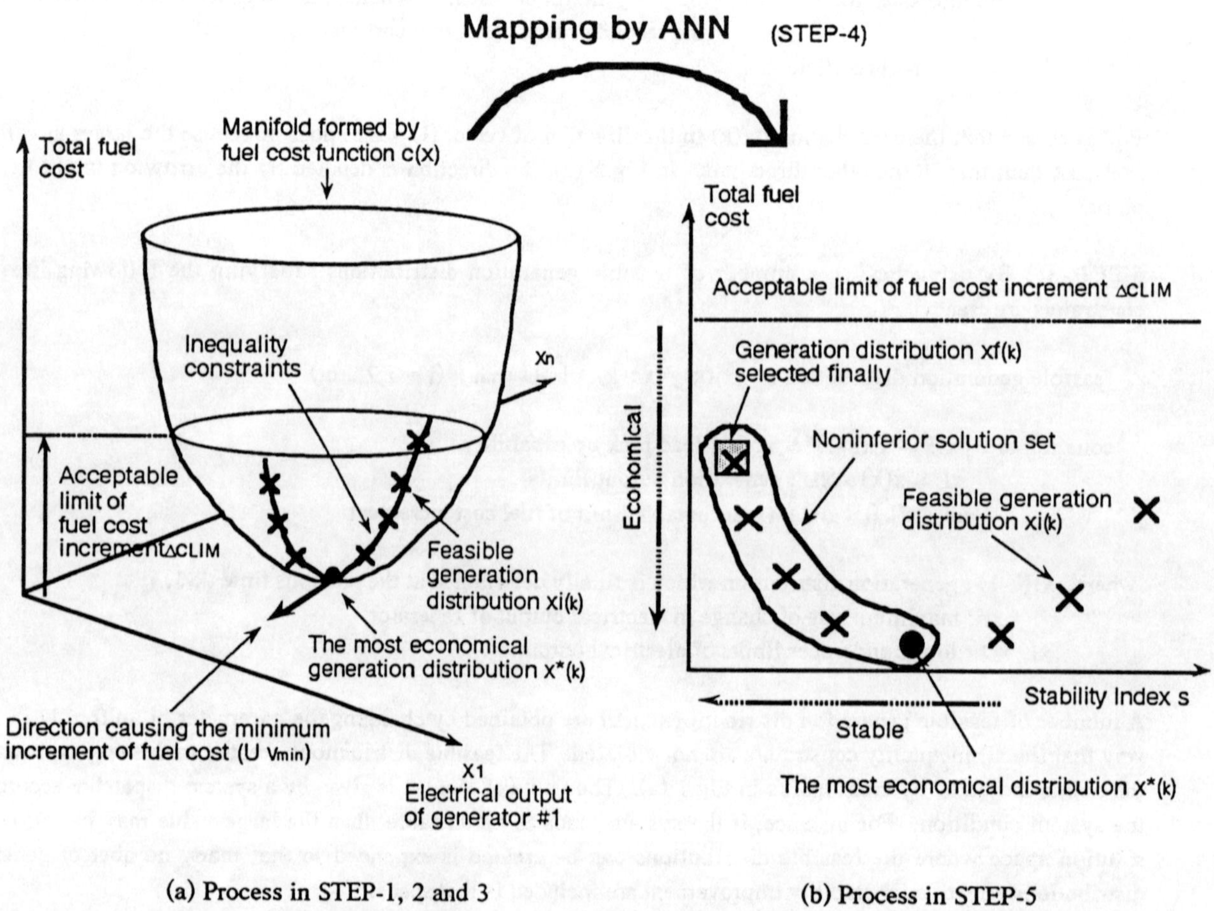

Fig.1 Searching process of the generation distribution xf(k) coordinating economy and stability

ANN FOR STABILITY EVALUATION

Generally, the dependency of system damping on the generation distribution is so complex that it will be difficult to identify the causal relation between them by using the mathematical technique such as sensitivity analysis of eigenvalue. To overcome this difficulty, ANN is applied for identifying the complex relation because it has the interpolative ability and can represent the relation with strong nonlinearity if the enough training data sets are given. The ANN used in the proposed algorithm is shown in Fig.2. Each input signal is the electrical output of generator. The forecasted load L(k) is also inputted in order to distinguish the system loading level. The output signal, namely, the teaching signal is the stability index sji corresponding to the inputted generation distribution and forecasted load. The number of the indices depends on that of electromechanical modes which should be monitored for evaluating the system damping. The training of the ANN is done in off-line through the following three steps. First, the transmission lines suitable for monitoring the modes are selected by the coherency based analysis. Second, for the preparation of training data, the power flow fluctuations of the selected lines are measured from the actual power system or obtained from the simulation in the various system conditions, and then the stability indices are extracted from those fluctuations. Simultaneously, the corresponding generation distributions and loads levels are acquired. Finally, the relation between generation distribution and stability indices is learned by the ANN with the error-back-propagation algorithm.

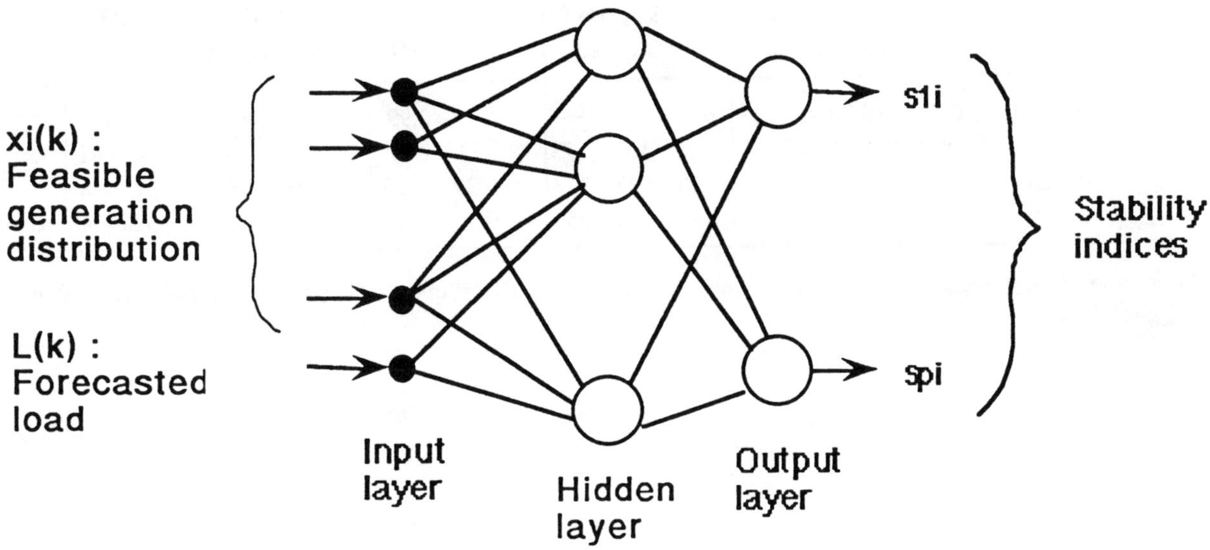

Fig.2 Configuration, input and output signals of the used ANN

TEST AND ITS RESULTS

The simulation of a three-machine one-load bus system model is done in order to verify the effectiveness of the proposed generation dispatch algorithm. It is assumed that each machine is a thermal generating unit of which the fuel cost characteristic is represented by quadratic polynomial form. And also, PSS is installed in it. Table 1 shows the generation dispatch schedule obtained by the proposed algorithm. For the comparison, the schedule by the conventional economic dispatch algorithm based on the equal incremental cost criterion is also shown. Here, the generation dispatch is assumed to be carried out with respect to a sequence of the forecasted load L(k) every 5 minutes. In this study, the acceptable limit of fuel cost increment ΔcLIM is set to 0.5% of the total fuel cost at the most economical generation distribution x*(k). The used ANN has 4, 20 and 2 neuron units in the input, hidden and output layers, respectively. The input signals are the electrical outputs of three generators and the forecasted load L(k). The output signals are the stability indices s1i and s2i corresponding to the oscillatory modes of 1.4Hz and 2.1Hz of the model system, which are extracted from the active power flow of the line between unit #1 and #3 because the both modes can be simultaneously observed on that line flow. In order to relate the generation distribution with the stability indices by the ANN, 92 sets of training data are prepared through the simulation in the various generation distributions at the load levels 500 and 550[MW]. Therefore, any distributions at the load levels between 500 and 550[MW] are untrained data. For the training, 50000 presentation of the data sets is done.

The result implies that the proposed algorithm realize the generation dispatch schedule effective for enhancing the damping on the modes within the smaller increment of fuel cost than the specified ΔcLIM. Especially, the damping on 2.1Hz mode is rapidly improved during the scheduling period of time compared with the conventional approach. The rotor speed deviations of unit #3 in Fig.3 also shows the improvement of the damping on that mode.

Table 1 Generation dispatch schedules obtained by the proposed algorithm and the conventional one

Time tk [min]	Total load L(k) [MW]	Conventional Dispatch x*(k) [MW]			Proposed Dispatch xf(k) [MW]			Real parts of eigenvalues of electromechanical modes				Total fuel cost [R]	
								CGD		PGD		CGD	PGD
		x1	x2	x3	x1	x2	x3	1.4Hz	2.1Hz	1.4Hz	2.1Hz		
0 - 5	500	215	152	133	—	—	—	-0.36	-0.99	—	—	297482	—
5 - 10	520	219	156	145	208	139	173	-0.37	-1.07	-0.34	-1.24	307441	307519
10 - 15	540	222	160	158	200	127	213	-0.37	-1.15	-0.33	-1.47	317418	317727
15 - 20	550	224	162	164	190	110	250	-0.38	-1.19	-0.31	-1.68	322414	323163

* [R] : fictitious money, CGD : conventional dispatch algorithm, PGD : proposed dispatch algorithm.

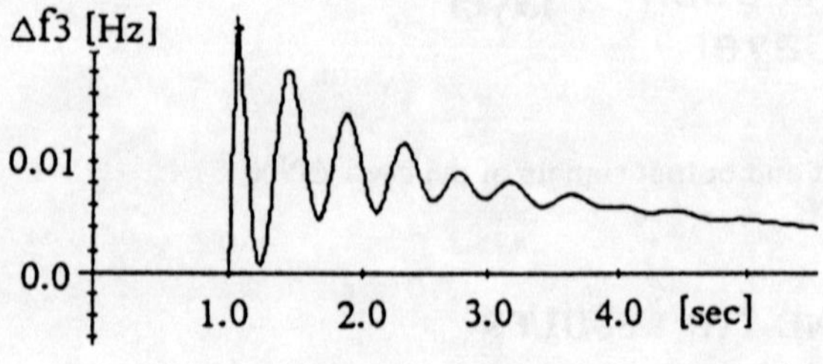

(a) Conventional generation dispatch

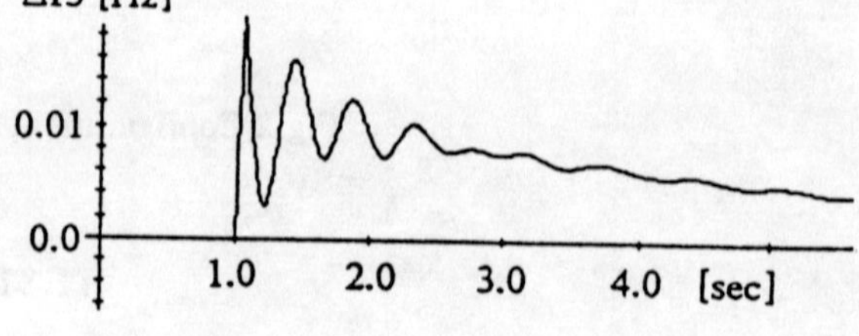

(b) Proposed generation dispatch

Fig.3 Rotor speed deviations of unit #3 following 50[MW] load reduction during 0.06[sec] at load level 550 [MW]

CONCLUSIONS

The ANN based generation dispatch algorithm for coordinating system economy and stability is proposed. The concept of the algorithm is based on the competitive nature of economy and stability in the generation dispatch scheduling. The ANN is applied for evaluating the contribution of generation distribution to system damping, and plays a role of the interface between fuel cost evaluation and stability assessment. The result of the simulation study shows that the proposed algorithm is useful for the coordination of economy and stability in power system operation.

REFERENCES

[1] C.T.Leondes ed., "Analysis and Control System Techniques for Electric Power Systems", Control and Dynamic Systems, Advances in Theory and Applications, Academic Press, Vol.41, part 1 of 4, pp359-461, 1991.

[2] H.Saitoh, J.Toyoda & Y.Kobayashi, "A New Index Extracted from Line Flow Fluctuation to Evaluate Power System Damping", IEEE Tr. on PS, Vol.6, No.4, pp1473-1479, Nov. 1991.

I. Fault Diagnosis

IMPULSE TEST FAULT DIAGNOSIS ON POWER TRANSFORMERS USING KOHONEN'S SELF-ORGANIZING NEURAL NETWORK

Thomas Baumann *Alain Germond* *Daniel Tschudi*

Ecole Polytechnique Fédérale de Lausanne
Laboratoire de Réseaux Electriques
CH-1015 Lausanne, Switzerland
Fax: + 41 21 693 46 62

ABB Sécheron Ltd.
CH-1211 Geneva, Switzerland
Fax: + 41 22 738 73 05

ABSTRACT

The paper presents a method based on Kohonen's self-organizing neural network to perform an impulse test fault diagnosis on power transformers. Digital signals obtained during transformer impulse testing are to be classified. In fact, if the model of a transformer is well known, a classifier could be trained off-line with all possible signals of simulated faults that can occur in a transformer. This would be useful to detect and locate failures in transformers during acceptance tests. Results of a classification of 28 (27 faulty and 1 faultless) different impulse test signals, derived from a simulated model of a power transformer, are presented in this paper.

Keywords: Transformer Diagnosis, Transformer Impulse Testing, Artificial Neural Networks, Self-Organization.

INTRODUCTION

Failure diagnosis during transformer impulse testing is usually a process which requires an experienced evaluator. Test signals are compared with similar signals which have been acquired earlier. A new technique to monitor this test is based on a deconvoluted frequency signature , the transadmittance of the transformer [3]. The magnitude of this characteristic is sensitive to failures of the apparatus during the test and can thus be used to locate them. The phase of the transadmittance shows similar changes too but is often noise-corrupted and is therefore not considered.
However, experience with human interpretation of the transadmittances has not yet been built up, that is why in this paper a classifier is proposed to perform failure location during the impulse test. The classifier can not only store recorded test data but also use data which is provided by simulations.
Many algorithms are known for signal classification and recognition. The advantage of artificial neural networks as classifiers is their tolerance to slightly modified inputs or noise. Furthermore, a hardware implementation of such networks will be possible soon, allowing thus fast training and classification of data.
In the domain of artificial neural networks the backpropagation learning algorithm for multi-layered networks is the most applied for the classification problem. This algorithm is called supervised, because the artificial neural network must, when a input is presented, minimize the error between the generated and the desired output in its learning phase.
An unsupervised algorithm, where the neural network organizes itself only in function of the inputs, presents the advantage of conserving the topology of the input vector space and of an automatic feature extraction. This enables an understanding of the classification process.
The Kohonen network, an example of an unsupervised neural network, can be applied to the classification, as showed in previous work. Kohonen himself showed the capabilities of his neural network for speech recognition [1]. Vercauteren et al. used the self-organizing Kohonen maps to do classification of brain tumours [2].

PROBLEM DESCRIPTION

Transformer impulse testing

Voltage and current signals acquired at a reduced (generally 50% of the maximum test voltage, also called Basic Insulation Level, BIL) and at full test level are used to detect failures during transformer impulse testing. The digital monitoring technique used today for this test is known for already 5 years [3]. It consists of digitally recording the input voltage, measured through a HV-divider, and the output current at a selected terminal (usually the ground or neutral terminal) measured at a coaxial shunt.

Reprinted with permission of the Organizing Committee of the Third Symposium on Expert Systems Application to power Systems, April 1-5 1991, Tokyo & Kobe, Japan

The transformer's transadmittance T(jw) (called transfer function) is then deconvoluted :

$$T(jw) = \frac{F\ (i(t))}{F\ (u(t))} = \frac{I(jw)}{U(jw)}$$

where F denotes the Fourier transform.
This frequency signature shows characteristic changes related to the type and the location of an impulse failure. The comparison between the two T(jw) calculated during the reduced (no faults are expected) and the full level test, can thus detect a possible failure and eventually reveal its location.

Fault location during impulse test

The classical approach to locate a failure which has occurred during the impulse test is to interpret the current or voltage records using a reasoning related to the concept of travelling waves. When the so-called series capacitance Cs (cf. Fig.1) in transformer windings is no longer negligible (e.g. interleaved windings) the transformer winding can no longer be regarded as a transmission line and therefore a more accurate method has to be found.
This paper deals with a new approach to the problem. Several characteristic transfer functions shall be classified to form a database for future recognition and by that means location of the impulse fault.
The first step, presented in the following paragraphs, consists of an examination of the possibilities and limits of Kohonen's neural network as a classifier.

CREATION OF THE DATABASE

Model description

At present we do not yet have at our disposal a database of measurements on a real world transformer. To evaluate the performance of an artificial neural network as classifier and to develop the methodology of using it for transformer test diagnosis, a database resulting from simulation on a six section single winding model of a transformer (as in [4]) is used.
The equivalent circuit of such a winding as well as the diagrams of the connected measuring equipment and the impulse generator are depicted in Fig.1. The circuit is composed of six identical sections, the mutual inductances being not represented in the figure. In the simulation, the impulse generator is modelled by a charged capacitance connected to a resistive circuit which helps shaping the input waveform. The input voltage U_{in} and output current I_{out} (the shunt is modelled by a low-ohmic resistor) are then calculated for a specified time duration after the switch-on of the generator.

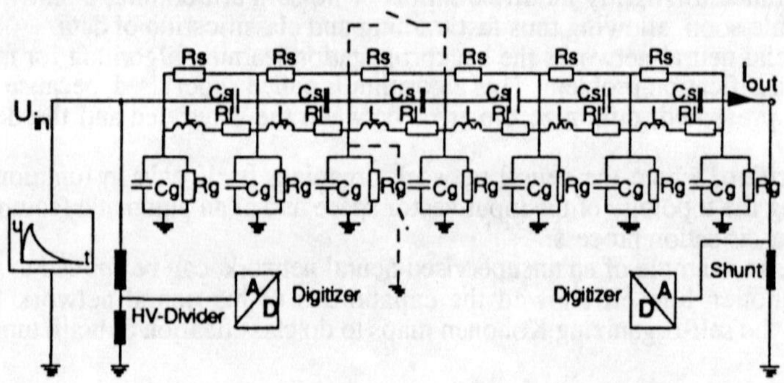

Fig. 1 Model of a six section single winding model of a transformer

Failures during this "impulse test" are simulated, bridging one or several consecutive sections (The dashed lines in fig.1 show two of these possible failures). The instant of switch-on of these short-circuits is determined through an analysis of the differential voltage between the nodes to be bridged. It corresponds to the 90%, 80%, or 70% voltage-level before the maximum differential voltage (100 % level) is reached. Thus, three families, T_{90}, T_{80} and T_{70} of current-voltage pairs result, representing each the 27 possible failures (7 failures to ground, 6 failures bridging one section, 5 failures bridging two etc...).
The corresponding transfer functions are deconvoluted then, their magnitude forming the database used for classification. Fig.2 shows the shape of these frequency characteristics for T_{90}, the row on top showing the T(jw) of failures to ground (nodes from left to right corresponding to figures from left to right), the second row depicting the T(jw) of single section failures etc... and the transfer function at the right lower corner representing the "healthy" transformer.

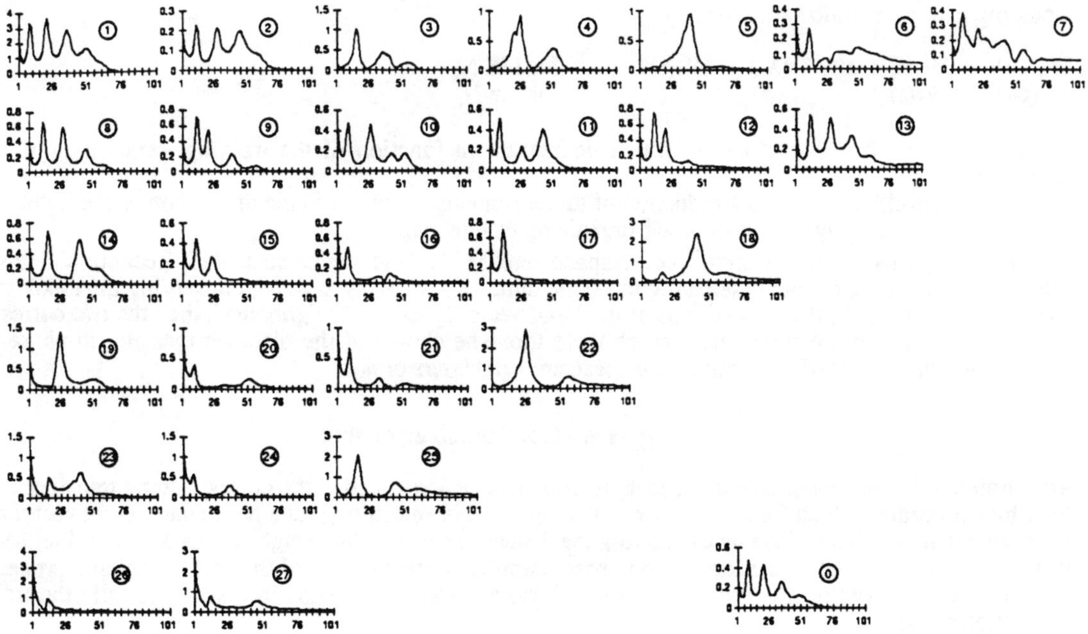

Fig. 2 Transadmittances of 27 transformer faults and of the faultless transformer.

SELF-ORGANIZING FEATURE MAPS FOR CLASSIFICATION

Training of the self-organizing feature maps

A detailed description of the algorithm can be found in [5] and [6]. Consider a two-dimensional lattice with M^2 neurons (also called units) and N input neurons. Every input neuron is connected through a synaptic weight w_{ij} ($i = 1 .. M^2, j = 1 .. N$) to every neuron of the map (cf Fig.3).

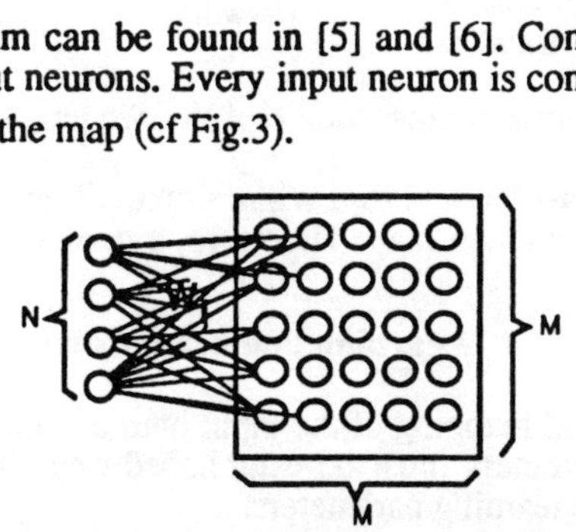

Fig.3 Schematic representation of a Kohonen map

A singular learning step can be summarized by the two following phases:

Phase 1: Find the best-matching unit c
An input vector $X = [x_j ; j = 1..N]$ is presented to the map and the distance between this input vector and all the weight vectors W_i ($i = 1 .. M^2$) is calculated. The smallest distance defines then the "active" neuron called unit c. Possible choices for the distance measure are the Euclidian and Manhattan distance and the scalar product. By using the latter, its maximum must be found and the input vectors must be normalized. This phase can be summarized by:

$dist(W_c,X) = min(dist(W_i,X))$ for $i = 1..M^2$ max for scalar product.

Phase 2: Update the unit c and its neighbours
Unsupervised learning proceeds by updating units within a neighbourhood N_c defined around unit c. The radius of neighbourhood decreases with the learning step until it reaches zero. If the radius is zero, only the unit c is updated.

Updating is done as follows:

$$\mathbf{W}_i(t+1) = \mathbf{W}_i(t) + a(t) * (\mathbf{X}(t) - \mathbf{W}_i(t)) \qquad \text{for } i \in N_c$$
$$\mathbf{W}_i(t+1) = \mathbf{W}_i(t) \qquad \text{for } i \notin N_c$$

$a(t)$ is the gain term ($0 < a(t) < 1$) which decreases in function of the training iteration step, as well as the neighbourhood.
Briefly, the algorithm performs a reduction of dimensionality. In our case the dimension of the input vector space is reduced to two, arranging vectors on a self-organizing feature map.
Kohonen maps partition the input vector space into M^2 sub-spaces, also called "receptive fields", which are determined by the set of inputs selecting the same unit. This mapping is done, conserving the topology of the input vectors. In other words, if two partitions of the input vector space are "neighbours", then the two corresponding units of the Kohonen map are topological neighbours too. The choice of the distance measure in phase 1 determines whether two sub-spaces of the input vector space are neighbours or not.

The classification algorithm

After training the self-organizing map, each neuron must be associated to the corresponding transformer fault. This is done by a procedure which finds the index of the input vector matching best a particular weight vector of a neuron of the Kohonen map. This is done by calculating the distances between the weight vector $\mathbf{W}_i$ stored in the neuron i and all the vectors contained in the training database. (similar to the phase 1 of the self-organizing process). Given the index of the input vector, the type of fault is determined too. Fig. 4 describes schematically the training and the association phase.

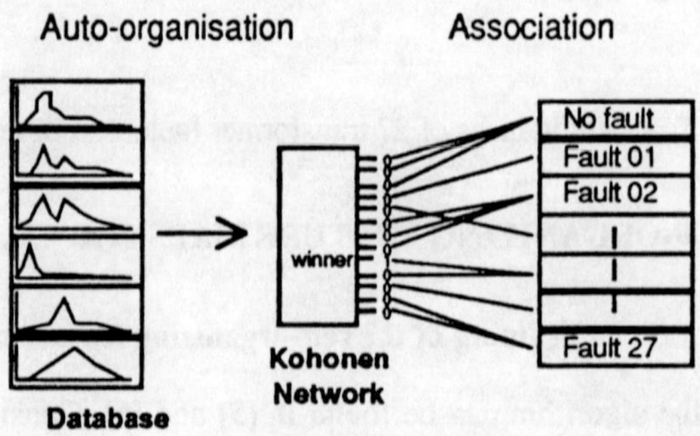

Fig. 4 Schematic representation of the two phases of the classification process

In other words, the Kohonen classifier acts as a winner-take-all network. The winner designates the unit matching best the input presented to the Kohonen map. Therefore, a certain fault is associated to the corresponding weight vector of a neuron.

Learning and network parameters

The main problem to perform good mapping of the input onto the Kohonen map is the determination of learning and network parameters. These parameters must be established experimentally by preliminary simulations. We can distinguish two network and three learning parameters:

Number of neurons of Kohonen map:
Generally, Kohonen's self-organizing neural network can be applied in two different manners to classification problems. If the number of input vectors in the database exceeds the number of units in the map, the network will cluster similar faults into one unit. This will be the case if we have a large database available, resulting from impulse tests on several transformers. The same grouping effect can happen if two inputs are very close (in the sense of the chosen distance measure).
Otherwise, the neural network will work as a generalizer network (as the Backpropagation net). In this case, the input vectors will be distributed on a part of the units of the Kohonen map, and the remaining units will represent an interpolation of the surrounding units. In order to obtain a well organized generalizing network, the learning parameters must be properly chosen, especially the neighbourhood.
Our simulations showed that these two modes normally appear mixed if the number of neurons and input vectors do not differ significantly.

Number of dimensions of the input vector:
Generally, the number of components of the input vector is determined by the database, but we took it as small as feasible due to the approximatively linear dependence on the simulation time.

In our case the input vectors are composed of the sampled magnitude of transadmittances (transfer function) derived from simulated transformer, the significant part of the sampled data being 6 - 80 kHz. Beyond 80 kHz the magnitudes are close to zero. With the sampling rate of nearly 1 kHz the input vector is composed of 75 components.

Initial learning gain a and neighbourhood N_c and their variation in time:
The choice of the learning gain and the neighbourhood is dominant for a topologically correct organization of a Kohonen map. Our simulations showed that the initial learning gain a should be between 0.2 and 0.5 and the initial radius of the neighbourhood must be about the half of the maximal dimension of the Kohonen map.
The decreasing of the two parameters with the discrete-time index is very important. A slow decreasing leads to better organization but needs also more simulation time. In all our simulations both parameters decrease with the inverse of the number of iteration steps. Further we define a (as in [7]), being also dependent on the distance between the best-matching unit and the unit to be adapted. We use the following formula:

$$a = a_0 * \exp\left(- \frac{(U_c - U_a)^2}{R^2} \right)$$

where U_c is the best matching unit and U_a the unit to be adapted, R indicating the radius of the neighbourhood.

Choice of the distance measure:
Previously we mentioned "topologically correct organization of a Kohonen map". What does it mean? The answer is very subjective because it depends on the choice of the distance measure. Kohonen [5] himself proposes the scalar product and the Euclidian distance to calculate the similarity between the input and the weight vectors.
Due to high computation time of the scalar product, we used the Euclidian distance for our simulations.

Pre-processing:
Generally, intelligent pre-processing of the input vectors can strongly improve the classification results with artificial neural networks. The objective of pre-processing is to render the input vectors more specific and there are no limits for the definition of an algorithm. The most used are:

Scaling of input vectors:
- Normalization to a constant value
- Scale to the maximal value of a vector

Scaling of input vector components (Component scaling [8]):

$$z_i = \frac{x_i - m_i}{s_i} \qquad \text{for } i = 1..N$$

where m_i is the mean and s_i is the standard deviation of all the components x_i. This pre-processing is especially adequate, if the components of input vectors do not have the same scale and variance, but are equally important.
In our case pre-processing starts with the choice of the magnitude of the transfer function which presents the advantage to be independent of the input signal. The phase of the transfer function is very sensitive to noise and is therefore not considered for classification.
Preliminary simulations showed that scaling of input vectors to the maximal value suited best the problem. A change of the maximal amplitude of the frequency characteristic does not indicate a failure but a partial discharge [3] and is thus insignificant for the classification.

The Kohonen map simulator

Unfortunately a Kohonen map realized on hardware that could manipulate nets of the order of 400 neurons and 75 inputs does not yet exist. Therefore we were forced to simulate it on a conventional computer. The program, realized on Apollo workstations, allows to set interactively all learning and net parameters described above and extracts the simulation results.
Several laboratories around the world carry out research on hardware implementation of Kohonen networks. Due to parallel architecture the training and the recall phase could be done in much less time, in order that a classification result can be extracted e.g. after less than 100 µs for a network with dimensions N=100 and $M^2 = 20*20$ [9]. On a 4500 Apollo workstation one step takes about 60 ms for the same network. The whole training phase implies approximatively 30000 steps, therefore it takes about half an hour.
The purpose of this paper is the study of the application of Kohonen networks to diagnosis of transformer failures using the pre-processed digital test signals. Our application does not necessitate such a speed - a simulation on a workstation is sufficient. Nevertheless, due to its high processing speed, it could be of great interest to apply that type of classifier to the domain of monitoring, where dangerous situations must be detected very fast.

RESULTS AND DISCUSSION

Classification of signals derived from simulated impulse tests on faulty and faultless transformers is interesting for two different reasons. Firstly, the Kohonen classifier could be used to detect faults during acceptance tests and secondly, due to the topology conserving characteristic of Kohonen maps they could be used to find possible hidden relations between the different faults. This subject will not be discussed in this paper.

Kohonen maps used as a classifier

Performing classification, the Kohonen maps must be fairly large in order to avoid the grouping effect, where several input vectors are clustered into a single neuron. This effect can lead to ambiguous situations.
The performance is tested as follows: The classifier is built up with one part of the database, and the other part (the test base) is then used to evaluate the classification results. The classification error is evaluated by presenting all the vectors of the test base to the trained Kohonen net. It is expected that a particular failure will be classified in the same region where the corresponding failures of the training-base are mapped (cf. Fig. 5). The error consists then of the number of misclassifications with respect to the number of vectors contained in the test base.
Table 1 summarizes the classification error in function of the number of neurons and the type of training- and test base chosen. The numbers 1,2 and 3 correspond to the three families T_{90}, T_{80} and T_{70} of the database described above.

Case	M*M	Training base	Test base	Classification error
1	20*20	1 and 3	2	1.20%
2	15*15	1 and 3	2	0.90%
3	12*12	1 and 3	2	1.80%
4	20*20	2	1 and 3	10.70%
5	15*15	2	1 and 3	7.10%
6	20*20	1 and 2	3	4.80%
7	15*15	1 and 2	3	7.10%

Table 1 Classification error in function of the number of neurons and the training- and test base

Three different types of simulations were performed. In the first, the families 1 and 3 are used to train the Kohonen networks with dimension M^2= 20*20, 15*15 and 12*12. Family 2 is presented to the trained classifier to evaluate the classification performance. In case 1 three out of nine simulations, where each simulation is a training with different initial random weight vectors, showed a single classification error, which figured out to be three times the failure 02 (corresponding to a fault to ground, cf Fig. 2). In the other six simulations the classification of the test base was faultless.
In the second case, the classification result is even better as in the previous case, which may be due to statistical reasons, the number of simulations per case being rather small (<10). If the number of neurons is reduced to 12*12 only 1 of 2 simulations perform a 100% correct classification of the test base.
In case 4 and 5 the training- and the test base were exchanged and then tested. The classification error is much higher, because the network has more difficulties to generalize, being trained with a smaller database.
The third type of simulation is similar to the first. Actually, in the case 1, 2 and 3 we expected the classifier to perform a interpolation between the first and the third family. Now the classifier should extrapolate the third family, which seems to be more difficult than the interpolation. The classification error is 4.8 % for a 20*20 Kohonen network.
In Fig. 5 an example of a 15*15 Kohonen network is represented, showing the classification result of a successful simulation (case 2). This map shows which type of transformer fault is stored in each neuron. Each neuron is represented by a three digit number, where the first digit denotes the family and the two other digits the fault number (cf. Fig. 3). Only faults of family 1 and 3 are represented, because family 2 is not used for training. The neurons which store the exact type of fault, are marked by a square or a triangle. The other neurons are interpolating neurons, which store an average of their surrounding neurons. These interpolating neurons are very important for the generalizing characteristic of the Kohonen classifier. This feature can be examined by presenting the test base to the classifier, which is the family 2 in our case. The neurons which have been activated by the test base are marked with a circle.
Two neurons, which do regroup all three families, can be found. This happens when the three corresponding vectors are very similar. In all other cases the family 1 and 3 are stored in separated neurons. If the test base is presented, a vector corresponding to a certain fault must find its best-matching neuron. It has two possibilities: Either it choses a "triangle or square neuron", which is the case for fault type 08, 16, 17, 20, 24 etc., or a interpolating neuron is activated as for fault type 10, 11, 13, 21, etc. A case, where an interpolating neuron is activated is most interesting, because it demonstrates the generalizing property of the Kohonen classifier.

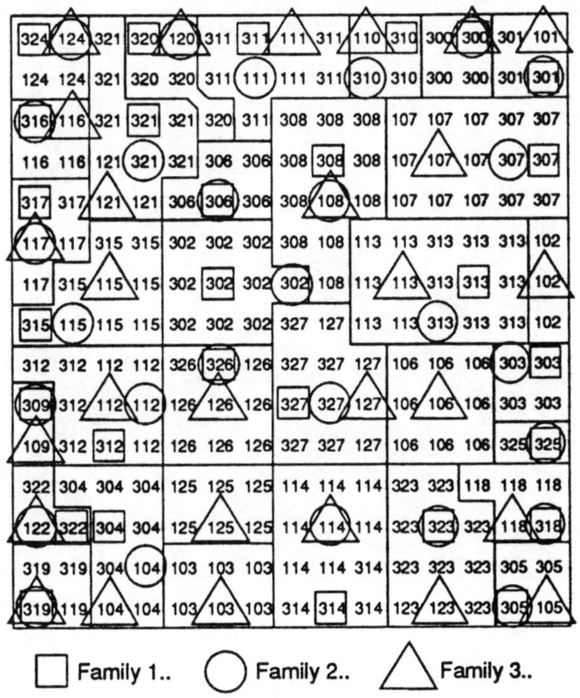

Fig. 5 Classification results after training of a 15 x 15 Kohonen network

CONCLUSION

In this paper we show that Kohonen's neural network can successfully be applied to the classification of signals resulting from impulse tests on power transformers, whereby the choice of learning and network parameters c an strongly influence the classification results. After training, the artificial neural network is able to generalize and recognize unknown pattern with an classification error of less than 5% (cf. Table 1).
At present all our results are based on transformer simulations in order to create the training- as well as the testing-base. Future work will focus on the testing of the classifier using real power transformer data.

ACKNOWLEDGEMENT

The research described in this paper was supported by the Swiss National Science Foundation, Project No. 20-26557.89

REFERENCES

[1] Kohonen, T., The "Neural" Phonetic Typewriter, IEEE Computer, pp. 11-22, March, 1988.

[2] Vercauteren, L. et al., The Classification of Brain Tumours by a Topological Map, Proceedings of International Neural Network Conference, Paris, France, pp.387-391, July 9-13, 1990.

[3] Malewski, R., Poulin, B., Impulse Testing of Power Transformers Using the Transfer Function Method, IEEE Transactions on Power Delivery, Vol. 3, No. 2, pp. 476-483, April, 1988.

[4] Degeneff, R.C., A General Method for Determining Resonances in Transformer Windings, IEEE Transactions on Power Apparatus and Systems, Vol. PAS-69, No. 2, pp. 423-430, March/April, 1977.

[5] Kohonen, T., Self-Organization and Associative Memory, 2nd Edition, Springer-Verlag, New-York, 1988.

[6] Niebur, D., Germond, A., Unsupervised Neural Net Classification of Power System Static Security States, Third Symposium on Expert Systems Application to Power Systems, Tokyo, April 1-5, 1991.

[7] Ritter, H. and Schulten, K., Kohonen's Self-Organizing Maps: Exploring their Computational Capabilities, Proc. IEEE First International Conference on Neural Networks, vol. 1, pp. 109 - 116, San Diego, Cal., July, 1988.

[8] Deichsel, G., Trampisch, H.J., Clusteranalyse und Diskriminanzanalyse, Gustav Fischer Verlag, Stuttgart, New York, 1985.

[9] Blayo, F., Marchal, P., Extension of Cellular Automata to Neural Computation, International Joint Conference on Neural Networks, Washington, June, 1989.

A CASE STUDY OF NEURAL NETWORK APPLICATION: POWER EQUIPMENT FAILURE DIAGNOSIS

Yaping Du, Fei Wang, T.C. Cheng
University Southern California
Los Angeles, CA 90089-0271
Tel: (213) 740-8535 E-mail: ydu@aludra.usc.edu

Abstract

This paper presents a successful application of neural network-based pattern classification system to the failure diagnosis of underground oil switches. Trained and tested with actual data collected from 67 samples, the two-layer neural network classifies the oil switch status significantly better than the current threshold classification method. The study shows that the key to a successful neural network-based failure diagnosis is the physical understanding of the given problem and the proper incorporation of the knowledge in training. The obtained principle also applies to other diagnosis problems.

INTRODUCTION

Equipment failures are a major cause for power system outages. The detection and diagnosis of the failure conditions for power equipment is very important for the overall system reliability, and therefore has been attracting lots of research attention. The diagnosis procedure is essentially a decision making procedure, i.e., to determine whether a particular piece of equipment should fail based on certain screening methods, and suggest corresponding preventive measures. A good diagnosis scheme, like a good doctor, should neither miss an early failing symptom, nor should it generate false alarms.

The difficulty for failure diagnosis lies in the fact that an equipment failure is a complex event, which can be a resultant of many factors and whose early symptoms can appear in various forms. In the case of oil switches widely used in underground power distribution systems, a simple oil sampling test reveals that parameters such as moisture content, dissipation factor and dielectric strength can all *partially* tell the condition of an oil switch in service, though it is believed that the complete picture should be depicted considering all these parameter together. Currently, a typical utility practice is to set a separate threshold for each parameter when screening underground oil switches using oil sample testing. If the test result for any parameter exceeds its threshold, the oil switch is classified as failing. Based on this simple method, an 80% diagnosis success rate has been achieved, with a judgment often tending to be conservative. Another shortcoming of the current practice is its incapability of utilizing the data with less obvious effects. These data include the oil switch age and the test date.

In order to devise better oil switch diagnosis criterion, extra efforts are needed to study its failure mechanism. USC's work is a good start [1]. In the meantime, acknowledging that diagnosis is a pattern classification or recognition problem, pattern recognition techniques can be used to find the combined relation (mapping) between the tested parameters (input features) and the oil switch status (output patterns). In recent years, artificial neural network has been used successfully for pattern recognition. Using sample data for training, a neural network (NN) can learn and memorize patterns, and predict the class of a new sample based on the previous knowledge. Compared with some statistical pattern recognition methods and expert systems, a neural network requires no explicit knowledge about probability distribution or models, and can handle numerical data better [2-8]. Neural network method is especially suitable for failure diagnosis, where no reliable models and little explicit knowledge are available. Works have already been attempted in the area of power equipment failure diagnosis [4-5]. Since neural network approach is a model-free and problem-dependent method, each neural network based system must be designed and trained according to its own specification. The success of a neural network pattern classification system depends primarily on the designer's physical understanding to the problem, as will be demonstrated in the present study.

In this paper, the principle of neural network based pattern classification will be reviewed. The formulation of the oil switch failure diagnosis as a neural network pattern recognition problem will be presented, and the input parameter selection (feature extraction) and output pattern representation will be discussed. Training and testing procedures will be designed and implemented using the actual data collected from oil switches in the field. The results will be compared with those using the current utility diagnosis practice based on threshold classification methods.

NN-BASED PATTERN RECOGNITION

Neural Network

Inspired by studying human learning mechanisms, the artificial neural network model was proposed as early as in the 1940's. Its applications became widely spread in pattern recognition, control, signal processing, etc. during the 1980's. A neural network itself is a structure of highly connected nodes called neurons, representing a preliminary human neural system [9]. In a multilayer feedforward perception model [9] used in pattern recognition, each node is connected with all nodes in the next lower layer through weighted connections. The input data are propagated forward through nodes and their connections in the hidden layer(s) to the output nodes.

The system error for a given pair of input-output data is defined by the mean squared error, illustrated as

$$J = \frac{1}{2M} \sum_{k=1}^{M} (t_k - o_k)^2 \tag{1}$$

where t_k and o_k are the desired output and neural network output. M is the number of outputs of the neural network. In the learning process, the system error is back-propagated through the network to update all weights, by using the popular generalized delta rule [9, 10, 11]. This is an iterative learning process, since weights are modified once when a sample is present. And it will end if J is less than a desired value.

Neural Network-Based Pattern Recognition

From the geometric standpoint, pattern recognition is nothing but partition of the input space into subregion(s) for each output pattern. For the neural network-based pattern recognition, the partition, or the mapping between input space (spanned by input features) and class space (spanned by output patterns) is implemented through a neural network. Since a neural network has a high degree of connections among nodes and nonlinear activation functions, it can handle complex decision-region problems. The complex mapping can be learned with the learning algorithm—the generalized delta rule through training data, then inherent patterns from input data are memorized with adjustable weights. The learning algorithm is iterative, so that, the NN-based classifier can be viewed as an evolving machine. The more the NN classifier learns from the world, the better it performs. After the learning process, the NN classifier is ready to predict the output class with the incoming sample data.

A pattern recognition problem consists of two basic steps: feature extraction or selection and classification. The two steps are closely related by the fact that a better feature extraction or selection will result in an easier classification. The neural network-based system can be designed to test a significant feature by comparing the system error J of a class output defined in (1). If J drops down significantly with iteration number when a tested feature is added into feature space, it indicates a significant correlation of the feature. Otherwise, the feature might be uncorrelated or there might be no significant features available.

OIL SWITCH DIAGNOSIS SYSTEM DESIGN

The status of an oil switch is characterized by the degree of its internal tank corrosion. Thus, the output patterns or classes for the present diagnosis system are:

- Class (I) — Non-corrosive condition
- Class (II) — Corrosive condition

The task of the neural network diagnosis system is to determine whether a given switch belongs to Classes (I) or (II) by using measurable features, without dismantling the switch or disrupting its service. The system design work involves two steps: selection of features and selection of NN architecture.

Feature Parameters

The feature selection design involves the appropriate selection of input parameters that: a) will reflect the status of equipment; b) can be obtained externally through routine maintenance check-up or from historic design and operating data. For underground oil switches, the corrosion degree was found closely correlated with the condition of the oil [1]. The parameters that characterize the oil condition are important features for the present neural network, including:

Moisture Content *(ppm)*: The moisture content in the oil has been observed the most relevant indicator of the oil condition. The current utility practice screens out a switch if the moisture level in its oil is above certain threshold.

Power Factor: This parameter, indicating losses of the oil , is closely related to the moisture content and also reflects the oil degradation due to aging. A threshold value is also set in the current switch screening.

Dielectric Strength *(kV/cm)*: This important indicator for oil aging condition is also checked during the maintenance test.

The other relevant parameters include the service age of the oil and the date on which a switch is tested. The possible effect of the age is obvious. The possible effect of the test date was found through screening a utility's testing results on its oil switches. It was observed that the oil tested in summer tends to have higher moisture contents, higher power factors and lower dielectric strength than that tested in winter.

It should be noted that the condition of an oil switch has a monotonous relation with all the feature input parameters. Physically, a switch by itself should be in a worse condition when its oil gets older, and has higher moisture content, higher loss, and lower dielectric strength. The test date can be converted to a monotonous curve by defining summer months as the maximum and winter months as the minimum.

Architecture of Neural Networks

Since the class output varies with input features monotonously, the diagnosis of oil switches is a two-open-decision-region problem. A two-layer neural network with two class outputs is selected. As a result, each sample's output is represented by a two dimensional vector. The vectors for the desired classes (I) and (II) are assigned to be $(0.9, 0.1)$ and $(0.1, 0.9)$. As explained previously, the five input features are moisture, power factor, dielectric strength, age and test date. In the classification problem, each input feature can fall into two possible regions, and a 32-node hidden layer will be used.

EXPERIMENTS AND RESULTS

For the experiment, feature data are collected from 67 oil switches. Table 1 categorize all the oil switches according to different manufacturers and switch types. Among the 67 samples used for neural network construction, 70% of them are used for training and 30% for testing.

Table 1: Categories of the Sample Oil Switches

Manufacturer	A		B	C
Type	BURD	RAM	RAM	BURD
Number	21	18	14	14

The original data are normalized to be in the range $[-1, 1]$. A new sample will be classified into (I) when the geometric distance between its output and the center of class (I) is shorter than that between its output and the center of class (II). [see Figures 2].

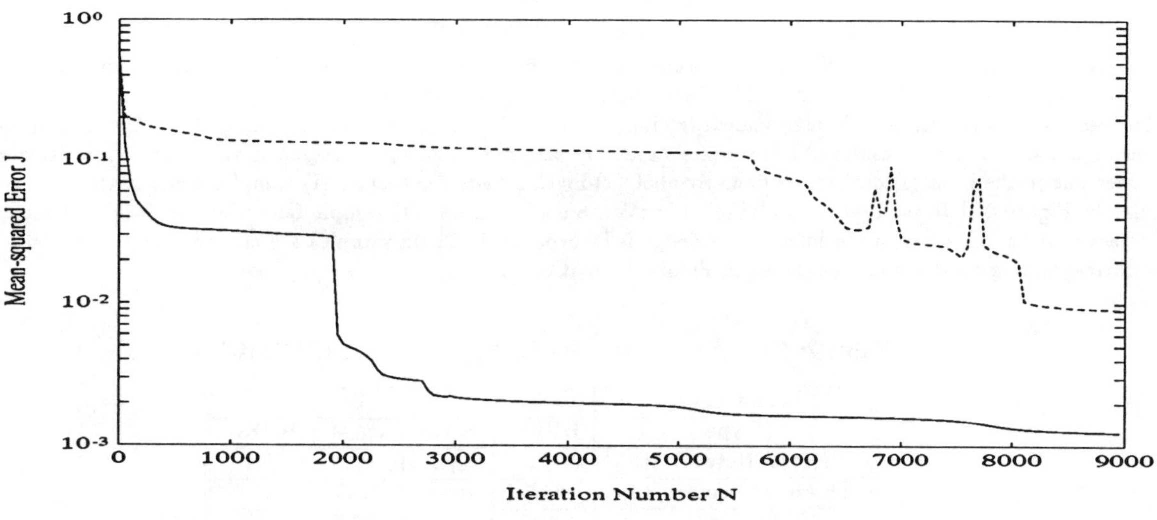

Figure 1: Learning Curves with Three and Five Input Features

Training

Since the training algorithm is iterative, samples are presented sequentially to update weights. In one cycle or iteration, each sample is used once. This process stops when the system error J in one cycle is sufficiently small, indicating that the neural network has memorized all the patterns implicitly stored in the training samples.

feature testing: Figure 1 demonstrates the significance of related features with all 67 samples. the dashed learning curve is for neural network with three features: moisture, power factor and dielectric strength, while the solid curve is for the neural network with two more features: age and test date. experiments show that the neural network with five features can be successfully trained, while three features alone are insufficient to build a neural network with good performance.

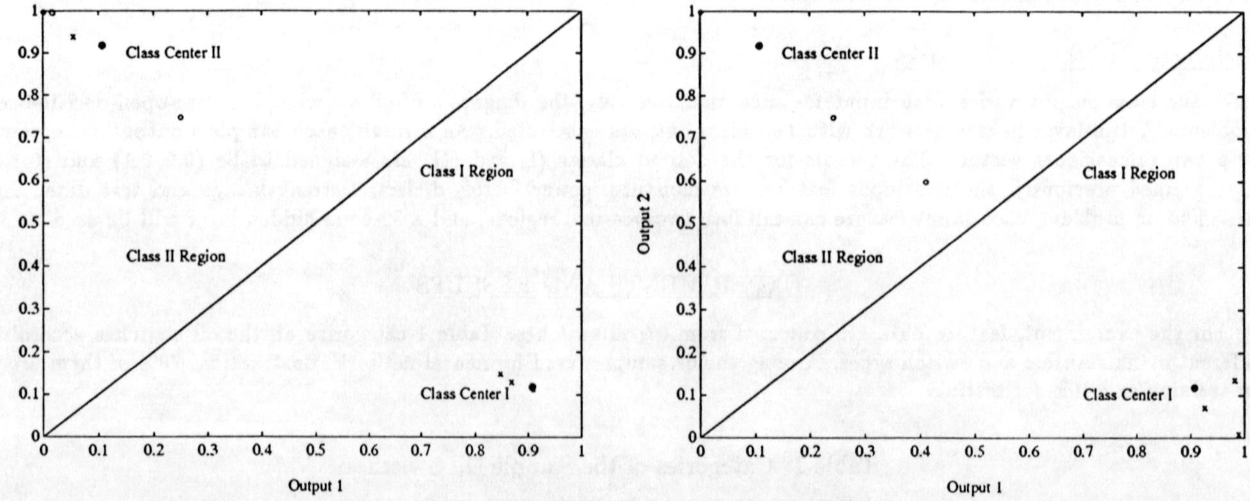

Figure 2: Classification Results: Without Human Knowledge Incorporation (left one) and With Human Knowledge Incorporation (right one)

Knowledge Incorporation: Like a human brain, a neural network will be smarter if more patterns (knowledge) are input to it. The samples for training not only come from actual data collection, but also should include the generally established knowledge for the given problem. The following knowledge is used in the learning process:

1. An oil switch will not corrode with an extremely low moisture content;
2. An oil switch will definitely corrode with a very high moisture content.

The knowledge is linguistic data that can be translated into numerical data as additional samples for training. A better physical understanding to the problem will lead to a better neural network classifier.

Figures 2 shows the effect of human knowledge incorporation for neural network learning. Both figures display, in a two-dimensional space, the test results of six samples randomly selected. The upper region is the class (II) decision region, and the lower one is the class (I) decision region. Symbol (x)-mark stands for a class (I) sample, while circle (o) for a class (II) sample. In Figure 2 (left one), where no linguistic data are used, a class (II) sample falls into the class (I) region, indicating a misclassification. However, when human knowledge is incorporated, all six samples are classified correctly and located near their corresponding class centers, as shown in Figure 2 (right one).

Table 2: Classification Results for Four Types of Oil Switches

Manufacturer	A		B	C
Type	BURD	RAM	RAM	BURD
Neural Network ID	#1	#2	#3	#4
Threshold Success Rate	84%	100%	75%	75%
NN Success Rate	100%	100%	100%	100%

Testing

Four neural networks corresponding to four categories of oil switches listed in Table 1 are trained with training samples and human knowledge data. The training error is chosen to be less than 2×10^{-3}. The testing results using these four neural networks are tabulated in Tables 2. For comparison, the results based on a threshold classification currently used by utilities are also given. The total success rate for the neural network classification is 19/19 (100%), while it is 16/19 (84%) for the threshold classification.

CONCLUSIONS

The following conclusions are reached through the application of neural network in the failure diagnosis of oil switches:

1. A neural network-based pattern classification system can be used for underground oil switch failure diagnosis. It performs better than the current threshold classification method.
2. A two-layer neural network with five inputs and two outputs is selected, for all the feature parameters have monotonous relationships with the output patterns.
3. The neural network approach is superior because the threshold classification bases its prediction on individual feature, while the neural network can handle collectively the complex nonlinear problems involving many implicitly correlated parameters.
4. The key to the success of the neural network-based diagnosis is the physical understanding to the problem and the incorporation of human knowledge in training neural networks and selecting features.
5. The neural network-based system has the capability of testing the significance of a feature.

References

[1] C.Y. Li, T.C. Cheng, F. Wang, E.M. Worland, "The Effect of Moisture on the Breakdown of Aged Oil In Underground Load Interrupted Switches", To appear on IEEE Trans. on Electrical Insulation.

[2] M.A. El-Sharkawi, R.J. Marks II, "Proc. of the First International Forum on Application of Neural Networks to Power Systems", Seattle, Washington, July 23-26, 1991.

[3] K.S. Swarup, H.S. Chandrasekharaiah, "Fault detection and Diagnosis of Power Systems Using Artificial Neural Networks", *idib*, pp102.

[4] Mo-yuen Chow, Griff L. Bilbro, Sui Oi Yee, "Application of Learning Theory to a Single Phase Induction Motor Incipient Fault Detection Artificial Neural Network", *idib*, pp98.

[5] Hiromi Ogi, Hideo Tanaka, Yoshiakira Akimoto, Yoshio Izui, "Fault Diagnosis for GIS Using an Artificial Neural Network", *idib*, pp112.

[6] K. Feser, U. Braun, "Application of Neural Networks in Numerical Busbar Protection Systems (NBPS)", *idib*, pp117.

[7] S.A. Khaparde, P.B. Kale, S.H. Agarwal, "Application of Neural Network in Protective Relaying of Transmission Lines", *idib*, pp122.

[8] Dejan J. Sobajic, Yoh-Han Pao, "Artificial Neural Net-Based Dynamic Security Assessment for Electric Power Systems", IEEE Trans. on Power Systems-4, No.1, pp220, 1989.

[9] Robert Hecht-Nielsen, "Neurocomputing", Addison-Wesley Publishing Company, Inc., New York, 1990.

[10] Yoh-Han Pao, "Adaptive Pattern Recognition and Neural Networks", Addison-Wesley Publishing Company, Inc., New York, 1989.

[11] R.P. Lippmann, "An Introduction to Computing with Neural Net", IEEE ASSP Magazine, April 1987, pp4-22.

INTEGRATING NEURAL NETWORKS WITH INFLUENCE DIAGRAMS FOR POWER PLANT MONITORING AND DIAGNOSTICS

Alice Agogino, Ming-Lei Tseng and Punit Jain
Department of Mechanical Engineering
University of California at Berkeley, CA 94720

ABSTRACT

The focus of this research is the integration of Probabilistic Neural Networks (PNN) with influence diagrams for power plant monitoring and diagnostics. Influence diagrams are used to provide a symbolic representation of the process system model. The conditional independence encoded in the influence diagram is utilized in defining the structure of the neural network, including the number and type of connections required. Neural networks are then used as a learning mechanism for estimating and updating conditional probabilities from a set of training data. This data may come directly from sensors associated with on-line monitoring systems, as was done on the prototype tested. Once the distributions are learned or updated off-line, influence diagrams are used to monitor new on-line data and answer both predefined and arbitrary diagnostic queries within its framework. This probabilistic framework allows the user (e.g., performance engineer or operator) to see not only the most likely failure causes, but also a probabilistic ranking of them in a manner which takes sensor validation into account.

INTRODUCTION

One goal of automatic diagnostic and monitoring systems is to take corrective action in the event of an accident or other unforeseen occurrence. More importantly, if such systems can effectively and efficiently pinpoint the causes of an incipient failure and take timely corrective actions, then catastrophic events can be avoided. Such systems, however, are highly sensitive to the reliability of the sensors on which they are based. Thus the integrity of the sensors defines the ability of these systems to monitor operations and diagnose failure.

The system envisioned in this research uses influence diagrams to provide a symbolic representation of the process system model. The independence and conditional independence encoded in the influence diagram are used to define the structure of the neural network to reduce the complexity (ie. the number of neurons needed and their connections). Probabilistic neural networks (PNN) are then used as a learning mechanism for estimating conditional probabilities. Once the distributions are learned the influence diagram can be employed to answer any query within its framework . The motivation behind this approach is to exploit the strengths of both influence diagrams and neural networks. Influence diagrams have proven to be an intuitive representational scheme, a rigorous uncertainty management mechanism and an efficient computational architecture for complex decision and inference problems. Neural networks have proven to be an effective learning scheme and an efficient parallel processing mechanism for complex learning tasks. In this study influence diagrams are used to effectively reduce the system complexity, to flexibly make probabilistic inferences, and to provide intuitive explanations. Neural networks then add learning, generalization, continuous updating, and parallel processing of large amounts of sensed-data of the influence diagram architecture.

BACKGROUND RESEARCH

Specht[1988] proposed probabilistic neural networks that could estimate conditional probability distributions. If the underlying density is smooth and piecewise continuous, a family of Parzen estimators[Parzen,1962] can be shown to asymptotically approach the probability density functions under some conditions. The particular Parzen estimator used by Specht for estimating the conditional probability density of an input vector X, given category A is,

$$f(X/A)=\frac{1}{2\pi^{d/2}\,\sigma^{d}\,n}\sum_{i=1}^{n}\exp\left[\frac{-(X-X^{i})^{t}(X-X^{i})}{2\sigma^{2}}\right]$$

where X^i is the i-th training pattern from category A, d is the dimension of the input patterns, σ is the smoothing parameter, n is the total number of patterns. The basic idea of this type of estimator is as follows. By observing X^i belonging to category A, one can assert that f(X/A) takes a non zero value at the point X^i. Moreover under the continuity assumption of f(X/A), it is reasonable to infer that f(X/A) will assume non zero values in the immediate neighborhood of point X^i, and the further away a point X is from X^i, the less contribution toward f(X/A) from observation X^i it gets [Devijver and Kittler, 1982]. This estimator can consistently approximate any smooth density function under certain conditions. The organization of Specht's model is re-drawn in Fig. 1 .

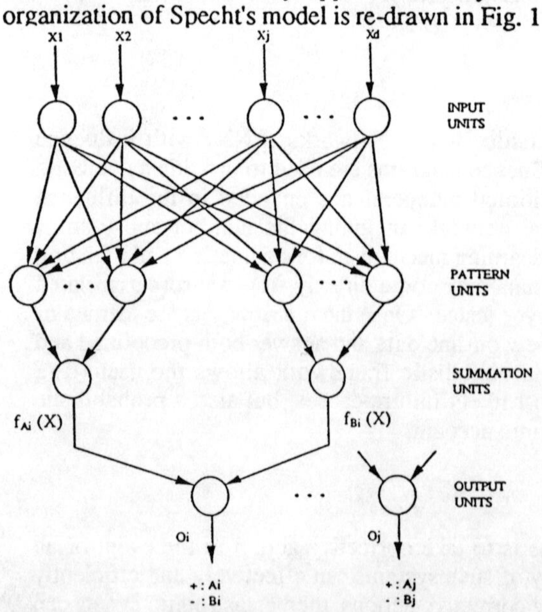

Fig.1 Specht's Probabilistic Neural Network

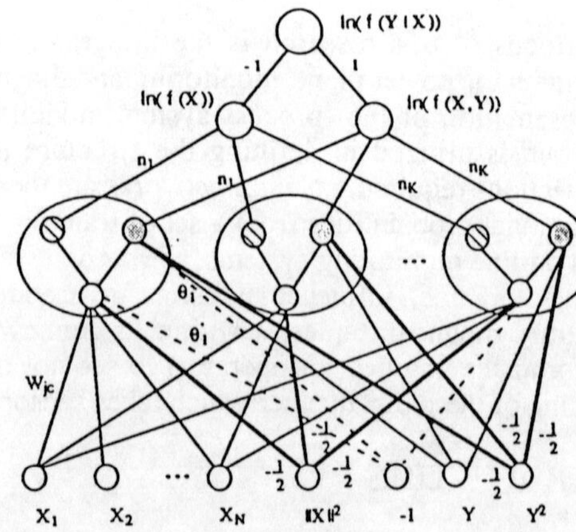

Fig. 2 Architecture of SOPNN

The original model requires as many pattern units as the training patterns. This requirement could cause indefinite growth of patterns units when one wants to update the probability density function as operating data accumulates. In order to overcome the drawback of unlimited storage, Specht proposed polynomial Adaline (or Padaline) paradigm, which was an approximation to his original model [Specht,1990]. This model too, however, proved to be ineffective in controlling the unlimited storage problem. In the next section a neural network model is proposed with the aim of resolving these limitations .

PROBABILITY ESTIMATION OF CONTINUOUS VARIABLES

In order to avoid the indefinite growth problem suffered by the Specht's model, it seemed desirable to use only a fixed amount of neurons to do the same job, but with as little loss in accuracy as possible. K-means clustering is used for this, i.e. a specific number of cluster means are used to encode the information contained in the samples. The architecture of the Self Organizing Probabilistic Neural Network (SOPNN) is drawn in the Fig. 2 . The first layer is a self-organizing net for clustering, and the second layer is essentially a modified Specht's model for estimating the probability density function. The idea is to represent similar training patterns in the same cluster by its centroid and then use these clusters for probability function estimation. The cluster nodes are basically localized gaussian receptive fields. The training data are presented one by one, and only once, to the neural net and a deliberate overlapping of the localized receptive fields is used to compensate for the potential loss in accuracy due to sub-optimal clustering. This is attributed to the fact that a misclustered point often lies on the border of its nearby cluster, and its contribution to the overall response function is more or less retained through the summation of contribution from all nearby receptive fields.

By incorporating the ideas of localized receptive fields and K-means clustering, the SOPNN has shown its effectiveness as a probability density function estimator. As an approximation to Specht's model the SOPNN requires much less neurons while retaining all the desirable properties without much sacrifice in accuracy. Given enough number of units, it can effectively estimate non- gaussian as well as multimodal distributions. The SOPNN is of fixed size and very general, not being limited to pre-specified parametric or unimodal distributions. The learning is incremental and one pass. Only new data need to be trained so that there is no need to store the entire training set. It is self-organized. We don't have to provide target clusters and then train it with the slow back-propagation algorithm.

APPLICATION TO SENSOR VALIDATION DATA FROM ELECTRIC POWER PLANT

The Specht's model, the Padaline and the SOPNN models were tested on a set of real sensor data obtained from the feedwater systems of an electric power plant. The outlet temperature of a particular heater was measured along with the corresponding power load. For clarity the dataset was divided into three regions namely: high load, low load1 and low load 2 regions. The scatter plot for low load region 1 is given in Fig 3. The resulting distributions obtained by the three models are plotted in Fig 4. The distribution in the low load 1 region looks like a combination of two normal distributions.

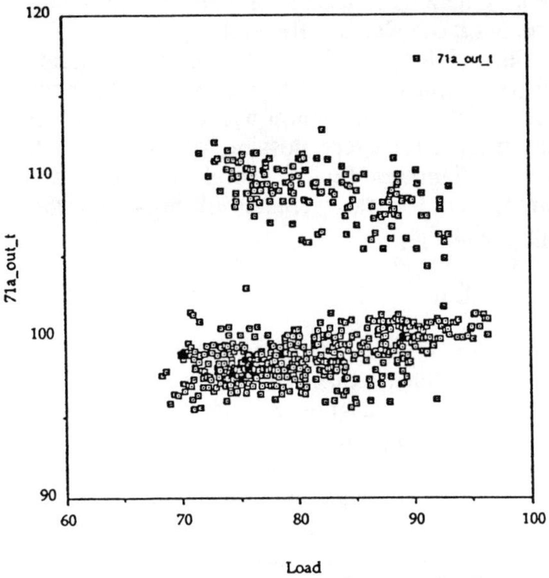

Fig. 3 Output Temperature of 71a vs load

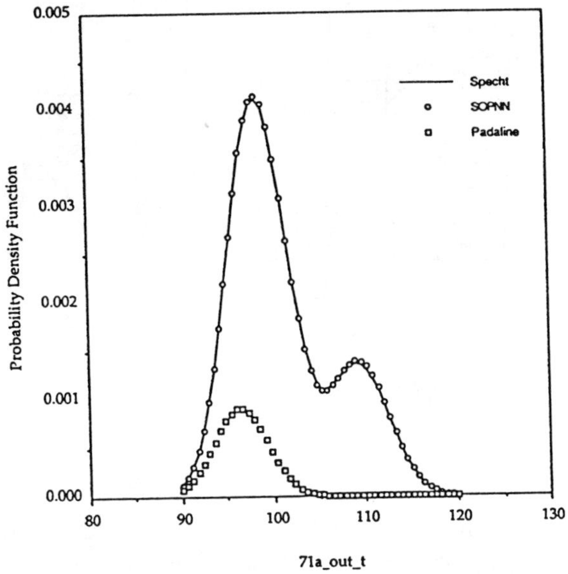

Fig. 4 Density estimation for power plant data

As indicated in the plots, the SOPNN performed essentially as well as the Specht's model but with less neurons. In contrast, the Padaline model does not perform as well.

INTEGRATING PNNs WITH INFLUENCE DIAGRAMS

In this section we focus on integrating probabilistic neural networks with influence diagrams for multiple sensor monitoring and diagnostic systems. It is our thesis that one should be able to come up with better artificial intelligence systems by combining the economy, efficiency and flexibility of influence diagrams with the learning ability, noise accommodation, and the massively parallel computation of neural networks.

Influence diagrams can be constructed in three ways : (1)encoding from human experts, (2) inferring from first principles, and (3)inducing from examples. Though human experts may be good at identifying critical variables, and providing essential information between the variables, they are limited in providing estimates for a large number of probabilities. It is a good idea to supplement probability assessments with data wherever possible. If enough representative data is available, it is possible to obtain the estimate for the probability density function. We have proposed the SOPNN to learn the probability distributions from data. Our proposed integration approach is as follows: An influence diagram is first constructed by the aforementioned methods. This system not only decomposes, in effect, a complex system or process subsystems but also provides the inference engine for probabilistic reasoning and decision making. Each family in the diagram defines the structure of a corresponding SOPNN. A family in this study is defined as a set of nodes consisting of a non-leaf node along with its immediate predecessors, i.e., for each set of conditional probabilities in the influence diagram there is a corresponding family. These networks are used in parallel to estimate probabilistic distributions from available data. The probabilistic distributions obtained are fed back to the influence diagram for probabilistic reasoning and decision making under uncertainty, as shown in Fig 5.

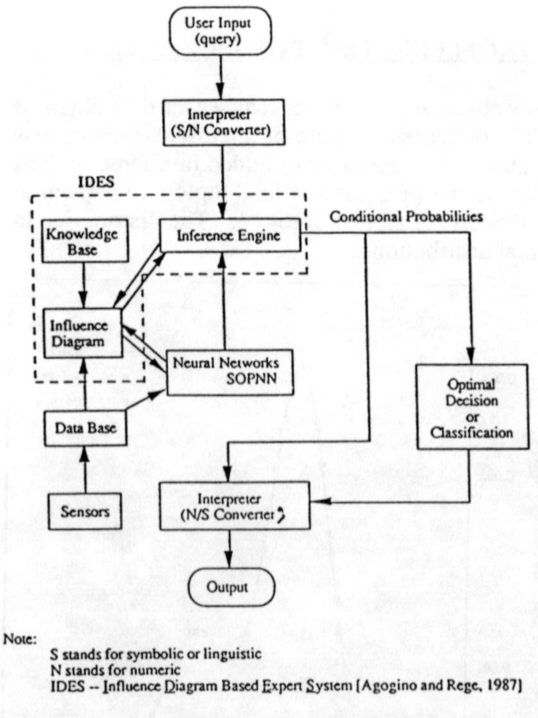

Fig. 5 Proposed Integrated System

The proposed integrated system can be used to reduce the system complexity in two ways : representation and inference. The conditional independence encoded in the influence diagram reduces the number of events to be considered, and consequently reduces the number of neurons needed in the SOPNN. Without influence diagrams one would have to consider all possible events, and this results in a very large number of neurons needed. Not only can influence diagrams be used to effectively represent a system, but it can also be efficiently applied to make probabilistic inferences, i.e. answer arbitrary queries within the influence diagram framework. Without influence diagrams one neural network will have to be constructed for every possible query, but with influence diagrams fewer neural networks are required; one for each cluster of nodes in the influence diagram.

SUMMARY

In this study we are interested in combining the strengths of influence diagrams and neural networks for multiple sensor diagnostic systems. We believe that by combining the economy, efficiency and flexibility of influence diagrams with the learning ability, noise accommodation and parallel computation of neural networks one should be able to come up with better artificial intelligence systems. To improve previously proposed models a Self Organizing Probabilistic Neural Network was designed to efficiently estimate continuous distributions. The effectiveness of each model has been tested through several examples using power plant data. The integrated system has been shown to reduce the complexity in both representation and inference. Since we have independently constructed only those PNNs for which the training data are available, our integrated system can better utilize partial data. By the same argument it can incorporate subjective information as well as empirical data, and is thus more flexible.

ACKNOWLEDGEMENTS

The authors are indebted to The Electric Power Research Institute (EPRI) and Sargent & Lundy Engineers for their technical support throughout this research. This research was supported by EPRI and Sargent & Lundy under contract RP2923-05.

REFERENCES

1. Devijver, P. A. and J. Kittler, "Pattern Recognition: A Statistical Approach", Prentice Hall, 1982.
2. Nadi, F, A.M. Agogino and D. Hodges, "Use Of Influeence Diagrams and Neural Networks in Modeling Semiconductor Manufacturing Processes", **IEEE Transactions on Semiconductor Manufacturing**, Vol. 4, No, 1, Feb 1991, pp. 52-58.
3. Parzen, E., "On Estimation of Probability Density Function and Mode," **Annals of Mathematical Statistics**, v33, pp 1065-1076, September, 1962.
4. Rege, A. and A.M. Agogino, "Topological Framework for Representing and Solving Probabilistic Inference Problems in Expert Systems, " **IEEE Systems, Man, and Cybernetics**, Vol. 18(3), May 1988, pp. 402-414.
5. Specht, D.F., "Probabilistic Neural Networks for Classification, Mapping, or Associative Memory," **IEEE International Conf. on Neural Networks**, vI, san Diego, CA, July 24-27, 1988, pp525-532.
6. Specht D.F, " Probabilistic Neural Networks and the Polynomial Adaline as Complementary Techniques for Classification," **IEEE Transactions on Neural Networks**, v1, pp. 111-121, March1990.

Use of Neural Networks in Optimizing RPV Bolting Procedures

by

William L. Biach
Biach Industries Inc.
Cranford, NJ 07016

Abstract- *The routine bolting and unbolting of the reactor pressure vessel in a nuclear power plant is performed through sequential application of a set of tools. This process produces numerous and cyclical loading of the studs. Variances in the results arise from mechanical and operator irregularities. Due to stress corrosion cracking and exposure to borated water it is desirable to minimize the number of cycles on the studs to maintain stud integrity* [2]. *By adjusting the process to reflect the actual response of the system we can optimize the procedure. A representational model would be too complex and wouldn't be practical in real time. The approach suggested here would be to adjust a strategy based on the current pattern of stud elongations after each step using artificial neural networks (ANN). There are three applications in this problem. First, classification of elongation patterns to identify systematic irregularities and problems. Second, to suggest the optimum strategy for continuing the process, given the state of elongations at any time. Third, to provide a model for the response of the flange when tightening a stud. Artificial neural networks have been successfully applied to problems of this type* [1],[3]

1. BACKGROUND

The reactor pressure vessel (RPV) head is a bolted, gasketed cover. It is usually about 15 feet in diameter with 54 to 64 studs.

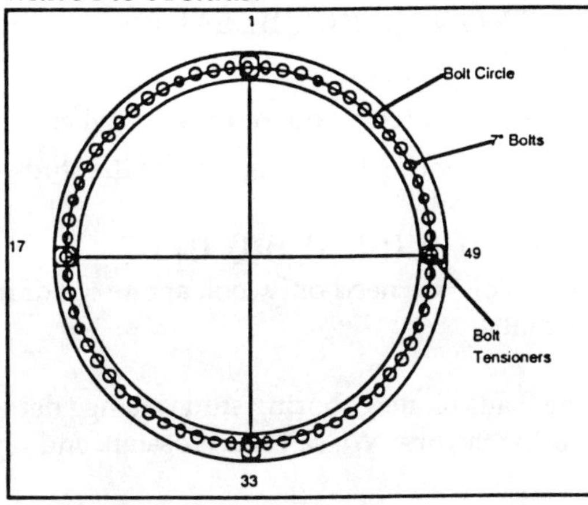

Figure #1. A top view of a typical Reactor Pressure Vessel bolt circle. This one has 64 studs numbered sequentially counter-clockwise. There also are 4 bolt tensioners symmetrically placed around the flange.

Each stud is between 6 and 7 inches in diameter and about 6 feet long. These studs require 1500 klbs to 2500 klbs of load to tighten them. This procedure is done with hydraulic stud tensioners which hydraulically stretch these studs to the desired strain and tighten the nuts to maintain the strain. Usually 3 or 4 units are used in a symmetric distribution around the flange (see fig. #1) and are moved from one stud to another always trying to maintain as symmetric a pattern as possible. Usually on the first pass (a pass being the application of the tooling to all of the studs) they are brought up to 75% of the desired strain to compress the gasket and pre-load the studs. Then another pass is done to bring them closer to final strain. Then they are adjusted until they all are within ±2%.

There is crosstalk in the system due to the elastic interaction between neighboring studs [2]. As studs are loaded they take away some of the burden from the adjacent studs depending on the loading of the other studs, the design of the flange, and the design and status of the gasket.

Inherent in the structure of the system is variation in the results of each loading of the stud. There is variation in the way the gasket responds. As the studs are cycled and subjected to borated water, stress corrosion cracking and other sources of performance degradation, the studs respond differently from the norm. The tools have a degree of inaccuracy and operators using them will get different results[2]. Therefore although the process is reasonably well controlled, the results have an unacceptable degree of variation. The following table shows results from field tests [5]. This chart shows the residual elongation of one quadrant after each pass. The target elongation on the passes was .030", .045", and .048 respectively.

TABLE #1

Stud #	1	2	3	4	5	6	7	8	9	10	11	12	13	14	15	16
1st pass	.003	.025	.016	.033	.018	.029	.022	.029	.004	.025	.017	.024	.013	.028	.023	.031
2nd pass	.034	.044	.039	.051	.042	.046	.039	.045	.035	.043	.037	.040	.033	.046	.042	.048
3rd pass	.044	.045	.045	.044	.048	.048	.048	.042	.046	.046	.045	.042	.042	.046	.044	.046

Studs 1,17,33, 1 and 49 were always done simultaneously. Notice the variation.

TABLE #2

Stud #	1	17	33	49
1st pass	.003	.002	.005	.002
2nd pass	.034	.034	.038	.035
3rd pass	.044	.044	.045	.046

2. MODELING THE PROBLEM

Let each application of the tools be considered an operator , $\mathbf{P}_{i,j}$ (where i is the step in the pass, j is the pass), on the vector space $\mathbf{E}$ representing the elongation of the studs. Starting with all of the studs without load,

$$\mathbf{E} = [0\ 0\ 0\ 0\ 0\ 0\ 0\ 0\ 0\ 0\ 0\ 0\ 0\ 0\ 0\ 0]$$

(Assuming 64 studs on the vessel and four tools, we need only look at one quadrant (or 16 studs), keeping in mind crosstalk between quadrants)

As load is taken by each stud, the loads on neighboring studs change depending on their proximity to the stud being loaded and other factors. We call this crosstalk and represent it symbolically as follows:

$$\mathbf{C}_{i,j} = [f_{l,m}(i,j)]$$

$f_{l,m}(i,j)$ is the influence a change in stud l has on the elongation of stud m for any i and j. f is a function of i and j because the influences change as the gasket is loaded and the flange reacts to different load distribution. This function is generally a bell shaped curve, dropping off quickly beyond two studs. For our example lets assume that for adjacent studs, their elongation is reduced by 60% of the increase in elongation of the stud being loaded. For studs two away from the one being tightened, the elongation is reduced by 15% of the increase. Of course they can't be reduced below 0.

$$f_{l,m}(i,1) = 1 \quad \text{for } m = l$$

$$f_{l,m}(i,1) = -.6 \quad \text{for } m = l \pm 1$$

$$f_{l,m}(i,1) = -.15 \quad \text{for } m = l \pm 2$$

$$f_{l,m}(i,1) = 0 \quad \text{for all other } m$$

For one quadrant $\mathbf{C}_{i,j}$ might look like:

$$\begin{bmatrix}
1 & -.6 & -.15 & 0 & 0 & 0 & 0 & 0 & 0 & 0 & 0 & 0 & 0 & 0 & -.15 & -.6 \\
-.6 & 1 & -.6 & -.15 & 0 & 0 & 0 & 0 & 0 & 0 & 0 & 0 & 0 & 0 & 0 & -.15 \\
-.15 & -.6 & 1 & -.6 & -.15 & 0 & 0 & 0 & 0 & 0 & 0 & 0 & 0 & 0 & 0 & 0 \\
0 & -.15 & -.6 & 1 & -.6 & -.15 & 0 & 0 & 0 & 0 & 0 & 0 & 0 & 0 & 0 & 0 \\
0 & 0 & -.15 & -.6 & 1 & -.6 & -.15 & 0 & 0 & 0 & 0 & 0 & 0 & 0 & 0 & 0 \\
0 & 0 & 0 & -.15 & -.6 & 1 & -.6 & -.15 & 0 & 0 & 0 & 0 & 0 & 0 & 0 & 0 \\
0 & 0 & 0 & 0 & -.15 & -.6 & 1 & -.6 & -.15 & 0 & 0 & 0 & 0 & 0 & 0 & 0 \\
0 & 0 & 0 & 0 & 0 & -.15 & -.6 & 1 & -.6 & -.15 & 0 & 0 & 0 & 0 & 0 & 0 \\
0 & 0 & 0 & 0 & 0 & 0 & -.15 & -.6 & 1 & -.6 & -.15 & 0 & 0 & 0 & 0 & 0 \\
0 & 0 & 0 & 0 & 0 & 0 & 0 & -.15 & -.6 & 1 & -.6 & -.15 & 0 & 0 & 0 & 0 \\
0 & 0 & 0 & 0 & 0 & 0 & 0 & 0 & -.15 & -.6 & 1 & -.6 & -.15 & 0 & 0 & 0 \\
0 & 0 & 0 & 0 & 0 & 0 & 0 & 0 & 0 & -.15 & -.6 & 1 & -.6 & -.15 & 0 & 0 \\
0 & 0 & 0 & 0 & 0 & 0 & 0 & 0 & 0 & 0 & -.15 & -.6 & 1 & -.6 & -.15 & 0 \\
0 & 0 & 0 & 0 & 0 & 0 & 0 & 0 & 0 & 0 & 0 & -.15 & -.6 & 1 & -.6 & -.15 \\
-.15 & 0 & 0 & 0 & 0 & 0 & 0 & 0 & 0 & 0 & 0 & 0 & -.15 & -.6 & 1 & -.6 \\
-.6 & -.15 & 0 & 0 & 0 & 0 & 0 & 0 & 0 & 0 & 0 & 0 & 0 & -.15 & -.6 & 1
\end{bmatrix}$$

$\mathbf{E}_{i,j} = \mathbf{E}_{i-1,j}\mathbf{C}_{i,j}\mathbf{T}_{i,j}$ is a function describing how the elongation vector changes on step i, with $\mathbf{T}_{i,j}$ the target amount of elongation. Therefore $\mathbf{P}_{i,j} = \mathbf{C}_{i,j}\mathbf{T}_{i,j}$. Let the notation $\mathbf{P}_{q \to r, s \to t} = \prod_{\substack{i=q,r \\ j=s,t}} \mathbf{P}_{i,j}$. Upon subsequent applications of $\mathbf{P}$ changing $\mathbf{T}$ to maintain symmetric loading

$$\mathbf{T}_{1,1} = \begin{bmatrix} 1\\0\\0\\0\\0\\0\\0\\0\\0\\0\\0\\0\\0\\0\\0\\0 \end{bmatrix} \quad \mathbf{T}_{2,1} = \begin{bmatrix} 0\\0\\0\\0\\0\\0\\0\\0\\1\\0\\0\\0\\0\\0\\0\\0 \end{bmatrix} \quad \mathbf{T}_{3,1} = \begin{bmatrix} 0\\0\\0\\0\\1\\0\\0\\0\\0\\0\\0\\0\\0\\0\\0\\0 \end{bmatrix} \quad \mathbf{T}_{4,1} = \begin{bmatrix} 0\\0\\0\\0\\0\\0\\0\\0\\0\\0\\0\\0\\1\\0\\0\\0 \end{bmatrix} \quad \mathbf{T}_{5,1} = \begin{bmatrix} 0\\0\\0\\0\\0\\0\\1\\0\\0\\0\\0\\0\\0\\0\\0\\0 \end{bmatrix} \cdots$$

we get the following results:

$$\mathbf{EP}_{1,1} = [1\ 0\ 0\ 0\ 0\ 0\ 0\ 0\ 0\ 0\ 0\ 0\ 0\ 0\ 0\ 0]$$

$$\mathbf{EP}_{1 \to 2,1} = [1\ 0\ 0\ 0\ 0\ 0\ 0\ 0\ 1\ 0\ 0\ 0\ 0\ 0\ 0\ 0]$$

$$\mathbf{EP}_{1 \to 5,1} = [1\ 0\ 0\ 0\ .4\ 0\ 1\ 0\ .4\ 0\ 0\ 0\ 1\ 0\ 0\ 0]$$

$$\mathbf{EP}_{1 \to 16,1} = [0\ 1\ 0\ .85\ 0\ .5\ 0\ .5\ 0\ 1\ 0\ 0\ .5\ 0\ .85\ 0]$$

3. Application #1- TROUBLESHOOTING WITH NEURAL NETWORKS

The first area of application is in troubleshooting the system. The above results were under theoretical, ideal conditions. $\mathbf{T}_{i,j}$ had the same magnitude for each operation and $\mathbf{C}_{i,j}$ was simple, predictable and consistent throughout the pass. Although the system itself produces results in which the studs vary from 0% to 100% of the desired elongation the pattern has some structure to it. When an irregularity in the process occurs, such as a defective stud, an irregularity appears in the pattern. It is not always easy to establish a rule by which such irregularities can be highlighted or distinguished from acceptable variations.

Figures 2 and 3 show patterns after the completion of the first pass in which there are anomalies. These particular problems could be highlighted by sorting the data. Identification is also simplified since the irregularities are presented isolated from other variances.

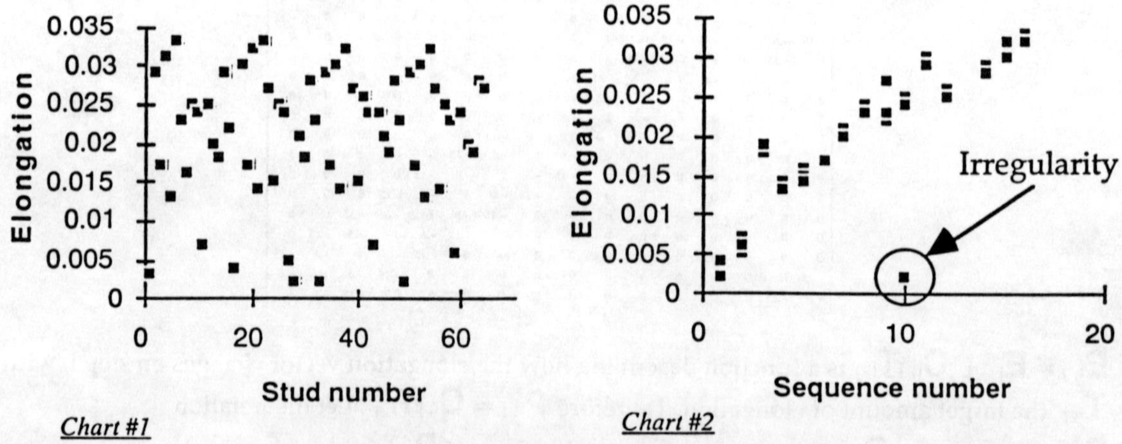

FIGURE 2. Chart #1 shows a plot of stud elongation after the first pass versus the stud number. The particular irregularity in this data can be seen by sorting the data into the sequence of tightening (what we refer to as i in the equations) as in Chart #2. (The elongations have been modified to illustrate the various scenarios).

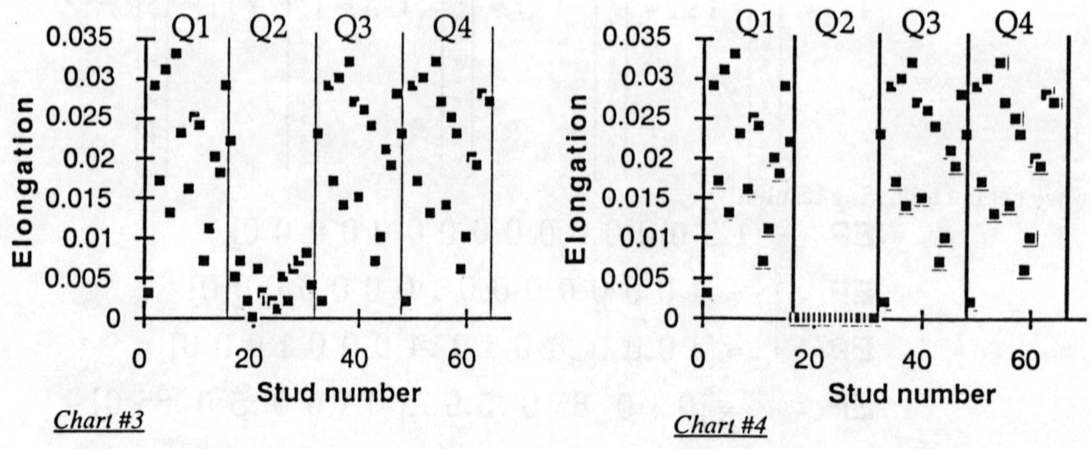

FIGURE 3. Chart #3 show a typical pattern when an operator in quadrant 2 isn't trained as opposed to Chart #4 where the equipment in quadrant 2 is defective. The main distinguishing characteristic is the uniformity of results in Chart #4 quadrant 2.

It would be desirable to develop a simple pattern recognition/classification model to identify these patterns. However, since these patterns often occur simultaneously, it would have to filter out the different effects. As a troubleshooting tool for the identification of correctable problems.this would have tremendous implications.

4. Application #2- OPTIMIZING THE PROCESS

Optimizing the process is a problem of choosing the best stud to tighten next and how much it should be tightened. In the simplest case we are trying to minimize the number of individual stud tightenings or application of the stud tightening equipment. An extension to the problem would be to minimize the tool transit time of the tooling between tightenings.

This problem is similar to the Traveling Salesman Problem. In place of distances, or cost function, we would use the number of additional tightenings resulting from following one stud by another. In other words, if we were to tighten stud **A**, how many additional tightenings would result from following **A** with **B**. There are characteristics which distinguish this from the TSP. First, you can return to a node, although this usually doesn't occur until after all of the studs have been tightened once. Second, the cost of **A-B** is not fixed, it depends on the state before **A**. Third, **A-B <>B-A**.

We would want to develop an ANN which would process the current state of the studs and suggest the optimum sequence before taking the next step. Self organizing networks like the Hopfield net and the Kohonen net have been used for this problem [3]. The strategy would be reevaluated after each step. This requires that the network could stabilize fast enough (i.e. the time it takes to disconnect and raise the tools for repositioning, appox. 10 sec). Since each tightening changes only a discreet number of studs we may be able to speed the convergence. The problem of local minimums remains an issue since it could lead to a trivial short term strategy which would result in a disastrous long term one.

5. Application #3- MODELING THE RESPONSE OF THE SYSTEM

As we stated earlier, when tightening a stud the resulting elongation of the stud and its neighbors is dependent on the state of the system before tightening, and the flange and gasket characteristics. This problem is complex to solve in an analytic manner. requiring finite element analysis. This is impractical in real time. However, it is a fairly typical problem for a learning network like a back-propagation net. We would have six inputs, the elongation of the stud, the elongation of its four neighbors, and the amount the stud will be elongated. The resulting elongation of the four neighbors would be the four outputs. The number of hidden nodes needs to be determined.

Developing this network would be crucial to the implementation of the optimization model of section 4. Initial training data could be developed from finite element analysis. We would have this model refine itself on each flange as data accumulates. However there is enough general similarity of response among flanges that a generic set of weights could be used to intelligently prime the net.

6. CONCLUSION

Research into the application of ANN to this problem is only in its formative stages. The problem is fairly well understood. However we have not sufficiently evaluated the features of various architectures to determine which to study further. There seems to be three very different, yet very appropriate applications for ANNs.

This problem will become more critical as there is more need for equipment to operate in an operator free environment (e.g.. fusion reactors). In some of our examples, there were correctable problems. The system may, in the future, have to compensate for problems rather than correct them.

It is also typical of many problems in procedure engineering where a process is completed in a series of steps. Although it is usually possible to determine the state of the system after each step, it is not usually clear what the next step should be.

REFERENCES

[1] Hopfield, J. J., & Tank, D. W. (1985). Neural computation of decisions in optimization problems. *Biological Cybernetics*, 52, 141-152

[2] *Maintenance and Technology Session on Bolting and Bolted Connections*, November 1985, EPRI

[3] Ritter, Helga, Martinez, Thomas, Schulten, Klaus, *Neural Computation and Self-Organizing Maps*, (1992), Addison-Wesley

[4] *Study of Bolting Problems, Tools and Practices in the Nuclear Industry*, August 1980, EPRI

[5] Documentation of the RPV bolting procedure, Utility's name withheld at their request

International Neural Network Society
1993

Facts About INNS

What Is INNS?

- INNS is a professional and scientific association which is dedicated to promoting scientific understanding about models of brain and behavioral processes and exploring the application of neural modelling concepts to technological problems.

Who Can Join?

- Any individual with an interest in learning about or contributing to this dynamic scientific field is invited to join INNS. INNS' 3,000+ members represent a diversity of disciplines, including computer science, neuroscience, psychology, engineering, physics, mathematics, business, and law to mention just a few.

- INNS members span the globe. The Society has active members on all continents except Antarctica.

- Students are especially welcome. INNS gives full-time students special discounts on membership dues and other benefits.

What Does My Membership Include?

- One year subscription to the journal Neural Networks.

- Discounts on registration fees for the annual World Congress on Neural Networks. Congresses are currently planned for 1993 in Portland and 1994 in San Diego.

- One-year subscription to the INNS newsletter, Above Threshold.

- Free Special Interest Groups (SIGs) focused on aspects on neural networks and based worldwide.

- Discounts on subscriptions to other scientific publications related to the field.

- Opportunities to participate in special professional education and development workshops on neural networks.

- Use of an on-line electronic message/bulletin board system devoted to the exchange of information about neural networks.

How Do I Join INNS?

- Complete the membership application and return it to INNS at the address above.

- Questions? Call 202-466-4667 or fax 202-466-2888 (both numbers are in the United States) or contact us via e-mail: 70727.3265@compuserve.com

INTERNATIONAL NEURAL NETWORK SOCIETY

Membership Application

Check one: ☐ **Regular** (US$60) ☐ **Student** (US$40)

Students - Please verify status by providing:

Expected Degree: __________ Graduation Date: __________

Faculty Signature: ______________________________

Dr./Mr./Ms.: ______________________________

Affiliation: ______________________________

Mailing Address:

____

____

____

____

Telephone: ______________________________

Fax number: ______________________________

E-mail: ______________________________

Check your areas of interest in neural networks:

- ☐ Biomedical applications
- ☐ Electronics/VLSI
- ☐ Expert Networks
- ☐ Finance/Economics
- ☐ Geology/Geophysics
- ☐ Higher Level Cognitive Processes
- ☐ Mathematics/Theory
- ☐ Mental Function & Dysfunction
- ☐ My Local City/State/Country SIG
- ☐ Neuroscience
- ☐ Optics
- ☐ Philosophy/Evolution
- ☐ Power Engineering
- ☐ Pulsed Networks
- ☐ Robotics/Automation
- ☐ Speech
- ☐ Standards
- ☐ Vision/Motion Analysis

Mailing List: INNS rents its mailing list to earn additional income. If you do not want your name used, check below:

☐ I DO NOT authorize INNS to use my name/address for non-neural network mailings.

Membership Fee Payment: I would like to pay for my membership with:

☐ Check (drawn in US funds)

☐ MasterCard ☐ Visa

Account number: ______________________________

Expiration Date ______________

Signature: ______________________________